The Kyoto Protocol

Research by the Energy and Environmental Programme of the Royal Institute of International Affairs is supported by generous contributions of finance and technical advice from the following organizations:

Amerada Hess ● Ashland Oil ● Blue Circle Industries ● British Gas
British Nuclear Fuels ● British Petroleum ● Eastern Electricity ● ENI
Enron ● Enterprise Oil ● Esso/Exxon ● LASMO ● Mitsubishi Fuels
Mobil Services ● Osaka Gas ● PowerGen ● Ruhrgas ● Saudi Aramco
Shell ● Statoil ● Tokyo Electric Power ● Texaco ● Veba Oil

This book also draws upon specific projects and meetings during the period of the Kyoto negotiations, which were supported by the following organizations:

New Energy Development Organization, Japan
The Environment Agency, Japan
MITI/Global Industrial and Social Progress Research Institute, Japan
UN Conference on Trade and Development
Department of Environment, Sport and Territories, Australia
Ministry of Environment, France
The Royal Ministry of Foreign Affairs, Norway
Department of Environment, Transport and the Regions, UK
World Wide Fund for Nature, UK
Department of State, USA
Environmental Protection Agency, USA
Department of Energy, USA

The Kyoto Protocol

A Guide and Assessment

Michael Grubb
with Christiaan Vrolijk and Duncan Brack

Additional contributions from Tim Forsyth,
John Lanchbery and Fanny Missfeldt

THE ROYAL INSTITUTE OF
INTERNATIONAL AFFAIRS
Energy and Environmental Programme

First published in the UK in 1999 by
Royal Institute of International Affairs, 10 St. James's Square, London SW1Y 4LE
(Charity Registration No. 208 223)
and Earthscan Publications Ltd, 120 Pentonville Road, London N1 9JN

Distributed in North America by
The Brookings Institution, 1775 Massachusetts Avenue NW,
Washington DC 20035-2188

Reprinted 2001

ISBN: 1 85383 580 3 paperback
     1 85383 581 1 hardback

1002545156

Earthscan Publications Limited is an editorially independent subsidiary of Kogan Page Limited
and publishes in association with the WWF-UK and the International Institute for Environment and
Development.

Typesetting by Koinonia, Manchester
Printed and bound by Biddles Ltd, *www.biddles.co.uk*
Original cover design by Visible Edge
Cover by Yvonne Booth

To all those who worked so long and so hard
for an effective agreement – and were still left standing
after those sleepless nights in Kyoto

'For, in the final analysis, our most basic common link is that we all inhabit this small planet. We all breathe the same air. We all cherish our children's future. And we are all mortal.' – John F. Kennedy, Commencement Address at the American University in Washington, DC, 10 June 1963

'The whole history of international environmental action has been of arriving at destinations which looked impossibly distant at the moment of departure.' – Tony Brenton, *The Greening of Machiavelli*

Additional citations

'The work of the Royal Institute of International Affairs on pressing international environmental issues has been ground-breaking, and their work on integrating economics and the environment is particularly valuable. This book's analysis of climate change and the challenges facing governments, business and the wider public is in that mould: a most welcome contribution.' – *Michael Meacher, Minister for the Environment, UK*

'Understanding how this complex and far-reaching agreement was negotiated is important to understanding its meaning. Analysing the commitments as signed, and the prospects, is equally important. This excellent book does both. All those interested in global responses to climate change would do well to study it carefully.' – *Michael Zammit Cutajar, Executive Secretary, UN Framework Convention on Climate Change*

'An extremely perceptive analysis of past events, the ongoing debate and future prospects, based on a sense of fairness and respect for equity on the part of all the actors in the arena ... Truly an outstanding work by a distinguished analyst and thinker.' – *Dr Rajendra Pachauri, Director, Tata Energy Research Institute, New Delhi*

'Michael Grubb's take on the debate is thoughtful and accessible. It's a provocative read for anyone with an interest in climate change.' – *Eileen Claussen, Executive Director, Pew Center, Washington*

'The outstanding merit of this book is its analysis of the complex and often hidden issues of the Kyoto Protocol. Michael Grubb and his colleagues emphasize that the road initiated in Kyoto still needs orientation, if it is to unleash its potential of stimulating technological change and economic progress world-wide.' – *Professor Stefan P. Schleicher, Chairman, Austrian Council on Climate Change*

Contents

Foreword by *H.E. Raúl A. Estrada-Oyuela* xiii

Preface xvii

Acknowledgments xx

About the authors xxii

Outline structure of Convention and Protocol xxiv

Glossary xxvi

Summary and conclusions xxxiii

Part I: The making of the Protocol

1 Analytic foundations: science, response options and the IPCC . . . 3

    1.1 Origins and the Intergovernmental Panel on Climate Change. . . . 3

    1.2 Scientific foundations and the First Assessment Report
        of the IPCC 5

    1.3 Findings of the IPCC Second Assessment Report. 7

    1.4 Key debates and implications of the Second Assessment
        Report 17

    1.5 Developments after the Second Assessment Report 21

    1.6 Conclusions 26

**2 Political and legal foundations: national perspectives and
the road to Kyoto 27**

    2.1 National interests, perspectives and negotiating groups 27

    2.2 The UN Framework Convention on Climate Change 36

    2.3 From Rio to the First Conference of Parties at Berlin 43

    2.4 The shifting political foundations 49

    2.5 Milestones to Kyoto. 52

3 Negotiating the Kyoto Protocol 61
   3.1 The negotiating process 62
   3.2 Policies and measures 65
   3.3 Defining emission targets: time scales, gas coverage and sinks . 68
   3.4 Assigning emission targets: differentiation and the 'EU
       bubble' 81
   3.5 International flexibility within Annex I: joint implementation
       and emissions trading 87
   3.6 International flexibility with developing countries: objections
       to JI and the emergence of the CDM 97
   3.7 Developing-country concerns and participation 103
   3.8 Extending commitments to new countries: evolution and
       voluntary accession 107
   3.9 Conclusions 111

4 The Kyoto Protocol 115
   4.1 Definition of Annex I commitments 115
   4.2 'Bubbling' and the EU redistribution of emission
       commitments 122
   4.3 Policies and measures 124
   4.4 Mechanisms for international transfer 128
   4.5 Additional issues relating to developing countries 138
   4.6 Compliance, future development and related issues:
       monitoring, reporting and review 142
   4.7 Conclusions 150

Part II: Analysis of commitments, mechanisms and prospects

5 Environmental and economic implications of the Kyoto
   commitments 155
   5.1 Environmental consequences of the Kyoto commitments 155
   5.2 Economic consequences of the Kyoto commitments 160
   5.3 The potential impact of flexibility within Annex I 171
   5.4 Potential impact of sinks and the CDM 186
   5.5 Conclusions: inflation, efficiency and competition between the
       mechanisms 191

6 Implementing international transfers under the Kyoto mechanisms **194**

6.1 General principles governing the creation and transfer of
emission credits and allowances 194

6.2 Acquisition and trade of emission credits from JI and the
CDM 198

6.3 Governing international emissions trading 206

6.4 Supplementarity, balancing and charges under the Kyoto
mechanisms 217

6.5 Conclusions 224

7 The clean development mechanism **226**

7.1 Introduction 226

7.2 A: additionality, and adaptation and administration charges... 227

7.3 B: bilateral or portfolio structure? 232

7.4 C: crediting, certification and the control of corruption 237

7.5 D: distribution of activities under the CDM. 238

7.6 E: eligibility criteria. 240

7.7 F : financing sources 242

7.8 G: governing the CDM 244

7.9 Conclusions 245

8 Prospects for the Kyoto regime **248**

8.1 The Buenos Aires 'Plan of Action' 248

8.2 Prospects for ratification and entry into force 253

8.3 Business and public involvement in the Kyoto regime 257

8.4 First-period developing-country commitments: accession,
expansion and inflation? 262

8.5 Longer-term approaches: an overview................... 265

8.6 Emissions convergence: any room for compromise? 269

8.7 Conclusions 275

Part III: Appendices

1. Text of the Kyoto Protocol 281
2. Key themes in economic debates: insights from the IPCC Second
   Assessment Report 304
3. Analysis of international trade in emission allowances 324
4. Further reading and sources of information 330

Index 335

Tables

3.1 Greenhouse gases in the Kyoto Protocol 73
3.2 LUCF emissions for Parties reporting with LUCF emissions
    greater than 5% of 1990 emissions 78
3.3 Total anthropogenic CO_2 emission changes, excluding LUCF,
    1990–96 82
4.1 Emissions and commitments in the Kyoto Protocol (from
    base year) 118
4.2 The internal distribution of the EU 'bubble' 123
5.1 Estimated costs of achieving Kyoto commitments from various
    economic models 164
5.2 Emission projections up to 2010 submitted by Parties
    (% from base year) 166
5.3 CO_2 emissions increases and required cutbacks by major
    Annex I groups: IEA projections (MtC) 168
5.4 Base assumptions in the ITEA analysis 173
5.5 Aggregate implications of Kyoto emission commitments 178
5.6 Traded volumes under the high case (no constraints:
    moderate trading) 179
5.7 Emission distribution in Annex I after trade (% relative to
    base year) 180
5.8 Impact of trading on the cost distribution for the Kyoto targets
    ('$/cap) 183
5.9 Potential sinks from anthropogenic LUCF activities 188
5.10 Estimates of the size of the CDM 190

Figures

1.1 Observed and modelled temperature trends, with and without aerosol corrections 10
1.2 Global average surface temperature, 1860–1998 22
1.3 Estimated and predicted global CO_2 absorption by vegetation ... 25
2.1 Global CO_2 emissions, per capita and per population 28
3.1 Trends in emissions of the Kyoto greenhouse gases, 1990–95 and 1995–6 83
5.1 Potential impact of the Kyoto Protocol on future emissions, concentrations, temperature and sea level 157
5.2 A view of optimal global CO_2 trajectories for stabilization at 450 ppm CO_2 and 550 ppm CO_2 159
5.3 Gap between Kyoto commitments and 1995 emission levels in principal Annex I groups 162
5.4 Standard marginal (a) and total abatement (b) cost curves for the ITEA model 172
5.5 US assigned amount and possible emissions under the Kyoto Protocol 181
6.1 Early crediting of projects 203
6.2 Accountability in international emissions trading 208
7.1 Portfolio/multilateral and bilateral structures for the CDM..... 233
8.1 Emissions convergence distributions under 450 ppm and 550 ppm 271
8.2 Kyoto commitments for industrialized countries............. 274

Boxes

1.1 A portfolio of actions 16
2.1 The distribution of global CO_2 emissions 28
2.2 Berlin decision on activities implemented jointly 46
2.3 The Berlin Mandate 48
2.4 Ministerial Declaration at the Second Conference of the Parties to the Climate Convention, 8–19 July 1996.................. 55
3.1 Progress on AIJ/JI projects before the Kyoto Protocol 101
4.1 Elements of policies and measures in the Kyoto Protocol...... 125
4.2 OECD proposal on emissions trading: the text that never made it .. 130

4.3 Joint implementation (Article 6) 132
4.4 The clean development mechanism (Article 12) 134
5.1 Countries and groups used for regional analysis of Annex I
commitments 161
5.2 Which way for Russian and Ukrainian greenhouse gas
emissions? 170
5.3 Abatement cost modelling in ITEA 172
6.1 Credit generation from joint implementation. 199
6.2 A 'review and assessment' approach to tackling 'hot air' trading 216
7.1 Bilateral versus portfolio approach to the CDM 235

Foreword

by H.E. Raúl A. Estrada-Oyuela

This excellent book is a much needed contribution to understanding the history, the meaning and the relevance of the Kyoto Protocol to the UN Framework Convention on Climate Change. Michael Grubb, together with Christiaan Vrolijk and Duncan Brack, has produced a most comprehensive analysis of a very complex and intricate set of commitments.

The Protocol concluded in December 1997, which may have a profound impact on industrial countries' production and consumption habits, is not a perfect international agreement, but it is the best compromise the international community was able to reach at that time. It was negotiated among a universal representation of the international community, including governments concerned with climate change and governments with totally different concerns. Some negotiators, particularly but not only the OPEC countries, were reluctant to agree measures or targets that might induce reductions in fossil-fuel consumption. Others, particularly but not only the USA, were trying to protect their life-styles by advocating a reduction in other countries' emissions. Developing countries were defending their right to make the same mistakes that industrial countries had committed before, although real economic development may not follow the old growth patterns; on the contrary, they are 'leap-frogging' between technologies, skipping many obsolete processes. Industrial countries in general behaved jealously, over-protecting their competitiveness in a context that revealed wide differences in their energy production profiles.

The authors carefully explore the Protocol's 'open tissues' – while it fills gaps left open in the Climate Convention, it has its own cavities. Governments represented at the negotiations were aware of these empty spaces, but had different ideas about how to tie them up and could not in all cases achieve consensus on one option. Those 'open tissues', constructive ambiguities on which the agreement was reached, can also be seen as systemic needs of the political process aimed at mitigating climate change. It is possible to dream of a complete treaty that fully covers and resolves all possible situations, but political and economic facts are stronger than dreams. Progress will be achieved through a sequence of steps, no single step being sufficient. Most probably, the most recent one – today the Kyoto Protocol – will call for additional action. The room for future commitments is located in the cavities of the 'open tissues'.

In order to have binding commitments in an international legal instrument, it was necessary to create flexibility mechanisms. The risk, clearly explained in this book, is that Parties to the Protocol could get entangled in these mechanisms without giving due consideration to climate change mitigation, which is the ultimate objective of the whole process. As in any other international joint venture, to find the right answers will require political will and wisdom. I am confident governments will eventually have both. When we arrived at Kyoto in December 1997 after two-and-a-half years negotiating, there were many sceptics, but I was optimistic. I even said that we diplomats are paid to be optimistic, which is a shorthand way of explaining that perseverance is a condition for diplomatic success, but it is not meant to be naïve.

Michael Grubb has followed the negotiations closely, with the twin advantages of being very well informed while having enough distance from the arena to keep it in perspective. His previous work on these matters, including analysis of different modalities of emissions trading, was familiar to the negotiators. I do not agree with every detail of the authors' assessments, but they are fair and well founded.

I share the concerns expressed in this book on the range of possibilities opened by the Clean Development Mechanism and the so-called 'hot air' or 'paper tons' included in 'assigned amounts' of two Annex B Parties. Misuse of the respective instruments may lead to an increase in global emissions by industrial countries. It is also true that the views of the USA were very carefully considered, not because it is the only remaining superpower but because it is the source of 37 per cent of the emissions from industrialized countries. That point was crystal clear to negotiators of the Convention in 1992 and of the Protocol in 1997; no effective agreement to mitigate climate change is possible without US participation. Nevertheless US views did not always determine the final product.

Developing countries' participation in mitigation efforts is not quantified by the Protocol, but it is clearly present in Art.10, as it is in Art.4.1 (b) of the Convention. However, their commitment to adopt policies and measures to mitigate climate change is conditional on the fulfilment of industrial countries' commitments to facilitate financial resources and the transfer of technology. Creating the option for voluntary quantified commitments to be undertaken by developing countries was a possibility in the negotiations, but it no longer exists in the Protocol. The source of the draft article on voluntary commitments may be traced to the original draft proposal of the Alliance of Small Island States (AOSIS), the Swiss proposal, 'Annex B' of the European Union proposal and the article on evolution in the US proposal. It was not an American creation, and at some points in the negotiations the US delegation made more difficult my work to gain acceptance of the draft article by developing countries.

The issue of developing countries' participation in the mitigation efforts should not became a vicious circle, blocking progress. Differentiation was established as the main criterion in the Convention, providing different treatment to industrial countries, economies in transition and developing countries. The Kyoto Protocol represented an important step in further differentiation. The possibility of deeper and more extensive differentiation opens up a promising field, provided each Party to the Convention honours the commitments already in force.

Eighty-four countries signed the Protocol in the UN Headquarters. These include all the industrialized countries covered by the quantified commitments except Hungary and Iceland, representing 99.4 per cent of the overall carbon dioxide emissions of industrial countries, and overall the initial signatories represent 88 per cent of global CO_2 emissions. That demonstrates the high level of acceptance of the text adopted at Kyoto. Under the Vienna Convention on the Law of Treaties, signatories of an international legal instrument, pending ratification, are obliged to refrain from acts which would defeat the object and purpose of a treaty. In the period until the Protocol enters into force, its purpose of reducing overall emissions from industrial countries is now protected by the relevant weight of the 84 signatories.

Reading this book may help negotiators to grasp what others think of the output of our work and also to dissect possible ways for further progress in understanding the mitigation of climate change.

Stanford, March 1999   Ambassador Raúl A. Estrada-Oyuela
Chairman, AGBM, and Negotiating Committee
of the Whole at Kyoto

Preface

Late in the morning of 11 December 1997, Chairman Raúl Estrada-Oyuela of Argentina brought down his gavel at the International Conference Centre at Kyoto and declared that the Kyoto Protocol to the UN Framework Convention on Climate Change was agreed unanimously. By then – 36 hours after the official deadline – a number of delegations had already left for home; many others were asleep, exhausted after the 12-day climax to more than 30 months of negotiations. The instrument that was adopted seeks to address the most challenging of global environmental problems: the threat posed by the steady accumulation of heat-trapping greenhouse gases, emitted largely from the prime economic engine of fossil fuel combustion.

In many respects, the Kyoto Protocol is without precedent in international affairs. Specific legal commitments capping emissions of such gases by each industrialized country constitute an achievement that many political sceptics had dismissed only months before as impossible. Yet these commitments are hedged around with such a complex of flexibilities that their exact meaning is in important ways opaque, and still disputed. The full implications are not yet apparent and to an important degree depend upon further negotiations that will determine how to implement the agreement as signed, and indeed whether and when it will be ratified by enough Parties to enter into force.

This book aims to explain and interpret the Kyoto Protocol: its background, provisions, meaning and prospects. It sets out the main debates and processes that led to the Protocol, provides a concise guide to its content, presents a numerical analysis of how the specific commitments may combine with the various flexibilities, and discusses key issues that remain to be addressed.

The book draws upon my own long involvement in, and experience of, climate change analysis and the political process, and that of others at

RIIA's Energy and Environmental Programme. In the aftermath of the Berlin conference that launched the Kyoto negotiations, the Programme hosted three annual workshops that brought together key negotiators, with analysts and various industrial and environmental non-governmental groups. I personally was closely involved in the debates surrounding the economic analysis of the problem and the development of emissions trading proposals, and attended many working meetings as well as the main negotiating sessions. Insights from these and related activities are reflected in these pages. This book also goes much further, in probing the agreement as signed and the issues that are yet to be resolved in its implementation.

Structure

The book comprises two main parts, plus substantive appendices.

Part I: The making of the Protocol sets out the pillars on which the Protocol rests, and describes the negotiating background and the content of the Protocol itself. Chapter 1 summarizes the intellectual pillars, in terms of the state of knowledge about the scientific and economic issues, and the processes by which the main conclusions have been accepted by governments. Chapter 2 summarizes the political and legal pillars: the main political interests and groupings, the international legal basis of the Protocol in the provisions of the UN Framework Convention on Climate Change, and the subsequent main political developments that led to Kyoto.

Chapter 3, the longest in the book, explains the main negotiating issues and the positions adopted by different Parties, and sketches briefly how these evolved under the pressures of negotiations. Chapter 4 describes the main provisions of the Protocol that emerged, interpreting their underlying meaning and drawing observations about the overall negotiating package and process.

Part II: Analysis of commitments, mechanisms and prospects provides a quantified analysis of the Protocol's numerical commitments and transfer mechanisms, and offers thoughts upon how these mechanisms and the Protocol overall might develop in the future. Chapter 5 provides a 'reality check', comparing the commitments against environmental objectives and actual emission trends and possibilities. It presents a numerical analysis and economic appraisal of the extraordinary impact of the Protocol's transfer mechanisms. Chapter 6 explores some of the critical issues concerning how the transfer aspects of these unique international mechanisms might be developed.

Chapter 7 provides an analysis of the Protocol's clean development mechanism, through which the commitments may attain a global reach. Chapter 8 then considers the Protocol's future, starting with the negotiating process launched by the follow-up conference at Buenos Aires, and going on to look at the core issues governing the prospects for the Protocol: the challenges of ratification, private-sector involvement, and the evolution of the regime to define future commitments and involve additional countries.

Part III consists of four Appendices. Appendix 1 contains the text of the Protocol itself. Appendix 2 describes some of the economic debates that set the context for the Protocol's development. Appendix 3 describes the detailed methodology and assumptions involved in our numerical analysis of the Protocol's commitments and international mechanisms. Finally, Appendix 4 details sources for additional reading and information.

March 1999   Michael Grubb

Acknowledgments

As with a river that has finally reached the sea, the origins of and contributions to this book are many and varied. Some are quite recent; indeed it is only a year since we agreed to write an extensive guide to and assessment of the Kyoto Protocol. However several components had by then already been largely completed, including the ITEA numerical analysis of emissions trading and commitments led by Christiaan Vrolijk (in Chapter 5), and the analysis with Duncan Brack of the IPCC's Second Assessment Report (in Chapter 1). Fanny Missfeldt had written a briefing paper on emission trends, projections and legislation in eastern Europe, and Tim Forsyth had written a briefing paper on Joint Implementation, both of these covering their topic in greater depth than could be incorporated fully in this volume. John Lanchbery agreed to help out on issues of reporting, monitoring, verification and compliance. We are deeply indebted to the contributing authors on this book.

Many others have helped in the latter stages of production, in particular the staff at EEP, Ben Coles, Nikki Kerrigan and Abigail Kellam, Margaret May of RIIA's publications department and of course the copy-editor Rosamund Howe and typesetter Peter Purdue.

Different parts of the book were presented at three different review meetings in June and December 1998 and January 1999, and we are grateful to all those who attended these meetings, offered thoughtful comments and pointed to major omissions.

Many people have contributed useful additional comments. Joanna Depledge and Cedric Philibert offered wide-ranging and thoughtful corrections and comments across the whole of the book. Peter Scott, Geoff Jenkins, Klaus Hasselman, Jo Haigh, Wolfgang Jung and Stephen Schneider reviewed Chapter 1 on the science and IPCC process. Yasuko Kawashima, Richard Kinley, Jonathan Pershing, Matti Vainio and Jose Maderia offered comments on aspects of the historical analysis and interpretation of the Protocol text in the rest of Part I. Jo Simons, Malik Amin Aslam, Jorgen Henningsen, Bill Hare, Aubrey Meyer and André Witthoeft-Muehlmann contributed varied comments in areas of Part II relating to their interests and expertise, and Richard Tol reviewed the Appendix on economic debates in the IPCC. To all these and many others we are much indebted. Of course they do not share any responsibility for the views expressed here, and any remaining errors and omissions are entirely the responsibility of the authors. The final stages of the book were overseen by RIIA's publications department under the skilful and patient eye of Margaret May.

The ultimate source of the book, however, lies much further back. RIIA's Energy and Environmental Programme has conducted work on climate change for ten years. We are deeply indebted to the companies and institutions that have sustained financial support for our work over the years and have respected and encouraged the intellectual freedom of RIIA's research, and to all the people from these companies, institutions and others who have contributed intellectually to RIIA's work on climate change. Their support enabled the appointment in 1988 of Michael Grubb to lead RIIA's research on environmental issues, and later the appointment of Duncan Brack to lead work on trade and environment, and Christiaan Vrolijk as a research assistant. In autumn of 1998, Michael stepped down as Head of EEP to allow more time for writing (this being the first product), and Duncan succeeded him as Head while Christiaan was appointed as RIIA's new Research Fellow on climate change. This book is thus also a tribute to the long-standing support of those who have helped us, and EEP, over the years.

London, March 1999   Michael Grubb
                                                Christiaan Vrolijk
                                                   Duncan Brack
                       Energy and Environmental Programme
                       Royal Institute of International Affairs

About the authors

Michael Grubb was Head of the Energy and Environmental Programme (EEP) at the Royal Institute of International Affairs (RIIA), Chatham House, from January 1993 until September 1998. He is well known for his work on the policy implications of climate change. He is an adviser to a number of international organizations and studies, particularly concerning economic and policy aspects of climate change, and has been a lead author on a number of studies for the Intergovernmental Panel on Climate Change.

Christiaan Vrolijk is a Research Fellow at RIIA studying international climate change issues and policy. Following almost two years as a research assistant, he took up the fellowship to develop a new project entitled 'Economic instruments in European electricity: levelling the carbon playing field'. He led development of RIIA's emissions trading model and has contributed to several briefing papers following the negotiations on climate change policy.

Duncan Brack has been Head of EEP since September 1998. His work focuses on the interaction between trade and environmental issues and he is currently completing a book on climate change policies and international trade. He is renowned for his work on CFC smuggling and on Multilateral Environmental Agreements. He is due to commence a new project looking at international environmental crime.

Additional contributors
Tim Forsyth was a researcher with EEP from March 1997 until August 1998. His work focused on technology transfer and foreign investment in Southeast Asia. He is now a Research Fellow at the Institute of Development Studies at the University of Sussex. He drafted the sections on technology transfer and the negotiations on joint implementation.

John Lanchbery is Climate Policy Adviser at the Royal Society for the Protection of Birds. Prior to that he was Director of Environmental Projects at the Verification Technology Information Centre (VERTIC). He has attended all the international climate negotiations since early 1992 and has written extensively on the subject. He drafted the sections on compliance, monitoring and verification.

Fanny Missfeldt was a researcher with EEP from March 1997 until June 1998. Her work focused on central and east European energy issues. She is now working on climate change issues in Eastern Europe at the Risoe National Laboratories in Denmark. She helped with data and drafting of material on the energy situation in Economies in transition.

Outline structure of Convention and Protocol

UN Framework Convention on Climate Change

Articles 2–3	Objective and principles
Article 4	Commitments
Articles 5 & 6	Research, education, public awareness
Articles 7–10	Establishment of the COP, Secretariat and Subsidiary Bodies
Articles 11 & 21	Financial mechanism and interim arrangement for GEF
Article 12	Communication of information related to implementation
Articles 13–14	Questions regarding implementation and dispute settlement
Articles 15–17	Amendments, Annex development and Protocols to the FCCC
Articles 18–26	Legal procedures
Annex I	Parties that take on the Article 4.2 commitments
Annex II	Parties that take on more financial commitments

Kyoto Protocol

Article 2	Policies and measures
Article 3	Quantified emission limitation and reduction commitments
Article 4	Joint fulfilment of commitments [bubbling]
Article 5	Methodological issues
Article 6	Transfer and acquisition of emission reduction units (joint implementation)
Articles 7–9	Communication of information, review of information and of the Protocol

Articles 10–11   Implementation of existing commitments and financial
                  mechanisms
Article 12   Clean development mechanism
Articles 13–16   Institutional issues (COP/MOP, Secretariat, Subsidiary
                  Bodies and Multilateral Consultative Process)
Article 17   Emissions trading
Articles 18–19   Non-compliance and dispute settlement
Articles 20–21   Amendment and Annex development procedures
Articles 22–28   Legal procedures, of which:
Article 25   Entry into force
Annex A   Greenhouse gases and sector/source categories
Annex B   Quantified emission limitation or reduction
                  commitments by Party

Glossary

AA   *Assigned amounts*, the official term in the Protocol defining the total allowed emissions for a Party over the commitment period 2008–12. **Emissions trading***, **JI** and the **CDM** could add to or subtract from this amount. More general terms used in the literature are allowance, cap, entitlement, permit and quota, but some of these are more appropriate at the corporate level.

AGBM   *Ad-hoc Group on the Berlin Mandate*, the principal negotiating body established by the Berlin Mandate to reach agreement in Kyoto on a 'protocol or other legal instrument'.

AIJ   *Activities implemented jointly.* Greenhouse gas reduction projects implemented in some **FCCC** Parties, which are funded from other Parties, but without any 'crediting' of emissions. Established by **COP**-1 as a pilot or test phase of **JI**.

Annex B   Annex listing initial national commitments under the Kyoto Protocol.

Annex I   The industrialized countries undertaking specific commitments under the **FCCC** and the Kyoto Protocol (almost synonymous with **Annex B** of the Kyoto Protocol, which has minor adjustments: see Section 4.1.1).

AOSIS   *Alliance Of Small Island States.* AOSIS, threatened by the rising sea levels, called for strong emission targets and a fast transition, and assistance for adapting to adverse affects of climate change.

---

* Words in bold type have a separate entry

Articles 2.3	Minimize adverse effects of climate change and mitigation measures on developing countries. (Articles in the Kyoto Protocol.)
Articles 4.8 and 3.14	Meeting the specific needs and concerns of developing-country parties arising from climate change and/or response measures. (Articles in the **FCCC**.)
Banking and 4.9	A system whereby emission allowances not used in any given compliance period can be saved for use in a future period. Banking of **assigned amounts** is possible under the Kyoto Protocol.
Base year	The **assigned amounts** are defined relative to emissions in the base year; for most **Annex I** countries this is 1990. Parties are allowed to choose 1990 or 1995 as the base year for the **industrial trace gases**. **EITs** are allowed different base years as defined under Decision 9/CP.2.
Baseline	Projection of emissions that would occur in the absence of an abatement project or policy intervention. The baseline is used to calculate the credits earned for a **JI** or **CDM** project, but details still have to be decided.
BAU	*Business-as-usual* projections of emissions, mostly used for the economy as a whole.
Borrowing	A system whereby emission allowances required by an emitter in any given period can be borrowed from the allowances available to it in a future period; the borrowing may carry an 'interest rate' or a 'penalty'. Borrowing is not incorporated in the Kyoto Protocol.
Bubble	An agreement by which Parties may redistribute their emissions commitments in ways that preserve their collective goal, allowed under Article 4 of the Protocol. Primarily intended for the EU and its member states, which have now signed the Kyoto Protocol together under a bubble agreement.
CDM	The *clean development mechanism* is defined in Article 12 of the Kyoto Protocol. It is similar in some ways to **JI**, but governs project investments in developing countries

	(Non-Annex I Parties) that generate **CERs** for **Annex I** Parties.
CEEC	Central and east European countries.
CER	*Certified emission reductions* can be earned with a **CDM** project in a developing country and be added to **assigned amounts** of an **Annex I** country.
CFCs	*Chlorofluorocarbons* are both ozone-depleting and greenhouse gases. Since the establishment of the Montreal Protocol they are being replaced rapidly with **HFCs**, which are also high-**GWP** greenhouse gases (**GWP** = 3800 and more). They are produced for solvents, refrigerants, aerosol spray propellants, foam packaging, etc.
CH_4	Methane is emitted from wet rice production, livestock, decay (e.g. landfill sites) and fossil fuel production. Methane **GWP** = 21 (over 100 years).
CO_2	*Carbon dioxide*, the main greenhouse gas, is released by burning fossil fuels, deforestation/land-use change and cement production.
Commitment period	Period for which commitments are adopted and defined under the Protocol. The first commitment period is 2008–12. Negotiations on commitments for a second period have to start by 2005.
COP	The *Conference of the Parties* to the **FCCC** is the supreme body of the Convention and meets each year. **COP**-1: Berlin (28 March–7 April 1995); COP-2: Geneva (8–19 July 1996); COP-3: Kyoto (1–10 December 1997); COP-4: Buenos Aires (2–13 November 1998); COP-5: Bonn (25 October–5 November 1999).
Credit	A unit added to the allowed emissions from a controlled entity, generally earned by investing in saving emissions from an uncontrolled source. The size of the credit is generally (though not necessarily) an estimate of the emissions saved relative to a **baseline** estimate of emissions from that source in the absence of that investment. In the Kyoto Protocol, **ERUs** and **CERs** are both emission credits.

CW	The *Committee of the Whole*, chaired by Ambassador Raúl Estrada-Oyuela, was the full negotiating body in Kyoto, preparing the Protocol for adoption by the **COP**.
CTI	*Climate Technology Initiative*, led by the **IEA**.
EITs	*Economies in transition*: the **CEEC** and **FSU** countries in transition to a market economy.
Emissions trading	The buying and selling of emission allowances. Article 17 of the Kyoto Protocol establishes trading of **AA** between **Annex B** Parties. It is expected that domestic and international schemes will be set up for industrial emissions trading. Some experience is gathered with the Montreal Protocol and (domestic) sulphur trading in the United States.
ERU	*Emission reduction units* can be earned with **JI** projects.
EST	*Environmentally sound technology.*
FCCC	See **UN FCCC**.
FDI	*Foreign direct investment.*
FSU	*Former Soviet Union.* The Russian Federation, Ukraine and Baltic countries are both in **Annex I** and **Annex B**, Belarus is only Annex I; in Buenos Aires Kazakhstan made the pledge to join Annex I at **COP-5**.
G77 (and China)	The *Group of 77* (and China) is the main negotiating group of developing countries, representing over 120 Parties in many international negotiations. The group includes countries with very different objectives, including **OPEC** and **AOSIS**.
GATT	*General Agreement on Tariffs and Trade.*
GDP	*Gross Domestic Product.*
GEF	The *Global Environment Facility* is entrusted with the operation of the financial mechanism; first on an interim basis (**FCCC** Article 21), and since Buenos Aires on a basis subject to four-year review (Decision 3/CP.4).
GHG	*Greenhouse gas*, or the basket of greenhouse gases controlled by the Kyoto Protocol.
GtC(e)	*Giga* (billion) *tonnes of carbon (equivalent).*

GWP   *Global warming potential*. Measure of the radiative (warming) impact of a molecule relative to carbon dioxide; 100-year global warming potentials are used for calculating the total **GHG** emissions under the Kyoto Protocol (in CO_2 equivalents).

HFCs   *Hydrofluorocarbons* replace the ozone-depleting **CFCs**, but have an extremely high **GWP** (140–11,700). Emissions are still low but growing fast.

Hot air   The excess emission allowances over the **BAU** emissions in the **commitment period** are called hot air. The existence of (too much) hot air could undermine the trading regime, or even the whole climate change regime. Hot air is believed to exist mainly in the **FSU** and some **CEEC** Parties. The extent will not be known until trends become clearer.

ICAO   *International Civil Aviation Organization*.

Industrial trace   The three sets of gases added to commitments in the closing
gases   stages of the Kyoto negotiations (**HFCs**, **PFCs** and **SF$_6$**).

IEA   *International Energy Agency*.

IMO   *International Maritime Organization*.

IPCC   The *Intergovernmental Panel on Climate Change* is the institution, established in 1988 by governments through **UNEP** and **WMO** to provide an authoritative assessment of the state of knowledge concerning climate change. The IPCC enlists several hundred scientists and other researchers from around the world to write reports that are subject to peer review globally, and publishes Policymakers' Summaries of the findings that are agreed line by line by the governments. The IPCC's First Assessment Report was completed in 1990, the Second Assessment Report in 1995. The Third Assessment Report is scheduled for 2001. The IPCC also publishes reports on special issues.

ISO   *International Standards Organization*.

ITEA   *International Trading of Emissions Allowances* model, developed at RIIA to provide a simple and transparent analysis of quantified commitments and **emissions trading**.

JI	*Joint implementation.* A term used widely and in many different ways to reflect ways of implementing commitments jointly. In the Kyoto Protocol, the term has become identified with the generation and transfer of **ERUs** under Article 6, in which investment in a project in one **Annex I** Party may generate a credit for the investing Party. Prior to Kyoto the term was also (and controversially) used in relation to possible projects in developing countries, now subsumed under the **CDM**.
JUSSCANNZ	Group of countries working together, tending to counterbalance the EU on the one hand and **G77** on the other. The group consists of Japan, the United States, Switzerland, Canada, Australia, Norway and New Zealand, though Norway and in particular Switzerland frequently stood somewhat apart from JUSSCANNZ positions.
LUCF	*Land-use change and forestry changes* are part of the regulated activities under the Kyoto Protocol, treated separately from other emissions and restricted to certain categories of human-induced changes.
MOP	*Meeting of the Parties* (to the Kyoto Protocol). The **COP** will also serve as the **MOP,** the supreme body for implementing the Kyoto Protocol after ratification and entry into force.
MtC(e)	*Mega (million) tonnes of carbon (equivalent).*
N_2O	*Nitrous oxide* (**GWP** = 210) emissions are mainly from fertilizers, fossil fuel burning, and land conversion for agriculture.
NGO	*Non-governmental organization.*
ODA	*Official Development Aid.*
OECD	*Organization for Economic Cooperation and Development.* In this book OECD is used to refer to those countries in **Annexes I** and II of the Convention, i.e. excluding Turkey and some **EITs** and newly industrialized countries that joined during the 1990s.
OPEC	*Organization of Petroleum Exporting Countries.*

Permit	A marketable instrument conferring the right to emit a quantified amount of a pollutant, e.g. one tonne of carbon per year. 'Permit' or 'quota' is the term which has tended to be used when industries or individual firms are the trading bodies.
PFCs	*Perfluorocarbon* (**GWP** between 6,500 and 9,200) emissions, mainly from aluminium, electronics and solvents industry, are low but growing.
QELRC	*Quantified emission limitation and reduction commitment.* The commitments adopted under Annex B of the Protocol. Before Kyoto QELRO was used (O = objective) as the general term for greenhouse gas targets and timetables.
Quota	See **Permit**.
SAR	*Second Assessment Report* (of the **IPCC**).
SBI	*Subsidiary Body for Implementation.*
SBSTA	*Subsidiary Body for Scientific and Technological Advice.*
SF_6	*Sulphur hexafluoride* (GWP is 23,900) emissions from electronics and insulation industry are low but growing.
Sinks	CO_2 absorbing ecosystems, such as young forests; see **LUCF**.
TAPs	*Technology Assessment Panels.*
Technology transfer	Technology transfer, established in Article 4 of the **FCCC** and Articles 10 and 11 of the Protocol, is one of the instruments for transfer of **Annex I** technology to the developing countries for their sustainable development.
Umbrella group	The Umbrella group, which emerged at Kyoto and afterwards, brings the **JUSSCANNZ** countries except Switzerland together with the Russian Federation and Ukraine.
UNCED	*UN Conference on Environment and Development.*
UNCTAD	*UN Conference on Trade and Development.*
UNEP	*UN Environment Programme.*
UN FCCC	*UN Framework Convention on Climate Change*, adopted in 1992 and entered into force in March 1994.
USIJI	*United States Initiative on Joint Implementation.*
WG	*Working group* (of the **IPCC**).
WMO	*World Meteorological Organization.*
WTO	*World Trade Organization.*

Summary and conclusions

The Kyoto Protocol to the UN Framework Convention on Climate Change (UN FCCC) represents a pinnacle of trends towards globalization in economic and environmental policy, and defines the basic structural elements upon which global efforts to tackle climate change in the twenty-first century will rest. Given extensive flexibilities in the agreement, the specific commitments of the Protocol are modest in terms of both environmental and economic impact: implementing the commitments themselves will neither halt global emissions growth, nor have a discernible impact on economic growth. Nevertheless they represent a fundamental change of course and a structure which, if ratified, implemented and expanded for subsequent periods, offers an effective international framework for tackling climate change. In several respects the Kyoto Protocol may prove to be the most profound and important global agreement of the late twentieth century.

Scientific basis and historical development

The Protocol rests upon approximately four decades of scientific efforts since the 1957 International Geophysical Year initiated the scientific networks of research and observation upon which theories of global warming have been refined. The First Assessment of the Intergovernmental Panel on Climate Change (IPCC), in 1990, provided sufficient consensus to underpin the UN FCCC, signed at Rio in 1992, which provides the legal and political foundations for international action. Emerging evidence of a 'discernible human influence' on global climate patterns aided the decision in 1995 to open new negotiations. The evidence for human-induced climate change with adverse impacts has continued to accumulate.

Subsequent US proposals for specific, binding quantified commitments, implemented with flexible mechanisms including international market instruments, defined the core structure of the Protocol, while pressure from the EU provided an impetus for stronger commitments than would otherwise have been possible. Renewed if fragile consensus among the developing countries to back strong industrialized-country commitments alongside measures to ameliorate adverse impacts removed the last hurdle in principle, but enormously divergent perceptions and interests remained. The Protocol was adopted in December 1997 under intense pressure from the deadline and desire for a meaningful agreement.

Specific industrialized-country commitments

The engine of the Protocol consists of specific limitations – assigned amounts (AA) – on the emissions of greenhouse gases from each industrialized country (listed in Annex I of the Convention) for the period 2008–12. The *aggregate* aim is to reduce these emissions to at least 5% below 1990 levels, which is roughly the level already achieved in 1995 due to reductions in the economies in transition (EITs). The headline commitments for all the OECD countries except possibly Australia would require strong action, particularly if applied to domestic CO_2 emissions in isolation. However, the collective Annex I goal is far weaker and the basic OECD commitments are modified by flexible mechanisms that serve both to weaken and redistribute the effort much more widely.

Source and sink flexibility is likely to mean that total industrial CO_2 emissions exceed the collective commitment by several percentage points (c. 5% with a wide margin of uncertainty):

- the inclusion of five other greenhouse gases (GHG), notably methane (CH_4), eases significantly the commitment as compared with CO_2 only for most industrialized countries other than Japan and Iceland; across Annex I these other gases account for about 20% of base year emissions;
- the inclusion of carbon 'sinks' – 'direct human-induced changes' in land use initially limited to 'afforestation, reforestation, and deforestation' – will be very important for some countries and may amount to a few per cent of Annex I emissions overall, within Annex I, depending on the detailed rules adopted.

International flexibility allows national emissions to diverge widely from the initial national commitments as a product of resource transfers and foreign investment. The three mechanisms for international emission transfers (joint implementation (JI) and emissions trading within Annex I, and the clean development mechanism (CDM) involving the developing countries) represent international market instruments that are unprecedented, controversial and poorly defined: much will hinge on specific rules to be developed through further negotiation. The different units will be fungible but subject to separate regimes of monitoring, certification, accountability etc. Another clause allows a group of Annex I Parties to redistribute their commitments (in a 'bubble') upon ratification; this was promoted for, and has been exercised by, the EU, and it is unlikely that other groups will use this provision in preference to trading.

Other aspects of the Protocol

The Protocol provides a menu of possible and mandated specific actions that will serve as hooks for future negotiations:

- a list of specific policies and measures will serve as a checklist for national reviews and peer pressure but is unlikely to gain a stronger role for some years;
- bunker fuels for international shipping and aviation, which are not included in national inventories, will come under rapidly increasing pressure for separate action;
- developing countries including oil-exporting countries will use key references in the Protocol to further debate on ways of minimizing any adverse impacts upon them.

Provisions concerning developing countries will support a significant expansion of their involvement in the process and capacity building, including further development of national inventories, programmes and reporting, and a new phase of negotiations on ways of promoting appropriate technology transfer. The loss of the proposed article on the adoption of voluntary commitments by developing countries – the *quid pro quo* for including emissions trading and the CDM – leaves the process by which new countries might be brought into the structure of specific commitments at least in the first period legally and procedurally unclear. This represents the only significant failure of core US objectives in the negotiations. In all other respects the structure of the Protocol is very much as sought by the US administration and represents a testament to the dominance of US power, and the continuing weakness of foreign policy in the EU, Japan and elsewhere, at the close of the twentieth century.

Quantitative implications of the international transfer mechanisms

The political and economic circumstances in which the Kyoto Protocol was negotiated led to some perverse emission allocations, and an overall pattern that may induce large-scale emissions trading, principally from the EITs to the United States, Canada and Japan. The allocations of Russia and Ukraine in particular could represent a surplus amounting to 5–10% of total Annex I assigned amounts. Transferring this surplus to other Parties would undermine the legitimacy of emissions trading and set dangerous precedents for the subsequent evolution of commitments and the trading system.

If emissions trading is wholly unrestricted and conducted purely to minimize costs in the first commitment period, the outcome could differ from the headline commitments to a far greater extent than the negotiators ever conceived. Under these conditions we estimate that energy-sector CO_2 emissions during the first commitment period could be:

- 25–30% above 1990 levels in Australia and some of the other smaller OECD countries;
- 10–15% above 1990 levels in the United States, Canada and Japan;
- around 1990 levels in the EU, with increases in most of the EU offset by reductions principally in Germany and the UK;
- 25–35% below 1990 levels in Russia and Ukraine, with wide variation across the other EITs.

These restrictions would not be trivial politically, but are clearly inadequate to lead efficiently towards atmospheric stabilization below a doubling of pre-industrial CO_2 levels, or to induce necessary technological changes. Indeed under some scenarios we find real resource abatement costs to be negligible, though Russia and Ukraine could still extract a price as effectively duopoly suppliers. The implied scale of West–East transfer of both emissions and finance would also raise potent political debate. Extensive use of sinks or the CDM would further reduce the level of effort required in Annex I, and would allow Annex I emissions to grow from the 5% reduction already achieved by 1995, violating the stated collective aim of the commitments.

If emission trends and further analysis do point towards such outcomes, debate will be reinforced about restricting the use of the transfer mechanisms. 'Supplementary constraints', to ensure adequate domestic action in countries that import assigned amounts, could take several forms. Quantified 'concrete ceilings' would disaggregate the market, and introduce opportunities for arbitrage and potential corruption across price differentials, and might not lead to any greater overall emission reductions. Efforts to make trading conditional upon specific policies would risk simply repeating the debates of the Protocol, with the same (ineffective) outcome. Other approaches could focus upon assessing the adequacy of domestic actions, either qualitatively, or quantified as a basis for determining allowed imports.

Excluding trading of surplus allocations ('hot air') would reduce national and aggregate emissions by several percentage points (as compared with no restrictions), and increase the value of the CDM (and even of trading for most parties within Annex I). Achieving this is politically and proce-durally problematic, but the most promising options could again be based on assessing the impact of domestic actions (this time in countries exporting assigned amounts), as an indicator for aggregate trading.

Excluding 'hot air' trading could also make room for the use of sinks and the CDM without violating the collective Annex I commitment, while shoring up the value of emission credits. Without any restrictions on Annex I trading, we find the value of CDM credits to be below \$5 billion annual-equivalent during the commitment period in most scenarios, and negligible in some.

Implementation architecture for international transfers

Emission credits from both the project mechanisms (ERUs and CERs, from Annex-I JI and the CDM respectively) must be certified *ex-post* on the basis of assessment project performance, but procedures for certain guarantees and credit estimation in advance will also be required. International trade considerations dictate that both ERUs and CERs must be fully tradable assets, and that early crediting, from shortly after the year 2000, should be available for JI projects within Annex I as well as under the CDM.

The project mechanisms necessarily and explicitly involve private entities. Emissions trading will also be most effective when conducted by private entities acting internationally. This requires either new legal procedures that transfer part of governmental obligations – assigned amounts – to private entities, or procedures for automating simultaneous transfers of private-sector permits alongside governmental assigned amounts.

Intergovernmental transfers must be conducted in units of assigned amounts, i.e. referred to the basket of six gases during the commitment period. Private-sector trading may be denominated in any subset of these gases and periods, and will probably first be enabled for CO_2 emissions from power generation and industrial processes.

To provide adequate incentives for compliance, units obtained from countries that do not comply with the system must be discounted accordingly – 'buyer liability' at the international level. Private entities will seek seller liability. New forms of insurance could make these objectives compatible by creating a market in insured emission units.

The possibility of applying charges to intergovernmental exchanges under all three of the international transfer mechanisms deserves serious consideration. Appropriate uses of revenues from such charges could include adaptation, regime implementation and technology development, and possibly also disaster relief.

Other aspects of implementation – monitoring, certification, etc. – are vital, and may take years to develop, particularly in less advanced Annex I countries.

The clean development mechanism

The CDM is the most unexpected and original of all Kyoto's innovations. It is a leap into *terra incognita* that is unlikely to work simply as a way to distribute abatement efforts globally at least cost. The difficulty of ensuring that crediting reflects real and additional emission reductions is compounded by the paradoxes that the most 'cost-effective' projects may be the least 'additional' and that strict project additionality would give perverse policy incentives. Other eligibility criteria will introduce several factors other than just greenhouse gas limitation.

Specific design of the CDM faces dilemmas of integrity and size. Rigorous rules and strong governance imply high administration costs, thereby deterring investment. Weak rules and governance may attract more investment at the risk of corrupting and discrediting the system, as well as undermining the criteria of sustainable development and the objectives of the Convention, since 'maximizing emission credits' under the CDM is equivalent to 'minimizing action within Annex I'. The combined lures of foreign investment and cheap emission reduction credits have created political pressures for rapid development of rules for the CDM that are potentially dangerous.

At best, the CDM could encourage large flows of foreign investment towards sustainable and low-greenhouse-gas development. At worst, it could become not just a source of spurious emission credits, but a sink for the intellectual as well as some physical resources of the developing world, and a distraction from the fundamental goals of sustainable development. However, for both good and bad, the CDM as generally conceived is likely to be constrained by the inherent limitations of credit-based, project-based systems, and for some countries it may be primarily a bridge towards adopting commitments and joining the international trading system.

However, the CDM may also evolve to perform a more fundamental and long-standing role. The defining purpose of the CDM should be to help direct foreign corporate investment towards goals of sustainable development in its many forms and interpretations, according in part to national preference. In this it should be complementary to the Global Environment Facility (GEF). Indeed, if appropriately designed, these institutions should correspond to the natural complementarity between the private and public sectors, at a global level.

Time scales and prospects for entry into force

The Protocol can legally enter force without US ratification but meaningful US domestic participation and adherence are essential to the regime. Although the US administration dominated design of the Kyoto Protocol, many in the US legislature remain relatively isolated from global realities and responsibilities, and reflect the deep resistance of the US body politic to emission restrictions. Domestic support for ratification will grow but continuing emission increases in the United States will make implementation more difficult and costly the longer it is deferred.

The Buenos Aires conference outcome focuses most key decisions before this upon the sixth Conference of Parties (COP), around the end of the year 2000. Many countries will seek ratification based on the outcome of this conference. The Protocol could enter into force during 2001–3 without the US, but serious action to stabilize US emissions or return them to former levels – the declared aim of the Convention commitments as ratified by the Bush administration – must be implemented during this period if the United States is subsequently to ratify the Kyoto Protocol. Implementation of existing Convention commitments and entry into force of the Protocol would plausibly establish leadership by the industrialized countries, and create a stronger basis for discussing subsequent global commitments.

Accession and expansion of commitments

Developing countries, as formally represented by the G77 and China, have firmly resisted efforts to open discussions on quantified emission commitments and blocked procedural routes to voluntary accession, fearing that individual countries would be pressured into joining by the might of the OECD. The status of countries seeking to join the trading system without undertaking full Annex I obligations remains unclear.

Assuming procedural issues can be addressed, accession of new countries individually is one route towards developing a global regime. This would tend further to inflate the trading system. This may be the price of developing an expanding regime; it could steadily undermine the G77, and maybe the Protocol itself.

Partly because of this, G77 opposition to any discussion of new commitments will ultimately become untenable. It offers no solution to the global problem, and no attractive alternative to those countries that may be tempted into the trading system. Especially if combined with meaningful industrialized-country implementation of Convention and Protocol commitments, this may force the G77 to adopt a more positive stance, predicated upon willingness to discuss quantified commitments if they are based upon an equitable distribution of allowed emissions, for example evolving towards comparable assigned amounts per capita. With the exception of those for Russia and Ukraine, almost all the current Protocol commitments already represent considerable per capita convergence.

Thus, one path may be for the Kyoto regime to evolve by accession and inflation, and loose development of allocation and trading rules to attract new entrants. This is as much as seems plausible for first period commitments. For subsequent periods an alternative might involve negotiations around principles of 'emissions convergence' towards more equitable long-term distribution of assigned amounts. A third approach could involve criteria for evolution of commitments based around threshold indicators but this may be less appealing for the G77. The scope for compromise between the different models is not obvious, but possibilities include the concept of a 'convergence corridor' towards per capita and perhaps other proposed indices of equity and performance.

Conclusions

The Kyoto Protocol is an extraordinary and unprecedented achievement in international affairs. Its inclusion of binding commitments with varied flexibility is more sophisticated than many political analysts considered possible. The many other elements – spanning global 'soft' commitments, related processes on technology transfer and financial mechanisms, policies and measures, minimization of adverse impacts, sinks, and compliance mechanisms – represent a complex package designed to gain and sustain global participation. The central structure of successive five-year commitment periods provides a natural basis for the evolution of the regime over time.

The transfer mechanisms potentially establish new markets based upon resources (emission credits and assigned amounts) that are purely human constructs – harnessing the greed of nations, and their industries, in the cause of climate protection. The incentives to expand ownership and access to these resources will be immense. Lax rules and wide scope for JI and the CDM would maximize the potential credit generation in EITs and developing countries respectively, but lessen the value of credits and action within the OECD. The major OECD importers may seek permissive rules on both the project mechanisms and trading, so as to maximize their access to cheap emission credits and to ease the struggle for ratification that may be long and difficult in some countries. However, the net result of all this would be massive inflation – resulting neither in significant resource transfers to anyone, nor in significant progress in tackling climate change. This may be restrained by competition between the mechanisms and regions, and by injecting a firmer sense of the overall constraints on emissions. Implementing the Kyoto Protocol without reference to long-term goals and trajectories would be like inviting Treasuries to print money without any conception of inflation.

Almost everything about the governance of these mechanisms remains to be determined. We have yet to develop a clear understanding, let alone consensus, about the international social, ethical and legal foundations upon which such instruments need to rest. Other aspects of the Protocol, including its provisions on policies and measures, minimization of adverse impacts, technology transfer, and compliance procedures, all require much further development to become effective.

Yet the Protocol has already made two fundamental achievements. First, its adoption has persuaded the private sector that the world will indeed start getting to grips with the problem of climate change, however slowly and ponderously, and indeed that there may be varied benefits accruing to those who move first. Second, it has advanced the debate on international economic instruments from whether to adopt them to how to implement them. The Kyoto Protocol could yet be destroyed: from within, by excessive national greed leading to uncontrollable inflation; or from without, if those who do not want any agreement, acting together with those for whom Kyoto is not good enough, can block ratification in key countries. Yet this is unlikely. Governments have already made major political investments to establish this nascent regime. It offers a solid basis, and there are no credible alternatives on offer. The world has started down an ambitious and efficient road to tackle its most daunting environmental problem, and there is no turning back.

Part I

The making of the Protocol

Chapter 1

Analytic foundations:
science, response options and the IPCC

The climate is changing. Ten years ago, that was a very controversial claim. Now it is widely accepted that changes are being observed, that human emissions of greenhouse gases are a cause, and that such emissions will lead to far greater climatic changes during the coming century. In 1990, governments began to negotiate on what to do about it, and in the mid-1990s they started to debate more substantive commitments.

The result of the frenzied international negotiations that followed was the Kyoto Protocol, which sets specific binding targets for emissions of greenhouse gases from industrialized countries, together with an array of complex mechanisms to give flexibility in how they are implemented and to assist global efforts towards more sustainable development.

The negotiation of the Kyoto Protocol followed decades of scientific research and many years of economic and policy analysis. This first chapter sets out that background, and summarizes more recent scientific developments. The other chapters in Part I of this book describe the political background and negotiations themselves, and the Protocol that emerged from them. Part II then examines the import of what was agreed.

1.1 Origins and the Intergovernmental Panel on Climate Change

The 'greenhouse effect' is not a new concern. As early as 1827 the French scientist Fourier suggested that the earth's atmosphere warms the surface by letting through high-energy solar radiation but trapping part of the longer-wave heat radiation coming back from the surface. This is caused by a number of 'greenhouse gases', notably carbon dioxide and water vapour. At the end of the nineteenth century the Swedish scientist Arrhenius postulated that the growing volume of carbon dioxide emitted by the factories of the Industrial Revolution was changing the composition of the atmosphere, increasing the proportion of greenhouse gases, and that this would cause the earth's surface temperature to rise.

The subject attracted little interest until the late 1950s, but in 1957 the International Geophysical Year provided the foundations for a global scientific community dedicated to understanding planetary processes and human influence on them, and established a network of monitoring stations. Observations immediately began to trace a steady rise in the concentration of carbon dioxide. A decade later, a study by the Massachusetts Institute of Technology (MIT) documented concerns about possible climate change, and by 1970 the Secretary General of the United Nations was sufficiently concerned to mention the possibility of a 'catastrophic warming effect' in his report on the environment.

The first World Climate Conference, in 1979, established the World Climate Research Programme that helped to stimulate and focus further research. During the 1980s the UN Environment Programme (UNEP) and the World Meteorological Organization (WMO) convened a series of international scientific workshops around which coalesced a tentative scientific consensus on the nature of the problem. Driven also by rising popular concern about environmental issues, during 1988 a series of international meetings culminated in the establishment of the Intergovernmental Panel on Climate Change (IPCC), under the auspices of UNEP and the WMO. Although led initially mostly by the industrialized countries, all governments were invited to join and the IPCC has expanded over subsequent years to almost global participation. It is the reports of the Panel that provide the scientific underpinning for the diplomatic processes of the UN Framework Convention on Climate Change.[1]

The purpose of the IPCC is to provide authoritative assessments to governments of the state of knowledge concerning climate change. In the ten years since it was established, it has evolved into what is probably the most extensive and carefully constructed intergovernmental advisory process ever known in international relations. Governments nominate

---

[1] An analysis of the history and evolution of the IPCC is given in Shardul Agrawala, 'Context and Early Origins of the Intergovernmental Panel on Climate Change', *Climate Change*, 39, 1998, 605–20; and Shardul Agrawala, 'Structural and process history of the Intergovernmental Panel on Climate Change', in *Climate Change*, 39, 1998, 621–42.

experts with an established record of research and publication on relevant topics. From these, writing teams within each of three working groups are selected by the bureau of governmental representatives for their level and breadth of expertise, range of views and geographical balance. The texts they produce are subjected to widespread peer review involving hundreds of individuals and groups world-wide.

Each working group produces a Summary for Policymakers, and the IPCC plenary meeting agrees a Synthesis Report. Since this represents a statement of what governments officially accept as a balanced account of the state of knowledge and reasoned judgment, its precise wording is subject to intensive negotiation between governmental delegations. The full assessment reports remain, however, the responsibility of the appointed authors, and the IPCC itself is precluded from making policy recommendations. Its purpose is to establish the basis of internationally accepted knowledge upon which other forums can base their negotiations and conclusions.

There are many guides to the science of climate change, including a book by the chairman of the scientific working group of the IPCC, Sir John Houghton.[2] No attempt is made here to duplicate discussion of the various scientific debates. Rather, the main part of this chapter summarizes the findings of the IPCC that have been accepted by governments world-wide as the analytic basis upon which rest the world's efforts to address climate change. The final sections outline some of the core debates and criticisms, and more recent developments.

1.2 Scientific foundations and the First Assessment Report of the IPCC

The IPCC produced its first report in 1990. The findings of *Working Group I* (science) probably had the greatest impact, with its key conclusion that rising concentrations of carbon dioxide and other greenhouse gases in the atmosphere were caused by human activities and would cause global temperatures to rise, with accompanying climatic changes. As the IPCC sought to explain, certain scientific fundamentals are undisputed. Greenhouse gases trap heat near the earth's surface: the planet would be

---

[2] J.T. Houghton, *Global Warming: The Complete Briefing*, Oxford: Lion Publishing, 1996.

much colder without them. The concentration of greenhouse gases is rising, due primarily to human activities. That increase is likely to lead to general warming of the earth's surface, though with important regional variations. These are the simple fundamentals on which all could agree.

Beyond that it gets more complex, principally because of the huge variety of additional influences upon the global atmosphere and temperature. These include natural feedbacks – changing global temperature itself changes the water cycle and associated cloud and ice cover, for example – and manifold extraneous influences, both natural (ocean current oscillations, sunspot changes) and human (sulphur and other emissions, and land-use changes). Disentangling these influences to predict the magnitude and consequences of human-induced climate change is a challenge of an altogether different order.

Nevertheless the IPCC, in its First Assessment Report conducted in little over a year, offered the central estimate that if greenhouse gas emissions continued to rise as projected, global average temperature would rise at around 0.3°C (±0.15°) per decade. This would be the fastest sustained, global rate seen for at least the past 10,000 years, and compares with a global average temperature difference of only 4–7°C between now and the last ice age. After a century or so, such warming would take the earth to temperatures beyond even the levels reached in the warm period before the last ice age, more than 100,000 years ago. The magnitude of the global warming seen to date was considered not inconsistent with predictions of climate models, but was rather lower and within the range of natural climate variability, so could not clearly be attributed to human activity.

Working Group II, which addressed the impacts of climate change, found widespread uncertainty and disagreement in its area. Sea-level rise (due primarily to thermal expansion of the oceans) and changes in rainfall patterns were expected to be major effects. The consequences for agriculture, wetlands, forest, coastlines and desertification could be significant, but the continuing high level of uncertainty, in particular about how climate would change locally, made it impossible to reach firm conclusions.

Working Group III, which reviewed potential responses to climate change, experienced substantial political disagreement over its deliberations. In the absence of any other policy forum, its report was subject to intense and highly politicized negotiations, in an atmosphere of widely diverging views over this relatively new issue. The outcome was a carefully hedged report whose one concrete recommendation was to start negotiations on a global agreement on climate change.

After considerable debate the assessment and recommendations were passed up to and accepted at the Second World Climate Conference in November 1990. The ministerial segment of that conference accepted the report and called upon the UN to open negotiations on what was to become the UN Framework Convention on Climate Change (UN FCCC) (explained in Chapter 2).

Meanwhile, the IPCC embarked upon an update of specific aspects of the science and emission scenarios, published in 1992. Its plenary meeting in November 1992 reorganized the Panel to reflect the new context, including a recognition that the IPCC would have input into the Convention process, and an acceptance (albeit reluctant by some parties) that it could be allowed to examine issues of international economics. The meeting launched the IPCC towards producing a set of special reports in 1994 and, in parallel, the full Second Assessment Report, finally accepted in late 1995 and published in June 1996.

The Second Assessment Report (SAR) marked a crucial stage in the progress of global action to combat climate change. It is this second report, more than any other, that set the context for the negotiation of the Kyoto Protocol. Until the Panel produces its third assessment report (due in 2001), the SAR remains the most authoritative summary of the science of climate change.

1.3 Findings of the IPCC Second Assessment Report

The report's main conclusions, as accepted by governments, were that:

- greenhouse gas concentrations have continued to increase as a result of human activities;
- global average temperature (and sea level) has risen, and recent years have been among the warmest since at least 1860;
- the ability of climate models to simulate observed events and trends has improved;

- the balance of evidence suggests a discernible human influence on global climate;
- on central emission projections, by the end of the twenty-first century global mean surface temperature is likely to rise by about 2°C, with a range of uncertainty of 1–3.5°C, and to continue rising for some decades thereafter even if greenhouse gas concentrations are stabilized by then;
- sea level would rise, with a mid-range estimate of 50 cm by 2100 (range 15–95 cm), and would continue rising for centuries thereafter;
- significant 'no regrets' opportunities are available in most countries to limit emissions of greenhouse gases (below levels that would otherwise be achieved) at no net cost;
- the potential risk of damage from climate change is enough to justify action beyond such 'no regrets' measures.

These headline findings were supported by almost 2,000 pages of analysis. The remainder of this section explains the key findings and provides a brief guide to some of the underlying debates in the IPCC and their implications.

1.3.1 The science of climate change

The science report[3] confirms that human activities are changing the atmospheric concentrations and distributions of greenhouse gases.

- *Carbon dioxide* had increased in concentration by nearly 30% from about 280 parts per million volume (ppmv) in pre-industrial times to 358 ppmv by 1994. It is the main anthropogenic greenhouse gas, pro-jected to account for about 70% of the 'radiative forcing' of climate over the next century. The 'carbon cycle' is complex: some emissions are reabsorbed rapidly but much remains in the atmosphere for a hundred years or more. The main anthropogenic sources are fossil fuel com-bustion and land conversion, e.g. deforestation.

---

[3] J.T. Houghton et al. (eds), *Climate Change 1995: The Science of Climate Change*, Cambridge: Cambridge University Press, 1996.

- *Methane* has more than doubled in concentration and on most criteria is the second most important anthropogenic greenhouse gas, but has a much shorter atmospheric lifetime than carbon dioxide. Main anthropogenic sources are agriculture, waste disposal, fossil fuel production and use; natural emissions from wetlands account for about 20% of the total methane emissions.
- *Nitrous oxide,* which has increased in concentration by 15%, is also significant and has a lifetime comparable with carbon dioxide. Main anthropogenic sources are agriculture and industry; natural sources are probably twice as large.
- *Halocarbons* (industrial chemicals such as chlorofluorocarbons (CFCs)) have increased rapidly but concentrations of key ozone-destroying gases such as CFC-12 are now scheduled to fall following their control under the Montreal Protocol. Some non-ozone-depleting CFC replacements, however, are greenhouse gases and may have a significant impact on climate change if their use grows.

Several other gases are involved, some indirectly through their impacts on atmospheric chemistry. One of the most important developments in the SAR was the improved understanding of the role of aerosols (particles, dust and very small droplets) that are present in the atmosphere as a result of processes both natural (e.g. dust storms, volcanic activity) and human (e.g. fossil fuel and biomass burning). By scattering and absorbing radiation, aerosols cool the surface regionally (e.g. around industrial areas and deserts). Improved modelling, particularly of sulphate aerosols resulting from fossil fuel combustion, has led to a much better fit between model predictions and the evidence of climate change (see Figure 1.1). Action taken to limit production of aerosols – such as the US 1990 Clean Air Act Amendments and the 1994 Second Sulphur Protocol in force across Europe – is expected to reduce these emissions (and hence their cooling effect, since they are removed from the atmosphere very quickly), and thereby uncover stronger signals of global warming. [4]

---

[4] The Second Sulphur Protocol is the 'Oslo Protocol' to the Convention on Long-Range Transboundary Air Pollution, UN Economic Commission for Europe, Geneva, 1979. The Protocol establishes differentiated commitments for deep cuts in sulphur emissions.

Figure 1.1: Observed and modelled temperature trends, with and without aerosol corrections

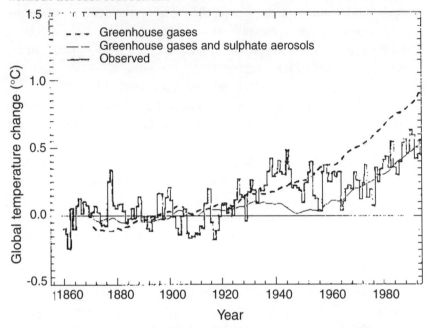

Source: J. T. Houghton et al., *Climate Change 1995: The Science of Climate Change*, Cambridge: Cambridge University Press, 1996, p. 33.

Most scenarios project both emissions and concentrations of all greenhouse gases to continue to rise steadily in the absence of controls. Even if global emissions of carbon dioxide were stabilized at current levels, its atmospheric concentration would still continue to grow; stabilization of *concentrations* would require emissions ultimately to drop well below current levels, particularly for the long-lived gases. Notably, global CO_2 emissions would have to be more than halved from current levels to stabilize concentrations, compared with projected doubling of global emissions

---

Despite its name, the UN-ECE region spans North America and the former Soviet Union as well as Europe itself, and signatories to the Oslo Protocol include Canada and Russia as well as the EU and other European countries. The United States adopted big reductions under the Clean Air Act Amendments. Other regional sulphur agreements are under consideration.

over the next few decades in the absence of controls. This underlines the sheer scale of the policy challenge.

The central issue of climate change is the extent to which growth in emissions of greenhouse gases affects climate. The IPCC had no difficulty agreeing that climate change could be *detected*:

- Global mean surface temperature has increased by 0.3–0.6°C since the late nineteenth century;
- Recent years have been among the warmest since 1860, despite the cooling effect of aerosols (including exceptional volcanic eruptions in the 1990s);
- Global sea level has risen by 10–25 cm since the late nineteenth century, and it is likely that much of the rise has been temperature-related;
- There have been changes in extreme weather events (heavy rainfall, hurricanes, etc.) in certain regions this century, though their statistical significance and relationship to wider climate change are still debated.

The core dispute was over whether such changes could be *attributed* to human activities. In what politically was the most important finding, the scientists concluded that 'the observed warming trend is unlikely to be entirely natural in origin'. The inclusion of aerosols and improved modelling (including of natural variability), and better data on the distribution of temperature changes, brought simulated changes in various dimensions much closer to those observed, and on the basis of such 'pattern matching' studies governments accepted that 'the balance of evidence suggests that there is a discernible human influence on global climate'.

For the mid-range scenarios developed by the IPCC, the models predict an increase in global mean surface air temperature of about 2°C by the year 2100 (in addition to what may have been induced so far), with a range of uncertainty of 1–3.5°C. For any figure within this range, the average rate of warming would be higher than any seen in the last 10,000 years. Even if stabilization of greenhouse gas concentrations were achieved (which is *not* assumed under these scenarios), temperature would continue to increase beyond this point because of the thermal inertia of the oceans; assuming stabilization in 2100, only 50–90% of the eventual temperature change would have occurred by then. Sea level would rise, with a mid-range estimate of 50 cm by 2100 (range 15–95 cm), and would continue rising for centuries thereafter.

The report was equivocal about the nature of climate change in a number of other areas of popular concern. One area was storm intensity, but the IPCC found 'little agreement between models on the changes in storminess that might occur in a warmer world'. Another fear was whether there was any chance of the West Antarctic ice sheet collapsing – raising sea levels by several metres. On this the IPCC said that 'recent theoretical work is equivocal' but the tone was reassuring. There was somewhat more confidence concerning daily temperature *ranges* (expected to decline) and some hydrological (water resource) impacts: warmer temperatures 'will lead to a more vigorous hydrological cycle' with various changes possible, of which one of the few relatively consistent ones is that 'most models produce a drier surface in summer in northern mid-latitudes'. Potentially the most disturbing findings concerned the big ocean currents (such as the Gulf Stream), which underlie many existing climatic patterns, and which evidence suggests would reduce in strength in the North Atlantic and 'could become less stable and more variable'.[5]

1.3.2 Impacts and adaptation

The IPCC's Working Group II was responsible for reviewing the state of knowledge concerning the impacts of climate change on physical and eco-logical systems, human health and socioeconomic sectors, and the technical and economic feasibility of a range of potential adaptation and mitigation strategies.[6]

The studies suggested extensive shifts in the composition and geographic distribution of ecosystems as individual species respond to the changing climate. Consequences would include:

---

[5] Houghton et al. (eds), *Climate Change 1995*, p. 213.

[6] Robert T. Watson et al. (eds), *Climate Change 1995: Impacts, Adaptations and Mitigation of Climate Change: Scientific-Technical Analyses*, Cambridge: Cambridge University Press, 1996.

- reduced biodiversity;
- major changes in the composition of an average one-third of forested areas;
- altered growing seasons and boundary shifts between grasslands, forests and shrublands;
- an increase in the temperature of deserts, and an increasing likelihood of irreversible desertification;
- the disappearance of one-third to one-half of existing glacier mass (with accompanying changes to snow and ice cover and river flow);
- changes in the productivity of lakes and rivers, with increases in flow variability and alterations in the distribution of wetlands;
- greater coastal erosion and flooding, with altered tidal ranges in rivers and bays;
- and changes in ocean circulation and behaviour and sea-ice cover altering biological productivity, with major impacts on marine ecosystems and heat and carbon storage capacity.

The report suggested that changes in the hydrological cycle could be of particular concern in regions where water availability is already low, such as the Middle East, North and East Africa and Central Asia. Global agricultural production was expected to be maintained (relative to projections in the absence of climate change), but with considerable regional changes in crop yields and productivity, particularly in the tropics and subtropics. Such variations could have serious consequences, since many of the world's poorest people, who are already vulnerable to agricultural failure, live in the areas which are most at risk and have the least ability to adapt or to import food. Global timber supplies may also be adversely affected by climate change.

Human infrastructure – settlements, industry, transport, etc. – tends to be less vulnerable to climate change than agricultural or natural ecosystems, but is most susceptible to sudden changes and increased frequency or intensity of extreme events. Some 46 million people are currently at risk of flooding owing to storm surges; a 50 cm sea-level rise would increase this number to about 92 million, and population growth would increase it even more. The impact of a 100 cm rise (expected during the twenty-second century on the central projections) was estimated to include land losses of 6% for the Netherlands, 17.5% for Bangladesh, and 80% for the Majuro Atoll in the Marshall Islands, with many nations, particularly small island states, losing capital value greater than 10% of GDP. Connected impacts include an increase in forced migration, and significant stress on the property insurance industry.

Wide-ranging and mostly adverse effects on human health, with significant loss of life, were also predicted. Direct effects include increases in mortality and illness owing to heat waves (partially offset by reduced cold-weather deaths), and death and injury from extreme weather events. Indirect effects, likely to be more important in the long term, include increases in the potential transmissions of vector-borne infectious diseases such as malaria, dengue and yellow fever; projections for the upper part of the IPCC-projected temperature range indicate that the geographical zone of potential malaria transmission would increase from about 45% of world population to about 60%. Reductions in supplies of fresh water could also have health consequences.

1.3.3 Analysis of responses in the Second Assessment Report

The IPCC is best known for its work on scientific issues, but the SAR also addressed response options. In this area, Working Group II examined a range of technologies to reduce emissions and enhance sinks of greenhouse gases, and concluded that significant reductions in net greenhouse gas emissions are technically possible and economically feasible. By the year 2100, the world's commercial energy equipment will in effect have been replaced at least twice, offering opportunities to change the energy system without premature retirement of capital stock. Energy efficiency gains of 10–30% above present levels are feasible at little or no net cost in many parts of the world; 50–60% gains are technically feasible. Fuel switching, from coal and oil to gas, and from fossil fuels to nuclear (if concerns over safety, waste disposal and proliferation are met) and renewable sources, will also reduce carbon dioxide emissions.

One of the chapters attempted to integrate this information by developing 'low emitting supply scenarios' showing that steady global emission reductions were possible – down from 6,000 million tonnes of carbon (6GtC) in 1990 to 4GtC in 2050 and 2GtC by 2100 – and that this might not cost much.[7] This depended, however, upon a number of assumptions, including technology developments over the next couple of decades. Biomass energy played an important role in these projections. The management of forests, agricultural lands and rangelands can also play an important role in reducing greenhouse gas emissions and enhancing carbon sinks.

A rather different approach was taken by the IPCC's Working Group III, which assessed the socioeconomics of impacts, adaptation and mitigation of climate change.[8] In practice the writing teams and the report were dominated by economics, with one chapter on equity and social aspects. Developing countries had expressed great hesitation about allowing the IPCC to address such potentially value-laden topics, and producing the report indeed proved a contentious task.

The working group concurred – hesitantly in the case of some economists[9] – that there are significant opportunities available in most countries for limiting greenhouse gas emissions at no net cost – the so-called 'no regrets' measures. These can follow from the lowering of market barriers and economic distortions (for example, through regulatory changes and reduced fossil fuel subsidies), and the reduction in other adverse impacts associated with CO_2 emissions, such as other polluting emissions, or traffic congestion

The working group also concluded that action beyond 'no regrets' measures is economically justified owing to 'the risk of aggregate net damage from climate change, consideration of risk aversion, and application of the precautionary principle'.

Many policy options are available and a prudent approach would use a portfolio of actions aimed at mitigation, adaptation and improving knowledge

---

[7] In the convoluted language finally negotiated, 'within the wide range of future energy prices, one or more of the variants could plausibly provide the demanded energy services at costs that are approximately the same as estimated future cost of current conventional energy'. Ibid.

[8] James P. Bruce et al. (eds), *Climate Change 1995: Economic and Social Dimensions of Climate Change*, Cambridge: Cambridge University Press, 1996.

[9] See discussion of 'bottom-up' and 'top-down' modelling, Appendix 2.

Box 1.1: A portfolio of actions

'A portfolio of actions ... that Policymakers could consider ... to implement low-cost and/or cost-effective measures'

- implementing energy efficiency measures including removing institutional barriers ...
- phasing out distortionary policies ... such as some subsidies and regulations, non-internalization of environmental costs, and distortions in transport pricing
- cost-effective fuel-switching measures ... such as renewables
- enhance sinks or reservoirs ... such as improving forest management and land use practices
- ... implementing measures and developing new techniques for reducing ... other greenhouse gas emissions
- encouraging forms of international cooperation ... such as coordinated carbon/energy taxes, actions implemented jointly (AIJ), and tradable quotas
- ... promoting ... national and international energy efficiency standards
- ... planning and implementation measures to adapt ...
- research aimed at better understanding the causes and impacts of, and adaptation to, climate change
- conduct technological research ... minimizing emissions ... and developing commercial non-fossil energy sources
- improved institutional arrangements, such as improved insurance arrangements, to share the risks of damages ...
- promoting voluntary actions to reduce greenhouse gas emissions
- education and training, information and advisory measures for sustainable development and consumption patterns ...

Source: IPCC Working Group III Policymakers' Summary

(see Box 1.1). An intensive debate on the costs and benefits of deferring abatement action resulted in acknowledgment that 'earlier mitigation action may increase flexibility in moving towards stabilization' and delaying responses was itself acknowledged to be a decision involving costs.[10]

The working group found that estimates of the cost of limiting carbon dioxide emissions from fossil fuels vary very widely, depending on such factors as choice of methodologies, underlying assumptions and policy instrument. 'Top-down' macroeconomic models simulate aggregate relationships such as those observed between energy consumption and GDP; these

---

[10] The report agreed that 'the choice of abatement paths involves balancing the economic risks of rapid abatement now (that premature capital stock retirement will later be proved unnecessary) against the corresponding risk of delay (that more rapid reduction will then be required, necessitating premature retirement of future capital stock)'. See Appendix 2 for more detail on these debates.

studies suggested that the cost of stabilizing carbon dioxide emissions at 1990 levels in the OECD over the next few decades lies between -0.5% and +2.0% of GDP (equivalent to a range between a gain of $60 billion and a loss of $240 billion) annually. 'Bottom-up' models focus on specific energy-using sectors of the economy and the scope for using better and more cost-effective technologies; these studies suggested that emissions might be reduced 10–30% below business-as-usual (BAU) trends within 20 to 30 years at no or negative net cost. In either case, flexible, cost-effective policies relying on economic incentives, as well as coordinated instruments, can considerably reduce mitigation or adaptation costs. A wide range of policy instruments is available, including carbon taxes, tradable permits, deposit-refund systems, subsidies, technology standards, performance standards, product bans, direct government investment and voluntary agreements.

Equity was acknowledged in the report as an important element in responses, both in its own right and as an essential requirement in gaining adherence to an international agreement. Various ethical bases in the literature were acknowledged, but the Summary for Policymakers expressed caution about the 'application to relations among states of principles originally developed to guide individual behaviour'. Nevertheless it was agreed that developing countries need support to enable them to participate fairly and effectively in climate change decision-making and implementation, and that the costs of climate change might well be borne inequitably unless this were addressed explicitly. Most proposals for a 'fair' distribution of emissions were found to centre on the idea of either equal per capita allocations or allocations based on incremental changes to national baseline emissions (current or projected). All the academics and philosophers in the world, however, could not have prescribed – or predicted – the distributions that were to emerge from the Kyoto Protocol (Chapter 4).

1.4 Key debates and implications of the Second Assessment Report

Not surprisingly, given the import of the Second Assessment Report, its evolution was marked by several important disputes. The most important and contested finding was that a 'discernible human influence' on climate could now be detected. The suggestion that human-induced climate change was being observed moves the issue from one of trusting scientific modelling projections to one of tangible evidence that the process is already and identifiably under way. As such, it had huge political impact; following reports of the draft findings in summer 1995, climate change received front-page coverage in the leading US newspapers for the first time for three years.

The conclusion was primarily based not upon observed global warming, but rather upon a collection of observed *trends* set against the much improved understanding and modelling of influences such as aerosols. The single most decisive evidence was based on a comparison of the *distribution* of temperature change in the atmosphere, vertically and across the surface, derived from models that include both carbon dioxide and aerosol effects. Other evidence was derived from comparisons with (disputed) estimates of natural variability, and patterns in the seasonal length and diurnal temperature ranges.

The conclusion that climate change was now being observed provoked vigorous discussion at the plenary meeting of Working Group I (Madrid, November 1995), which eventually agreed to the form of words noted above. However, some parts of the draft were loosely worded, and different parts carried different emphases on the degree of confidence and qualifications. This led to a major row when an industry group, the Global Climate Coalition, charged that the final editing had been politically influenced. The Working Group's chairman pointed out that the IPCC had been specifically requested to clarify the chapter's wording in the light of the discussions in Madrid; 40 scientific authors and contributors wrote in defence of the editorial changes, expressing full agreement with the chapter as published.[11]

In fact, the chapter itself remained confusingly opaque about the degree of confidence to be attached to human attribution, and exactly how far the world has gone along this road is a question of judgment and semantics.[12]

---

[11] For a detailed account see Chris Cragg, *Financial Times Energy Economist*, June 1996.
[12] The published concluding section noted that 'detection of human-induced change in the Earth's climate will be an evolutionary and not a revolutionary process. It is the gradual accumulation of evidence that will implicate anthropogenic emissions ... whilst there is

But this and the public squabble about editorial changes should not be allowed to obscure two key points. The first is that government representatives found the evidence presented in the draft chapter and authors' presentations sufficiently convincing to agree the wording presented in the Summary for Policymakers. Second, that key scientists were united in support of the chapter as published and the conclusions drawn from it – and fully expected their tentative conclusion to be confirmed by the steady accumulation of evidence. Subsequent developments, summarized below, suggest that they were right.

Though overshadowed by this debate, critical commentary greeted some other parts of the science report. Notably, criticism continued of the scenarios used for exploring future emissions, particularly of carbon and sulphur, with the World Energy Council repeating its complaint that data on emissions of both were implausibly constructed and (in the central scenarios) unrealistically high.[13] Others, however, warned that the central scenarios of CO_2 emissions might be too low. One result of this was a major (and ongoing) effort to develop a new set of emission scenarios in preparation for the IPCC's Third Assessment Report.

Greater controversy surrounded the chapter on energy supply options in Working Group II's report, which explored various technologies and scenarios for low-emitting energy systems. Several government representatives, and others, expressed scepticism over its conclusion that energy systems could be developed that would lead to steadily reducing global carbon dioxide emissions perhaps at little cost (see above). Furthermore, many critics were sceptical about the huge role given to biomass energy in some of the scenarios. Eventually, negotiations between government representatives and the lead authors resulted in the agreed wording, but many governments and others remained unconvinced by the chapter's optimistic tone and conclusions.

---

already initial evidence ... it is likely (if model predictions are correct) that this signal will emerge more and more convincingly with time ... any initial pronouncement will be questioned by some scientists ... and few would be willing to argue that *completely unambiguous* attribution ... has already occurred ... [but] the body of statistical evidence now points towards a discernible human influence on global climate.'

[13] *Climate Change 1995*, World Energy Council Report No. 5, London, March 1996.

Working Group III also produced several controversies. The most unpleasant concerned the economic valuation of the impacts of climate change – which necessarily subsumed the question of how to value impacts on human life. The available literature predominantly uses an approach that reflects 'willingness to pay' to avoid unwanted impacts. This is necessarily much lower in poorer countries, and developing-country delegates reacted furiously to the apparent implication that the lives of their citizens were somehow 'worth less' than those of the rich world. The economists involved rejected as inconsistent the proposition that risk of death should be valued equally between countries for climate damages, when it is so obviously not for other issues.[14] Delegates also expressed scepticism at the apparent confidence of the economic calculus, given the huge scientific and other uncertainties. In the end, governments accepted the chapter but amended the Summary for Policymakers into an ungainly text which implicitly criticized the underlying chapter – to the anger of the authors who dissociated themselves from it. The core finding that the impacts of projected climate change (for a 2°C rise) might equate to 'a few per cent' of global GNP was noticeably more vague and hedged than in the underlying chapter.

Despite the attention that this debate gained, other aspects of the economics report were probably more important. Governments accepted without great difficulty the significant potential for 'no regrets' measures and – with a little more debate – that action beyond this was justified. Efforts to quantify the costs of emissions abatement fell foul of the long-standing division, referred to above, between 'bottom-up' and 'top down' models; although the chapter marked some progress, the divide was never really bridged and reported results separately for each approach, while implicitly eliminating some of the most extreme economic claims. The SAR confirmed the existence of a huge range of options for limiting emissions, and advanced but did not resolve the economic debate on the real costs of implementing these options – rejecting the extreme claims that abatement would cripple economies, while also casting economic

---

[14] See Appendix 2. An overview of this and some related debates on the interface between economics and equity, some of which surfaced at various points in the WG-III debates, is given in M. Grubb, 'Seeking Fair Weather', *International Affairs*, Vol. 71, No. 3, July 1995.

scepticism on technologists' assertions that the problem can be fixed for free. These and other economic debates are considered more fully in Appendix 2 of this book.

Producing a text able to be approved by the wide range of governments represented in the IPCC was thus a difficult and politically contentious task. It resulted, however, in an intergovernmentally approved assessment that formed the key text setting the background for the negotiations on international agreement to limit emissions of greenhouse gases leading up to Kyoto. The conclusion of Working Group III – that responses should comprise a wide portfolio of actions including further research, preparation for adapting, and some degree of mitigation beyond 'no regrets' – stood as the cornerstone of a rational response. Quite *what* it implied for policy in detail depended on the judgment of policy-makers, industrialists and commentators on the report, and wider policy debate.

1.5 Developments after the Second Assessment Report

In the aftermath of the Second Assessment Report, of course, scientific debate and developments continued apace on many fronts.

In terms of basic trends, global temperatures continued to climb. Ground-based measurements showed that 1997 was the warmest year ever recorded – until the data for 1998 were collected. Stimulated in part by the unprecedented intensity of the El Niño change in Pacific Ocean currents, global temperatures in 1998 climbed further, reaching close to a full degree Celsius above the pre-industrial average (Figure 1.2). The reversal of El Niño towards the end of 1998 is expected to lead to lower global temperature in 1999, but scientists are now in no doubt about the underlying trend.

One reason for that confidence is the apparent resolution of the long-standing discrepancy between ground-based and satellite-based measurements, the latter having shown little sign of global warming. In 1998 it was shown that drift in the satellite orbits might explain much (though not all) of the discrepancy. Probably more important, however, the analysis pointed to the great technical complexity of deriving precise temperature estimates from satellite measurements due to the number of complicating factors and relative recency of the techniques.

Figure 1.2: Global average surface temperature, 1860–1998

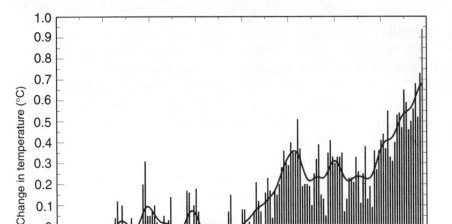

Source: Hadley Centre, *Climate Change and its Impacts,* Bracknell: UK Meteorological Office, 1998.
Note: The bars show the annual global average surface temperature; the smoothed curve represents the 5-year average.

There were important developments also in the debate about the confidence with which observed changes could be attributed to greenhouse gas emissions. Work on aerosols continued to improve understanding of the likely impact of sulphur, and highlighted that other aerosols, with different geographical distributions, probably also played a larger role than previously suspected. Climate models had also improved substantially. In particular, doubts had been raised about the fact that most climate models needed some arbitrary 'flux corrections' to enable them correctly to simulate present temperatures and prevent temperature drift in future simulations. With a much improved representation of ocean and atmospheric processes, especially higher ocean resolution and cloud feedbacks, several models (including the most recent release of the UK Meteorological

Office's Hadley Centre model) were able to simulate temperatures stably without any flux corrections. This increased confidence that they now captured the most important dynamics of energy exchange between atmosphere, land and oceans, and hence increased confidence in their predictions.

Putting these developments together also considerably improved simulation of recent global climate change, including the impact of greenhouse gases. The Hadley Centre concluded that 'comparison of model simulations and observations, based on new statistical techniques, indicates that human-made greenhouse gases have contributed substantially to global warming over the past 50 years, and this gives us confidence in prediction of the future'. Their revised model predicts that the business-as-usual emissions projected in the IPCC's central scenario would result in further warming of about 3°C over the next 100 years – slightly more than the central estimate in the IPCC's Second Assessment Report. On a more positive note, the model projects some slow-down, but no collapse, of the Gulf Stream that determines weather patterns in Europe and the eastern United States and Canada – though this continues to remain a point of divergence between the models.

In the late 1990s, therefore, the mainstream debate surrounding the science of climate change is moving beyond the question of whether human greenhouse gas emissions have caused and will cause the climate to change: most scientists would agree on both counts. The scientific question emerging with ever greater force concerns the likely impacts of such change.

That question has been given ever greater urgency by the series of weather-related disasters in the late 1990s. Unprecedented drought in Indonesia and some other parts of Southeast Asia led to agricultural and economic disruption well before the resulting Indonesian forest fires grabbed world headlines. On the other side of the world, Canada and Alaska too experienced unprecedented forest fires associated with heat waves and prolonged drought, while the Great Lakes area around Chicago suffered severe heat waves. Flooding of the Yangtse River in China left tens of thousands homeless, and hurricanes swamped Florida. Most devastating of all, Hurricane Mitch wrought such colossal damage in Central America that politicians claimed it had set the region back by a generation. The first global climate change negotiating conference after Kyoto, meeting in Buenos Aires in November 1998, sent a message of condolence and solidarity. Perhaps in their minds also was a degree of unease, lest it be thought that the devastation of Central America would somehow reflect on the slow and painful efforts in the negotiating chambers.

It is of course debatable whether these anomalies could be tied to global warming, given the considerable natural variability in extreme weather events. Almost certainly the anomalies were most directly associated with the unprecedentedly strong El Niño. That in turn begged the question of the relationship between human-induced warming and the intensity of El Niño, a topic on which the scientific community has not reached a consensus. The IPCC Second Assessment Report had stated that climate models do not show much consistent relationship between modelled climate change and the frequency of extreme events, and that there was no clear trend concerning the historical frequency of storms and their intensity – though the latter was contested by further research on hurricane records released late in 1998.[15]

The more direct consequence of these extreme events was that they showed just how vulnerable societies could be to them, even at a time when it is acknowledged that climate change is only just beginning. A Special Report by the IPCC in 1997 on regional impacts, based on the SAR, had highlighted various diverse vulnerabilities. Subsequent analysis seemed to add further to the possibilities. One of the leading models of global ecosystems predicted that the Brazilian rainforest would start to die back towards the middle of the twenty-first century as a result of the increased temperatures and reduced rainfall.[16] This was a symptom of trends with global implications. Although vegetation growth would be expected to absorb CO_2 during the first half of the century, under these stresses the

---

[15] M. Saunders et al., *Global Warming: The View in 1998*, London: Benfield Greig Hazard Research Centre, University College, 1988, cited in *Global Environmental Change Report*, Vol. XI, No. 2, Cutter Information Corp., 1999. One of the complicating factors is that storm intensity patterns appear to have varied on decadal time scales.

[16] These results were from the Institute of Terrestrial Ecology in Edinburgh: A. White, A. Friend and M. Cannell, 'Impacts of Climate Change on Natural Vegetation', in Hadley Centre, *Climate Change and its Impacts*, Bracknell: UK Meteorological Office, November 1998.

Figure 1.3: Estimated and predicted global CO$_2$ absorption by vegetation

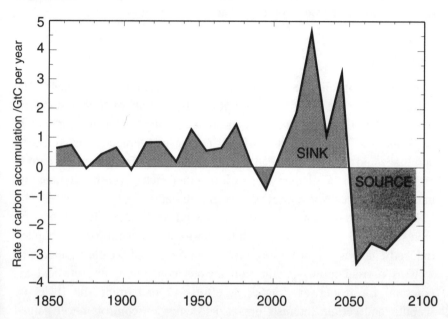

Source: Hadley Centre, *Climate Change and its Impacts*, Bracknell: UK Meteorological Office, November 1998.

ecosystem would turn into an additional emissions source from mid-century onwards, further exacerbating climate change itself (Figure 1.3). Related studies predicted shifting impacts on water resources and food supply, with broad changes particularly adverse in the developing world.

Projections of the regional impacts of climate change remain controversial, and doubtless these will form a key focus of attention in the IPCC's Third Assessment Report.

At the close of 1998, however, there was one other piece of scientific news from a different quarter, of a more positive if indirect nature. Global measurements showed that levels of ozone-destroying CFCs over Antarctica had started to fall, and that global ozone depletion appeared to be stabilizing, in response to the successive global agreements and domestic actions following the 1987 Montreal Protocol on ozone-depleting substances. Ozone depletion is of course a much simpler problem than climate change, with much quicker response possible. But it helped to show that the internecine complexities of international negotiations, and related domestic actions, could ultimately yield results.

1.6 Conclusions

The science of climate change rests on decades of assessment. The fundamental grounds for concern withstood the tests of time and critics: human emissions of greenhouse gases are changing the atmosphere in ways that trap heat near the surface, leading to average global warming and many and varied other changes. Many of these changes remain hard to predict in detail, particularly as various other factors – other human and natural influences – also affect the earth's climate.

With the completion of the IPCC's Second Assessment Report, governments accepted all this, along with the hard-fought conclusion that human-induced changes are becoming discernible in global weather patterns as well as myriad more specific conclusions concerning impacts and the potential for responses. Last-ditch efforts to undermine the IPCC by sceptics and certain interests opposed to emission controls never gained ground, and in the end backfired badly (see Chapter 2, Section 5). Subsequent scientific developments have strengthened further the scientific basis for alarm. Climate change is a serious problem, and by the mid-1990s governments were committed to doing something serious about it.

Thus was formed the intellectual background against which governments embarked upon some of the most important and complex global negotiations of the modern era: the effort to negotiate a regime to bring under control human interference with the global atmosphere and climate.

Chapter 2

Political and legal foundations: national perspectives and the road to Kyoto

As the issue of climate change began to develop from a purely scientific matter into a political issue during the late 1980s, countries began assessing their interests, objectives and concerns, and the steps that should be taken to develop an international response. This chapter summarizes the main national positions and groupings, and the international legal developments, that together serve to provide the political and legal foundations of the Kyoto Protocol.

2.1 National interests, perspectives and negotiating groups

Cooperation and action to limit climate change is complex because serious responses could reach deep into countries' economic and political interests. Carbon dioxide, the main contributor to projected climate change, comes predominantly from the use of fossil fuels and (especially in developing countries) deforestation. Energy use has been intimately related to economic development, and the fossil fuel industries comprise some of the largest and most powerful industries in the world. Even deforestation has complex causes and interests. Though there is a lot that can in principle be done to limit emissions, it is not easy.

Emissions themselves offer one easy measure of the relative significance of different countries in global efforts to contain climate change. In 1990, the year that the first IPCC report was accepted and negotiations towards a Convention initiated, the 'rich world' OECD countries accounted for more than half global fossil fuel CO_2 emissions and the central/east European countries, including the then USSR, for another 20%. The quarter of the world's population in the industrialized countries overall, therefore, accounted for about three-quarters of the fossil fuel CO_2 emissions (Box 2.1), and well over half the total global greenhouse

Box 2.1: The distribution of global CO$_2$ emissions

Countries differ hugely in their emissions of CO$_2$. Figure 2.1 shows the global distribution of CO$_2$ emissions in terms of three major indices: emissions per capita (height of each block); population (width of each block); and total emissions (product of population and emissions per capita = area of block).

Figure 2.1: Global CO$_2$ emissions, per capita and per population, 1995

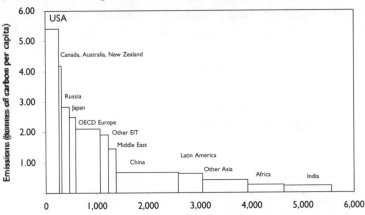

Source: Authors, data from IEA/OECD, 'CO$_2$ emissions from fuel combustion', Paris: IEA/OECD, 1997

This figure illustrates several dimensions of the policy debate. Per capita emissions in the industrialized countries are typically as much as ten times the average in developing countries, particularly Africa and the Indian subcontinent. This is one of the reasons why industrialized countries accepted the responsibility for leading climate change efforts: unless they can control their own high emissions there is little prospect of controlling emissions from developing countries that start from a very much lower base. There are also large differences among the industrialized countries, with per capita emissions in the EU and Japan at about half the levels in the United States and Australia.

At the same time, the large population of the developing countries indicates the huge potential for global emissions growth, if and as their emissions climb towards anything like levels in the industrialized world. There are also very large differences among the developing countries, with some of the Asian emerging industrial economies now with per capita emissions comparable with those in Europe and Japan.

emissions including deforestation and other gases.[1] Not surprisingly the main focus was upon these countries to act, and a principal dividing line was and remains the basic 'North–South' division, reinforced by the corresponding structure of all UN institutions.

There are, however, many differences within each of these groups. In addition to different past, present and future contributions, countries vary in the ease with which they can reduce emissions according to current efficiency, wealth and technological capabilities, as well as the role of domestic fossil fuel resources and access to non-fossil fuel resources. They also differ greatly in their economic strength and capacity to pay for responding. They may differ widely in their vulnerability to climate change: some, such as low-lying island states or countries in semi-arid areas, can expect to move even closer to the margin of existence. Countries also have more subtle differences on the issue, informed in part by cultural attitudes towards and experiences with nature, environment and technology.

The political fault-lines and groupings on the climate issue began to emerge clearly during the preliminary discussions in the late 1980s. With few exceptions, the general patterns that emerged have proved remarkably resilient.[2] This section outlines the interests and attitudes of the main groups in the negotiations.

2.1.1 The European Union

The European Community as an institution, and the European Union as a political grouping of (since 1995) 15 member countries, comprises the largest single political entity in the industrialized world in terms of both population and economic output. Many of its members (since 1995 including

---

[1] Detailed figures – which are particularly uncertain for non-energy sources in developing countries – depend on data sources and how the comparisons between different gases are made.

[2] Detailed analyses of economic interests and international political fault-lines at the time are given, among others, in M. Grubb et al., *Energy Policies and the Greenhouse Effect, Volume II: Country Studies* and *Technical Options*, London: RIIA, 1991. Because this section is about the political groupings and interests that helped to shape the Convention, and that have largely endured since, references are drawn largely from this set of national studies. For more recent appraisals see references later in this book.

the environmentally sensitive countries of Sweden, Austria and Finland) have called for action on climate change ever since the issue emerged politically. The northern European countries particularly have a relatively strong internationalist outlook and sense of global and environmental responsibility, and sympathy with the stance of developing countries. In addition, the Union overall and all its member countries except the UK are net energy importers.

In many European countries, environmental issues gained prominence during the 1980s, notably acid rain and ozone depletion. These were dismissed as scare-mongering at the beginning of the decade, and recognized as major threats justifying belated and sometimes expensive action by the end of the decade. This experience led to elaboration of environmental policy based on the 'precautionary principle'. Many of these countries also sense economic reasons for action, in terms of the benefits of both improved energy efficiency and technological leadership.[3] By 1990, many European countries had adopted indicative targets for limiting CO_2 emissions, and pressed for such responses across the industrialized world.

Of course, EU member countries vary according to certain economic and institutional factors; indeed some of the problems faced by the EU reflect those that could arise, on a larger scale, at the global level in the negotiation of a coordinated climate change strategy. There is a 'North–South' dimension, with four countries in a markedly less advanced development stage: the poorer 'Cohesion' countries do not want to bear the responsibility for past emissions of other EU countries, and they fear any constraint on energy consumption as an obstacle to the main aim of economic growth. Climate policy declarations in the EU have recognized this disparity, aware that emissions from these countries are likely to grow in the context of overall reductions, requiring bigger reductions from some other member states if the EU target is to be achieved. Overall, however, the EU has been at the forefront of pressure for stronger action on climate change.

---

[3] A view expounded most notably by the European Commission White Paper on Competitiveness and Employment, Luxembourg: EC, 1993.

2.1.2 The United States

The United States occupies a pivotal role in the global negotiations on climate change, not just because of its sheer political and economic power, but because of the weight of its emissions which comprise almost 25% of the global total. An agreement without the United States would be meaningless. Furthermore, as one of the world's highest per capita as well as absolute emitters, the global community tends to look to the United States as a bellwether of whether any action is being taken to modify emission trends.

In contrast to Europe, the United States has tended to be more hesitant about responses to the climate change issue overall and far more concerned about the economic consequences of CO_2 abatement. Apart from specific political circumstances, one source of resistance is the US low-price energy culture. The US economy grew on the back of seemingly limitless and cheap domestic fossil fuel resources, and it is the world's largest coal producer and the second largest producer of oil and natural gas, factors which combine to produce a huge resistance to measures that would limit or raise the costs of fossil fuels.[4] This feeds into the perception, widespread in the United States, that the costs of reducing CO_2 emissions would be very high, and hundreds of millions of dollars have been spent on economic analyses that support this perspective (in contrast to Europe, where more money has been spent on economic studies that support the opposite conclusion).[5] The relative isolationism of the United

---

[4] One analyst wrote starkly that 'the history of US energy demand and the existing resources, infrastructure and institutions make the US economy as dependent upon fossil fuels as a heroin addict is on the needle': Steve Raynor, 'The Greenhouse Effect in the US: The Legacy of Energy Abundance', in Grubb et al., *Energy Policies and the Greenhouse Effect, Vol. II*.

[5] The 1990 Economic Report of the President put the costs of reducing US CO_2 emissions by 20% at between $800 billion and $3.6 trillion, based on the modelling work of A.S. Manne and R.G. Richels, 'CO_2 Emission Limits: An Economic Cost Analysis for the USA', *Energy, the International Journal*, April 1990. A broad review of economic modelling studies is given in M. Grubb, J. Edmonds et al., 'The Economic Costs of Limiting Fossil Fuel CO_2 Emissions: A Survey and Analysis', *Annual Review of Energy and Environment*, Palo Alto: Annual Reviews Inc., 1993, and in Chapters 8 and 9 of the IPCC Second Assessment Report, Working Group III Report.

States, especially from the developing world, has also made it easier to divert attention from its domestic energy consumption by pointing to the role of other sources (such as deforestation and rice cultivation) and of developing-country growth.

However, the US position is by no means monolithic. The United States led the establishment of the IPCC and has remained at the forefront of scientific analysis of the issue. The credibility and political influence of the scientific sceptics declined after the publication of the first IPCC assessment (and was more or less buried by the IPCC's Second Assessment Report). In the Convention negotiations there was an emphasis on the contribution of existing environmental and energy policies (including the 1990 Clean Air Act Amendment and the National Energy Strategy).

There are also big structural divisions in the US political system. In particular, the US administration conducts international negotiations, and is more exposed to international realities than the domestic legislature. The election of the Clinton administration, with Vice-President Al Gore who had written a book on environmental concerns, led over time to a very different tone in the administration, with consequences that are traced in this chapter. However, whatever the administration agrees internationally has to be ratified by the legislature if it is to have any legal force, and the Congressional system in particular is heavily influenced by the interests of coal-producing states and oil and electricity companies. Congress thus exercises a virtual stranglehold on what can be implemented – and the administration has continually to negotiate with this in mind.

2.1.3 Japan and other OECD countries

The position of Japan is more ambiguous. Institutionally most closely allied to the United States, in 1990 Japan followed the European lead and adopted a delicately worded two-tier CO_2 emission target, and it has continued to seek a middling position. Its position reflects scarcely concealed intense internal debates. Japan is already one of the most energy-efficient countries in the world especially in its industrial sector, and so was and remains concerned about its ability to make additional reductions. Also, Japan has shown relatively little direct concern about possible climate change impacts, and has not traditionally been one of the more 'internationalist' countries.

But there are also strong contrary factors. With an economy wholly dependent upon imported fossil fuels, 'CO_2 constraints are perceived as an opportunity for Japan to revitalise energy conservation and other policies which are desirable in and of themselves'.[6] Japan has built technological confidence, and is seeking an international political profile to match its economic might. Barred by its constitution from contributing to international peacekeeping efforts, in the early 1990s debate about Japan's '*Kokusai Koken*' (international contribution) turned towards environmental issues as an area where the country could assume its rightful role as a mature and responsible international player, leading the world into a sustainable twenty-first century.[7]

Other OECD countries have varied attitudes. Canada and Australia share with the United States economies that are highly energy-intensive, and growing populations, making them concerned about the economic costs of emission constraints, although in Canada this is offset somewhat by a long-standing tradition of concern about international environmental issues. New Zealand, Norway and Switzerland, in sharp contrast, were among the lowest per capita CO_2 emitters in the OECD owing to their hydro-dominated economies; though the latter two in particular have long traditions of environmental sensitivity, this also makes the task of reducing emissions from historic levels more difficult than in the historically coal-based EU economies. Australia emerged with a very special position. Heavily dependent upon coal and heavy industry oriented towards export, predominantly with Asian neighbours that are not in the industrialized group of countries, the government has been exceptionally concerned about the potential adverse impacts of action to limit emissions.

Thus, the position of OECD countries presents a complex network of resource-based interests and historically acquired attitudes towards

---

[6] Akira Tanabe and Michael Grubb, 'The Greenhouse Effect in Japan: Burden or Opportunity?', in Grubb et al., *Energy Policies and the Greenhouse Effect, Vol. II*, p. 281.
[7] Yasuko Kawashima, 'A Comparative Analysis of the Decision-making Processes of Developed Countries toward CO_2 Emissions Reduction Targets', *International Environmental Affairs*, Vol. 9, No. 2, Spring 1997, pp. 95–126.

environment, technology and internationalism. However, the one thing that almost all the non-EU OECD countries have in common is greater difficulty than the EU in reducing emissions below 1990 levels. As that emerged as the focus of negotiations, these countries were forced together in a somewhat unnatural alliance – the JUSCANZ group, later expanding to JUSSCANNZ including Switzerland and Norway, that emerged as a serious counterweight to the EU in the Kyoto negotiations. (We use JUSSCANNZ in the text for most purposes, although Switzerland in particular frequently stood apart.)

2.1.4 Economies in transition

The countries of central and eastern Europe, including Russia and several states of the former Soviet Union (FSU), have since the inception of the climate change regime been in varying stages of economic and political transition. To a large extent, this has precluded them from taking a very active role in the negotiations. However, they have participated sufficiently to reflect and protect their main perceptions and interests, and these have become clearer over time.

Almost all the countries share the fact that by the late 1980s they had developed relatively energy-intensive economies – in some cases, grossly so – and that emissions have declined with the economic transition. The collapse of emissions since the late 1980s has made it easy for these countries to achieve the Convention's goal. Beyond that, there are important divergences. Most of the central European countries, and some of the FSU, have found themselves heavily dependent on imported energy, increasingly at or near world prices. Six of these countries are expected to join the European Union early in the twenty-first century, and several others hope to do so subsequently. They have consequently tended increasingly to align themselves with the EU position.

Russia itself, however, remains a huge and resource-intensive country that has developed closer political ties to the United States than to the EU. The specific developments in the Kyoto negotiations, described below, brought the United States and Russia closer together with a common interest and indeed, as analysed in Chapter 5, the outcome creates a potentially almost symbiotic interdependence between the two former superpower rivals in the area of climate change. In the final stages of the Kyoto negotiations and subsequently, Russia (and Ukraine) increasingly joined discussions with the JUSSCANNZ countries, forming (without Switzerland) what became known simply as the Umbrella group.

2.1.5 Developing countries and the G77

Institutionally, in the United Nations the developing countries seek to group together under the umbrella of the 'G77 plus China' (numbering more than 120 countries at full strength). However disparate their real interests and perspectives, these countries feel that their only source of strength lies in numbers and unity when faced with the might of the OECD. Since the inception of the climate regime, developing countries have indeed been united by the principle that any response action should be led by the industrialized world, but not by much else. In practice, at least three distinct groups can be readily identified.

The Alliance of Small Island States (AOSIS) emerged in the early days of debate about climate change. This comprises 42 states from the Pacific, Indian and Atlantic Oceans, some of which are only 2 metres above sea level at their highest point. They are thus extremely vulnerable to the impacts of climate change, particularly sea-level rise, some having their very existence as states threatened. In the negotiations they proved vocal and well organized, pushing for early action to reduce CO_2 emissions and halt deforestation, and for the resources to help them adapt to the changes that are now considered inevitable whatever action is taken.

In many ways at the opposite end of the spectrum was the Organization of Petroleum Exporting Countries (OPEC), created in the 1960s to represent the interests of developing-country oil exporters which felt that they were being exploited by Western governments and their multinational corporations. Having seized control of the oil markets in the 1970s, and watched that control slip from their fingers during the 1980s, these countries came to fear a further decline in their fortunes in the 1990s and beyond if and as efforts to limit CO_2 emissions depressed global oil consumption or, even more worrying, were used as a rationale to seize yet more oil rents through taxation. OPEC countries, usually led by Saudi Arabia, highlighted scientific uncertainties and argued – sometimes to the point where they were accused of obstructionism – that the emphasis of any action should be placed on research, sinks and others actions that would draw the pressure away from CO_2 emissions.

These two groups, while vocal, are relatively small. The rest of the developing countries form a third and much looser group with its own internal divisions, for example with discrete regional groups for Africa and Latin America. China and India (also often Brazil and Indonesia) often strike a more individual stance within the G77: China has more individual weight, and in fact is officially distinct from the G77; the developing countries negotiate as 'the G77 and China'. Their emphasis has been upon equity and development concerns, on the principle that 'the developed countries bear the main responsibility for the degradation of the global environment'.[8] Economic development is their primary focus, and they have argued centrally that any developing-country commitments should be purely dependent on the provision of financial resources and technology transfer by the North, and should not infringe developing countries' sovereignty over the use of their natural resources.

2.2 The UN Framework Convention on Climate Change[9]

Against this broad political background, countries met under UN auspices in February 1991 to begin negotiating a global Convention on Climate Change. The UN Framework Convention on Climate Change (UN FCCC) that emerged was signed at the Rio Earth Summit in June 1992; and the Kyoto Protocol and the way in which it developed subsequently cannot be understood without reference to the Convention on which it rests. This section describes the main elements of the Convention.

---

[8] Beijing Ministerial Declaration on Environment and Development, Beijing, 19 June 1991.
[9] The definitive guide to the negotiation and meaning of UN FCCC is Dan Bodansky, 'The UN Framework Convention on Climate Change: A commentary', *Yale Journal of International Law*, Vol. 18, 1993. This section is based upon the summary analysis in M. Grubb, *The Earth Summit Agreements:A Guide and Assessment*, London: RIIA/Earthscan, 1993.

The UN FCCC was born out of the collision between the explosion of environmental concerns in the late 1980s, including growing political and public awareness of the issue of climate change, and the older legacy of UN politics, particularly its North–South dimensions. The gulf between negotiators was enormous. Some were convinced about the seriousness and urgency of the problem and wanted an agreement which contained binding commitments to limit greenhouse gas emissions. Others recognized the problem, but wanted to tread more cautiously, agreeing first a general legal framework and leaving any binding measures until subsequent negotiations. Some developing-country participants even feared the whole issue was a conspiracy by Northern countries to impede their development, and their sole objective was to protect themselves, and if possible seek any access to finance and technology that might aid development.

With the Rio Earth Summit set as a deadline, the Convention was completed in fifteen months. Just weeks before Rio, the Chairman welded a mass of incoherent texts and conflicting political positions into a credible document that spelled out the core principles, institutions, procedures and initial commitments that, with limited further changes negotiated by the Parties, came to be agreed as the Framework Convention on Climate Change.

2.2.1 Elements of the Convention

Article 2 sets out the overall Objective of the regime:

> The ultimate objective of this Convention ... is to achieve ... stabilization of greenhouse gas concentrations in the atmosphere at a level that would prevent dangerous anthropogenic interference with the climate system. Such a level should be achieved within a time frame sufficient to allow ecosystems to adapt naturally to climate change, to ensure that food production is not threatened and to enable economic development to proceed in a sustainable manner.

The core objective is thus to contain the rate of change within safe limits and ultimately to stabilize the atmosphere at safe levels. Quite how the terms 'dangerous ... interference' and the requirement to 'allow ecosystems to adapt naturally ...' are interpreted is unclear, but given the gulf between the requirements of stabilization and the business-as-usual projections, it is likely to be a far from trivial undertaking.

The preamble and the principles to the Convention serve as a set of agreed facts and guiding principles, covering a wide spectrum: the underlying scientific concern; the primary role of industrialized-country emissions in the past; the principle that appropriate levels of response 'should reflect the environmental and developmental context'; the 'principle of Sovereignty of States in international cooperation'; the need for continued research and measures 'continually re-evaluated in the light of new findings'; the idea that 'various actions ... can be justified economically in their own right and can also help in solving other environmental problems'. Specific principles acknowledged included:

- the need to protect the climate system 'on the basis of equity and in accordance with [States'] common but differentiated responsibilities and respective capabilities. Accordingly, the developed country Parties should take the lead';[10]
- 'the specific needs and special circumstances of developing country Parties, especially those that are particularly vulnerable' (e.g. the low-lying island states and the primary fossil fuel exporters);
- the need for 'precautionary measures in the absence of full scientific certainty,' qualified by the need to be 'cost effective' and 'comprehensive', i.e. to take into account all sources and sinks, adaptation and all economic sectors;
- 'Parties have a right to and should promote sustainable development integrated with national development programmes', and they 'should cooperate to promote a supportive and open international economic

---

[10] When this principle was adopted, there was little dissent that developed countries should take the lead, but there were important disputes as to why. Developing countries argued that a major factor was the historical 'debt' arising from the much higher past emissions of industrialized countries, which have monopolized the available 'environmental space' of the planet. The United States argued that past emissions were a matter of history, and current generations could not be held accountable for this; the need for developed countries to take the lead was simply a reflection of their current relative wealth. This dispute is likely to resurface when attempts are made to reach agreements which require closer specification of the meaning of 'on the basis of equity'.

system. Measures taken ... should not constitute a means of arbitrary or unjustifiable discrimination or a disguised restriction on international trade.'

The impression conveyed is thus one of concern, tempered by the acknowledged uncertainties and the stated precedence of national economic interest and development. It clearly places the emphasis on developed-country action, and points towards an iterative approach which adapts to improving information and changing circumstances.

The central focus of debate was upon Article 4 (Commitments). This finally emerged as a long and convoluted article, deliberately ambiguous in places. The first extended paragraph commits all Parties to:

- 'Develop, periodically update, publish and make available to the Conference of the Parties, national inventories' of greenhouse gas emission sources and sinks;
- 'Formulate, implement, publish and regularly update national and, where appropriate, regional programmes containing measures to mitigate climate change ... and measures to facilitate adequate adaptation to climate change';
- 'Promote and cooperate in the development, application and diffusion, including transfer, of technologies, practices and processes';
- 'Cooperate in preparing for adaptation to the impacts of climate change; develop and elaborate appropriate and integrated plans for coastal zone management, water resources and agriculture, and for the protection and rehabilitation of areas, particularly in Africa, affected by drought and desertification, as well as flood';
- Various general aims including 'promote sustainable management', and 'promote and cooperate' in a wide range of other measures including conservation and enhancement of sinks and reservoirs of greenhouse gases; scientific, technological, socio-economic, etc. research and the open and prompt exchanges of relevant information; education, training and public awareness;
- 'Communicate to the Conference of Parties information related to implementation'.

This applied to all Parties. The second extended paragraph – Article 4.2 – commits 'developed country Parties and others included in Annex I' to more specific measures, namely the adoption of 'policies and measures that will demonstrate that developed countries are taking the lead in modifying longer-term trends in anthropogenic emissions consistent with the objective of the Convention'. Deliberately disjointed references in the first two paragraphs (a) and (b) suggested that this would be demonstrated by the indicative aim of returning their emissions of CO_2 and other greenhouse gases to 1990 levels by the year 2000, and this became the focus of attention in the years immediately after the Convention.[11]

With hindsight, however, probably the most important aspect of this convoluted commitment was that it locked the Parties into a process of providing detailed information on their emission programmes and prospects, and reviewing the adequacy of their efforts:

> The Conference of the Parties shall, at its first session, review the adequacy of subparagraphs (a) and (b) above ... [and] take appropriate action, which may include the adoption of amendments to the commitments ... A second review of subparagraphs (a) and (b) shall take place not later than 31 December 1998, and thereafter at regular intervals determined by the Conference of the Parties, until the objective of the Convention is met.

---

[11] Specifically: (a) 'Each of these Parties shall adopt ... policies and take corresponding measures on the mitigation of climate change, by limiting its anthropogenic emissions of greenhouse gases and protecting and enhancing its greenhouse gas sinks and reservoirs. These policies and measures will demonstrate that developed countries are taking the lead in modifying longer-term trends in anthropogenic emissions consistent with the objective of the Convention, recognizing that the return by the end of the present decade to earlier levels of anthropogenic emissions of carbon dioxide and other greenhouse gases not controlled by the Montreal Protocol would contribute to such modification.' This is qualified by 'taking into account' the need to recognize various differing circumstances, and to 'maintain strong and sustainable economic growth'. It states that 'Parties may implement such policies and measures jointly with other Parties'.
(b) 'Each of these Parties shall communicate, within six months of the entry into force of the Convention for it and periodically thereafter ... detailed information on its policies and measures referred to in subparagraph (a) above, as well as on its resulting projected anthropogenic emissions ... for the period referred to in subparagraph (a) [i.e. to the year 2000] with the aim of returning individually or jointly to their 1990 levels these anthropogenic emissions ... This information will be reviewed by the Conference of the Parties, at its first session and periodically thereafter, in accordance with Article 7.'

This section further requires each of these Parties to 'coordinate as appropriate with other such Parties, relevant economic and administrative instruments' and 'identify and periodically review policies and practices which encourage ... [greater emissions] than would otherwise occur' – a pointed reference to policies such as coal subsidies. Other parts of Article 4.2 required a review of the list of countries in the Annexes, and invited other countries to join the commitments in (a) and (b) – an implausible prospect for any country with growing emissions.

The following three paragraphs (4.3–4.5) established the financial responsibilities of countries in Annex II – essentially the OECD countries – to assist developing countries to meet their obligations under the Convention and to help particularly vulnerable countries to adapt to climate change, and 'the transfer of, or access to, environmentally sound technologies and know-how'. The other Parties in Annex I – the 'transition economies' of central and eastern Europe, and some of the FSU – did not have the same financial commitments and were to be allowed 'a certain degree of flexibility' in implementing their emission commitments (Article 4.6).

The other commitment that was to prove politically important was the requirement to try to meet the 'specific needs and concerns of developing country Parties arising from the adverse effects of climate change and/or the implementation of response measures', with a list that included, first, various categories of countries vulnerable to impacts, and then (h) 'countries whose economies are highly dependent on income generated from the production, process and export and/or consumption of fossil fuels and associated energy-intensive products'. The Conference of Parties 'may take actions, as appropriate, with respect to this paragraph'.

Subsequent Articles established the essential institutional machinery of the Convention. An annual Conference of Parties was charged both with sorting out all the issues which could not be resolved in the time span of the Convention negotiations, and with reviewing progress in the light of expanding knowledge and changing circumstances: it is the central body with the authority to determine what happens, when and how. To this end, the first meeting was determined to take place within a year of entry into force of the Convention, and every year thereafter unless otherwise decided by the Conference of Parties itself. The Secretariat would deal with day-to-day running of the Convention and associated efforts, transmission of reports, and preparation for meetings of the Conference of Parties. A Subsidiary Body for Scientific and Technological Advice (SBSTA) was to 'provide the Conference of the Parties and, as appropriate, its other subsidiary bodies with timely information and advice on scientific and technological matters ... This body ... shall comprise government representatives competent in the relevant field of expertise'; unlike the IPCC, this was to be essentially a government negotiating forum that would meet between the annual Conferences of Parties. A parallel Subsidiary Body for Implementation (SBI) would 'assist the Conference of Parties in the assessment and review', similarly composed of government representatives. SBI subsequently endorsed and agreed to oversee the Secretariat's role in reviewing National Communications, which includes making country visits and reviewing the actions being taken to implement the Convention.

Article 11, Financial Mechanism, addresses the thorny issue of control of finances, and 'defined' a 'mechanism for the provision of financial resources on a grant or concessional basis, including for the transfer of technology'. Its operation shall be 'entrusted to one or more existing international entities'. The Global Environment Facility, housed at the World Bank and overseen jointly with UNEP and the UN Development Programme, was accepted – reluctantly by many developing countries especially – as the 'interim' agency, though in reality it always seemed likely that it would maintain that status indefinitely.

The remaining Articles of the Convention mostly address the procedural machinery governing reporting and other aspects of implementation, and the provisions for Amendments and for Protocols. The Convention states that 'if all efforts at consensus have been exhausted ... amendment shall as a last resort be adopted by a three-fourths majority vote,' with further details regarding procedures for amendments.[12] By contrast, a skeletal

---

[12] An amendment so adopted '... shall enter into force for those Parties having accepted it on the ninetieth day'. Other Parties are not bound until they indicate acceptance. Thus, the rule of consensus may be breached, but with the greatest reluctance; any such breaches shall not be binding on the dissenting Parties (a caveat which is unavoidable, since countries cannot be bound to things they have not accepted).

article on Protocols states, in five short sentences, that Protocols may be adopted by the Conference of Parties, and should essentially be self-contained instruments bearing only upon the Parties to the Protocol concerned. There was no agreement on anything concerning rules of procedure or adoption. The sense that a Protocol to the Convention could be the really decisive step was already in the air, and countries did not want to give anything away.

In all this, the Convention was probably as successful as could reasonably be expected. It provides an international legal framework and set of principles which was acceptable to almost all the countries involved. It accepts that climate change is a serious problem and reassures developing countries that addressing it at present is primarily a responsibility of the industrialized countries. A step-by-step, evolving approach based on the expectation of improving information is clearly established, with considerable emphasis on the need both to report information and to re-evaluate commitments. There is also clear emphasis that information should be open to public scrutiny. Its most obvious limitation lay in the lack of any specific binding commitments regarding emissions; but it established the institutional and procedural basis upon which the next steps could be taken – as and when the international community was ready for them.

2.3 From Rio to the First Conference of Parties at Berlin

To have legal authority under the terms of international law, any international agreement needs to be ratified by the domestic legislature in participating countries. It then acquires the status of a legal commitment by the ratifying countries; and the agreement itself needs to acquire a minimum number of ratifications before it 'enters force' as a confirmed international legal agreement.

The US government, under the Bush administration, was among the first to ratify the Convention, and others accumulated quite rapidly. By December 1993 the UN FCCC had been ratified by 50 countries – the number required for entry into force, which occurred a statutory three months later. This was remarkably quick for a global treaty of this scope. Ratifications continued apace: by the time of Kyoto, the Convention had been ratified by 167 countries plus the European Community.[13] This represents almost global participation.

On other fronts, progress was more mixed. The aftermath of Rio brought a degree of exhaustion coupled with some backlash against environmental issues in general, and (especially in the United States) against climate change in particular. Scepticism concerning the science and scare stories about the economic impact of emission constraints were sometimes given great prominence, though for the most part the issue was just ignored.

But no one ignored proposals for carbon taxes. Clinton's 'Btu' tax proposal,[14] presented as an environmental initiative, crashed disastrously in 1994, probably contributing to big gains by the Republican Party in that year's Congressional elections. The EU's carbon tax proposal that was first presented around 1990 was rejected decisively by the UK in 1993, and the support was not really very deep in many other EU countries; it remained bogged down for years before finally being abandoned. Other initiatives made mixed progress.

However, the machinery related to the Convention process was in place. Governments proceeded to develop national plans, indicating emission projections and actions they were taking or planning to limit emissions. Since these were governmental communications they had to be agreed, at least in principle and often in practice, across various government departments. The constituency of officials who had to know something about climate change, and national plans to address it, grew steadily. And, with the Convention's entry into force secured, attention at the international level started to turn to the next steps.

The core of 'next steps' was the requirement to hold the first Conference of Parties to the Convention within one year of entry into force; and for that conference to review the adequacy of progress and to consider

---

[13] Since the Kyoto conference, the UNFCCC has been ratified by an additional eight countries. The Convention has received 176 instruments of ratification as of 7 October 1998.

[14] US energy is counted in Btus (British thermal units) and this is how the tax proposal quickly became known, leading one wit to remark that for the United States to raise revenue by taxing British energy was an excellent idea. The British government, however, ran into enough trouble trying to tax energy for itself.

next steps. The German government, a long-time proponent of stronger action on climate change, was to hold this critical conference in Berlin. Although the German government and a few others expressed hopes that substantive new commitments would be agreed in Berlin, this was never a realistic proposition. The final meeting of the Preparatory Committee, in New York in February 1995, failed even to agree on whether the current commitments should be formally described as 'inadequate' and it cast doubt on whether the Berlin meeting would be able to initiate negotiations on next steps. In the event, the Berlin conference proved far more significant than this backdrop suggested.

The institutional issues at Berlin proceeded quite smoothly, apart from a failure to agree formal Rules of Procedure – a situation that persisted throughout the Kyoto negotiations.[15] The budget was secured and it was agreed – after an ugly fight between the Swiss and the Germans –that the Secretariat would be moved from its interim location in Geneva to a long-term home in Bonn. The conference formally established the two subsidiary bodies under the Convention: one for scientific and technical advice (SBSTA), the other on implementation (SBI). The Global Environment Facility was approved as the continuing interim financing agency, now in a fully operative rather than a pilot phase. Thus the major institutional machinery was established and confirmed by the Berlin conference.

A more contentious area of debate concerned joint implementation (discussed in Chapter 3). In the end, the conference reached an enabling compromise for a pilot phase of 'activities implemented jointly', on an opt-in basis, without explicit crediting of emission reductions to the donor country (see Box 2.2). The Convention's subsidiary bodies were charged with a monitoring and accounting role, and the whole process was to be reviewed so as to take a 'conclusive decision on the pilot phase and the progression beyond that' by the end of the decade.

---

[15] Several oil-exporting countries held out for a formal requirement for consensus on major decisions. Other countries, fearing that this would give the oil exporters effective veto power over the whole process, sought some form of enhanced majority process. The impasse meant that the conference, and subsequent negotiations, were conducted under draft rules without voting, which placed more power in the hands of the chairman to judge when to push through decisions even if some countries were still unhappy with them.

Box 2.2: Berlin decision on activities implemented jointly

> The Parties agree:
> (a) 'to establish a pilot phase for activities implemented jointly among Annex I Parties
>   and, on a voluntary basis, with non-Annex I Parties who so request';
> (b) 'that activities implemented jointly should be compatible with and supportive of
>   national environment and development priorities and strategies'...;
> (c) 'that all activities implemented jointly under this pilot phase require prior
>   acceptance, approval or endorsement by the governments of the Parties
>   participating in the respective activities'.
>
> Such measures are required:
> (d) 'to result in real, measurable and long-term benefits that would not otherwise have
>   occurred, and
> (e) to involve genuine additional finance'.
>
> Most important,
> (f) 'no credits shall accrue to any party ... during the pilot phase from activities
>   implemented jointly'.

The most contentious and focal point of the negotiations at Berlin, however, concerned the adequacy of commitments and potential next steps. Although the backlash after Rio brought a period of popular scepticism in the media, the scientific evidence was in fact hardening in important respects, as set out in Chapter 1. It was also becoming clear that many OECD countries were not on course to fulfil the non-binding 'aim' in the Convention's commitments to return their emissions to 1990 levels by 2000. Furthermore, it was quite obvious that commitments that did not extend beyond 2000 were not adequate to address a global and long-term problem. Most countries had accepted the obvious in preparatory discussions, namely that the Convention commitments were not adequate.

It was, however, a highly politicized process. A few countries (notably oil-exporting countries) and powerful groups (notably some US industry groups) were opposed to the development of new commitments and had previously blocked agreement that commitments were inadequate. Furthermore, the US industry groups insisted that any new negotiations should be global, and include commitments for developing countries. The oil-exporting countries in turn played on the fear that new negotiations would involve developing countries in commitments – and that they would be adversely affected by industrialized-country emission reductions – in order to try to get the G77 group of developing countries to oppose any negotiations on new commitments. At the opposite end of the spectrum, AOSIS presented a proposal calling for a 20% cut in industrialized-country CO_2 emissions by 2005.

The political pressures to reach some kind of enabling agreement intensified in the run-up to Berlin and at the conference itself. After a week of internal dissent, the G77 group of developing countries split. A majority 'Green Group' – excluding the oil exporters – emerged, accepting the current commitments as inadequate and calling for the industrialized countries to accept binding emission reductions while rejecting firmly any process that might lead to commitments for developing countries. This then put pressure back on the OECD countries. They had already accepted that their commitments and performance were not adequate to the problem, and were hard put to present the failure to achieve their initial aim under the Convention as indicating the kind of leadership that could justify starting to involve developing countries in commitments.

It was finally agreed at Berlin that the current Convention commitments were inadequate, and consequently to 'begin a process to enable it to take appropriate action for the period beyond 2000, including the strengthening of the commitments of Annex 1 Parties', i.e. the industrialized world. Specifically (see Box 2.3), it was decided that the process should lead the industrialized world to:

- 'elaborate policies and measures'; and
- 'set quantified limitation and reduction objectives within specified timeframes, such as 2005, 2010 and 2020, for their anthropogenic emissions'.

These negotiations should not introduce new commitments for developing countries, and should be 'scheduled to ensure completion of the work as early as possible in 1997 with a view to adopting the results at the third session of the Conference of the Parties'. Thus were launched the intensive negotiations that finally culminated in Kyoto.

Box 2.3: The Berlin Mandate

Conference of Parties
First session, Berlin, 28 March–7 April 1995
Review of adequacy of Article 4, para 2(a) and (b) [industrialized-country emission commitments], including proposals related to a protocol and decisions on follow-up

The Conference of Parties, having reviewed [the commitments] and concluded that these are not adequate, agrees to begin a process to enable it to take appropriate action for the period beyond 2000, including the strengthening of the commitments of Annex I Parties ...

2.   The process will, *inter alia*:

(a)   Aim, as the priority in the process of strengthening the commitments in Article 4.2 (a) and (b) of the Convention, for developed country/other Parties included in Annex I, both
  • to elaborate policies and measures, as well as
  • to set quantified limitation and reduction objectives within specified time-frames, such as 2005, 2010 and 2020, for their anthropogenic emissions by sources and removals by sinks of greenhouse gases not controlled by the Montreal Protocol taking into account the differences in starting points and approaches, economic structures and resource bases, the need to maintain strong and sustainable economic growth ...

(b)   Not introduce any new commitments for Parties not included in Annex I, but reaffirm existing commitments in Article 4.1 and continue to advance the implementation of these commitments in order to achieve sustainable development ...

(c)   Take into account any result from the review ...

(d)   Consider, as provided in Article 4.2(e), the coordination among Annex I Parties, as appropriate, of relevant economic and administrative instruments ...

(e)   Provide for the exchange of experience on national activities in areas of interest, particularly those identified in the review and synthesis of available national communications; and

(f)   Provide for a review mechanism.

3.   The process will be carried out in the light of the best available scientific information and assessment on climate change and its impacts, as well as relevant technical, social and economic information, including, *inter alia*, IPCC reports ...

4.   The process will include in its early stages an analysis and assessment, to identify possible policies and measures for Annex I Parties ... [and] ... could identify environmental and economic impacts and the results that could be achieved with regard to time horizons such as 2005, 2010 and 2020.

5.   The protocol proposal of the Alliance of Small Island States [AOSIS – proposing a 20% reduction in industrialized-country emissions by 2005] ... should be included for consideration in the process.

6.   The process should begin without delay and be conducted as a matter of urgency ... to ensure completion of the work as early as possible in 1997 with a view to adopting the results at the third session of the Conference of the Parties.

2.4 The shifting political foundations

To those uninitiated in the rituals of international diplomacy, agreement to negotiate next steps looked like a rather pathetic result of the three years' discussion since the grand heights of Rio. Yet the Berlin conference was a watershed. The period after Rio had been one of scepticism, retrenchment and failure to implement the visionary but vague promises of Rio. Given the perceived lack of strong public concern and the strength of opposition from vested interests, many doubted whether governments had the stomach for a more serious round of negotiations. It is worth pausing to consider the political shifts that underlay the outcome of Berlin, and the coalitions that first became clear there.

2.4.1 `National positions

The most obvious shift surfaced prominently during the Berlin conference itself. The developing world had previously maintained a mostly unified position under the umbrella of the G77 group. However disparate their real interests and perspectives, these countries feel their only true strength comes from numbers and unity when faced with the might of the OECD. But the internal tensions at Berlin were too great for this to hold. The oil-exporting countries feared for their income. But some others – particularly the small island, low-lying states and those in danger of desertification – feared for their survival. The resulting politics in the G77 were not pretty; one can imagine the kind of response that Bangladesh or the East African states were inclined to give to OPEC pleas of impending poverty. From the wreckage of this untenable alliance emerged the majority view that action was indeed required – so long as it was confined to industrialized countries. For the first time, the vast majority of developing countries unambiguously accepted climate change as a serious problem that had to be addressed, and they called for strong action by those they held to be responsible.

The politics were also complex within the industrialized countries. The EU countries remained the most proactive. Their position was strengthened by the fact that they were the only OECD group whose emissions had actually stabilized since Rio, and the UK was now more firmly aligned to the mainstream EU position.[16] Many of the EU countries – and the EU as a group, now with the full backing of the UK – sought specific commitments on emission reductions below 1990 levels, on specified targets and timetables. They were also more sympathetic to the insistence by developing countries that it would be inequitable to expect them to make additional commitments at this stage, and that the industrialized world had yet to demonstrate that it was taking an adequate lead as required in the Convention.

The United States and others in the JUSSCANNZ group emerged as more cautious, particularly with regard to commitments to emission reductions below 1990 levels. The United States and Australia emerged as a hard core, opposing any reference to reductions below 1990 levels and insisting also that developing countries should be involved in new commitments.

Eventually these countries were presented with a choice by others in the OECD: either accept commitments to domestic emission reduction targets and timetables below 1990 levels, or drop insistence on new commitments for developing countries. They accepted the latter.

2.4.2 The industrial landscape

The OPEC oil-exporting countries had previously gained their negotiating strength not only from their role in the G77 but also from their links with Western (particularly US) industry. It is here that the second big development occurred. At the final preparatory meeting in New York, the industrial front seemed to be dominated by the concerns of the US coal and oil industries. Indeed, just when it appeared that the governments might recommend that current commitments should be declared to be inadequate at Berlin, key energy-exporter governments renewed their objections, apparently at the behest of industry lobbyists who rushed to the negotiating floor.

Such behaviour may have seemed like tactical sense, but it irritated governments and embarrassed more moderate industrial colleagues. As a

---

[16] This was due to a number of factors, including the centrist shift of Conservative Party policy on Europe after the fall of Mrs Thatcher, and the fact that UK emissions had declined as a result of introducing competition and greater efficiency in the electricity industry (at the expense of the coal industry).

result of these actions, in Berlin all non-governmental groups were barred from the negotiating floor. But it emerged that the industrial front, like the G77, had fractured anyway. At key meetings in the run-up to Berlin, the industrial hardliners failed to get the backing of wider industrial alliances, with other companies either seeing some opportunities or at least fearing that industry risked losing the relative goodwill it had achieved in the Rio process. Thus OPEC's industrial pillar in the West, and the stranglehold it sought to place, particularly on US negotiators, lost its force.

Indeed, at Berlin three disparate industrial groupings emerged clearly for the first time. At one end, the hardline group of predominantly US, predominantly coal, interests struggled to halt the tide towards new steps. At the other end of the spectrum, the 'sunrise' industries of renewable energy and energy efficiency, together with at least one major gas company, joined forces with the insurance industries in a call to support the AOSIS protocol. Stretched painfully and widely across the middle ground were the oil companies, some wedded to the coal position and others, with dominant interests downstream and in gas, suspecting that their long-term interests lay in going with the tide or at least in not being seen to stem it.

2.4.3 Environmental groups

Environmental NGOs continue to play an active role in the climate change process and, like their industrial counterparts, were frustrated with the relative lack of direct access to negotiators at Berlin. Nevertheless, as well as the traditional NGO role in public campaigns and in lobbying governments on the seriousness of the problems, Berlin saw a continuing growth in the political sophistication and scope of their actions. Immediately before Berlin, Greenpeace organized a conference on the concerns of the insurance industry, attracting high-level participation, which immediately injected an industrial counterpoint to the coal and oil industries' concerns. During the conference, NGOs also publicized the links between the US fossil fuel industries and oil-exporting countries, a conspiracy theme that was picked up by *Der Spiegel* to great effect. And NGOs probably played a significant role in persuading Indian and Brazilian delegates to make the moves that broke the G77 impasse and that led ultimately to the developing countries' 'green paper' and thence to the Mandate.

2.4.4 Scientific developments

Whether either of the key political developments was also influenced by scientific developments in the months before Berlin is unclear, but these may have played a part; certainly the continual evolution of the science played an important role in the process over the coming years. As the dust settled from the 1991 eruption of Mount Pinatubo, which had perturbed patterns of global climate and temperature, underlying trends seemed to be re-emerging. Just in the few months before Berlin fresh evidence was added. As discussed in Chapter 1, the inclusion of sulphates and other factors in the models has greatly improved the simulation of observed trends over the past century. Separate analyses of trends in temperature patterns and rainfall, conducted in the United States, the United Kingdom and Germany, concluded that the climate was changing, in ways that seemed consistent with mainstream simulations of expected climate change. And Antarctica saw the dramatic calving of an iceberg the size of Oxfordshire, and the first melting since records began of the Ross Channel ice sheet, in a region that has warmed rapidly.

None of this constituted proof, in a scientific or legal sense, that greenhouse gas emissions were changing the climate. However, by the time of Berlin, the confluence of climatic events was already shifting the balance of scientific opinion – and public perception – towards the view that climate change was not only a threat but was already being observed. This carried obvious political repercussions. Yet the resurgent concern and the political watershed around the Berlin conference could not answer the fundamental question: what actually could countries now agree to do?

2.5 Milestones to Kyoto

In the aftermath of the Berlin decision, the Parties established a negotiating body under the chairmanship of Argentinian Ambassador Raúl Estrada-Oyuela to oversee the negotiations. This became known as the ad-hoc group on the Berlin Mandate (AGBM: a somewhat more neutral acronym than the original suggestion of 'Berlin Mandate Working Group', which appeared altogether too German). Apart from this, however, uncertainty reigned as to how to start implementing the mandate for negotiations.

At the AGBM's first full session in the summer of 1995, the EU submitted a proposal to structure an agreement on specific policies and measures, divided into different categories. The fate of this proposal is outlined in Chapter 3, but at least it provided something tangible for negotiators to react against. The United States in particular reacted against it; but what that country actually supported remained unclear.

The question of emission targets was left for AOSIS to pick up. Before Berlin they had presented a Protocol proposal that including a legally binding 20% reduction in CO_2 emissions by 2005. Though generally recognized as a politically infeasible goal, this reaffirmation of the 'Toronto target' served as a rallying point for the NGO community and left hanging the question of 'if not this, then what?'

Possible answers to that question were hotly debated within the US administration throughout the year after Berlin. Upon coming into office, the Clinton–Gore team had appointed many environmentalists into the administration; bruised by political experience, they retained their basic if moderated aims and learnt the art of the possible. And there was a major environmental success story to show, in the form of US controls on acid rain, under which major emission reductions were implemented at far lower cost than originally predicted. Furthermore, the administration had scored a significant political victory over the industrial lobbyists in reaching an agreement at Berlin, and opinion polls showed consistently that the American public remained concerned about environmental issues and looked to the Democrats to do something about them. The administration had embraced the IPCC conclusions, but decisively rejected the EU approach to harmonizing specific policies and measures. What would follow?

Against this background, Senator Tim Wirth, Secretary for Global Affairs at the US State Department, arrived to head the US delegation at the second Conference of Parties (COP-2) in Geneva in July 1996. Set in the context of US politics, his speech was one of the strongest of the negotiations and a decisive turning point. He first rounded on the US industrial critics of the IPCC's Second Assessment Report:

> The IPCC's efforts ... serve as the foundation for international concern and the clear warnings about current trends are the basis for the sense of urgency within my government. We are not swayed by and strongly object to the recent allegations about the integrity of the IPCC's conclusions ... raised by naysayers and special interests bent on belittling, attacking and obfuscating climate change science ... let me make clear the US view: the science calls upon us to take urgent action.

He then went on to declare:

> Our analysis and consideration of this issue to date have led us to certain conclusions ... Sound policies pursued in the near term will allow us to avoid the prospect of truly draconian and economically disruptive policies in the future ... denial and delay will only make our economies vulnerable in the future.
>
> The US will seek market-based solutions that are flexible and cost-effective ... The US recommends that future negotiations focus on an agreement that sets a realistic, verifiable and binding medium-term emissions target ... met through maximum flexibility in the selection of implementation measures, including the use of reliable activities implemented jointly, and trading mechanisms around the world.

This was in fact the first time that any major Party had specifically called for quantified commitments adopted under the negotiations to be made binding. In a curious way, this reflected a long-standing US concern. The United States has long argued that because of its culture and internal political structures, it would be held accountable to any specific commitments made – a situation it contrasted with what it perceives as the tendency of many European countries to declare fine targets without any detailed plan to achieve them.

Anxious to rebut such assertions and pursue its long-standing goal of a strong climate agreement, the EU readily accepted this new emphasis on legally binding targets, while still pressing its list of policies and measures. Japan hesitated, but Russia and Australia were the only Annex I countries to object strongly. Within two days, the ministers in Geneva

Box 2.4: Ministerial Declaration at the Second Conference of the Parties to the Climate Convention, 8–19 July 1996

Ministerial Declaration

The Ministers and other Heads of Delegation ...

Recognize and endorse the Second Assessment Report of the IPCC as currently the most comprehensive and authoritative assessment ... and believe that [it provides] a scientific basis for urgently taking action ... the continued rise of greenhouse gas concentrations in the atmosphere will lead to dangerous interference in the climate system.

Reaffirm the existing commitments under the Convention, including those intended to demonstrate that Annex I Parties are taking the lead in modifying their longer-term trends ... and agree to strengthen the process ... for the regular review and implementation of present and future commitments... many of these Parties need to make additional efforts to overcome difficulties they face in achieving the aim of returning their emissions of greenhouse gases to 1990 levels by 2000.

Instruct their representatives to accelerate negotiations on the text of a legally-binding protocol or other legal instrument ... for adoption at the 3rd Conference of Parties. The outcome should fully encompass the remit of the Berlin Mandate.

---

produced a declaration endorsing the IPCC Second Assessment Report, reaffirming the terms of the Berlin Mandate with the words 'legally binding' inserted, and also highlighting the importance of a global effort on technology transfer (see Box 2.4). Though not adopted as an official conference document due to the objections of Russia, Australia and OPEC, this became the guiding statement upon which the negotiations entered a new and more deadly serious phase.

COP-2 was a turning point because it made it plain to the world that the US administration was fully committed to the negotiations, supported legally binding commitments, and had some new ideas about how they should be approached. There was a strong reaction from some quarters in the United States, and industry groups threatened that Tim Wirth's statement would make the 'irresponsibility' of the US position a major election issue. But they were wrong. Opinion polls showed consistently that the environment was a vote-winner for the Democrats, and it would clearly be against Republican interests to complain that the Democrats were proposing legally binding action to protect the planet. Many environmental groups in the United States were enthusiastic about their delegate's move, but other environmentalists remained sceptical, wanting to know just what the targets would be.

The negotiating process itself was slow to digest and react to the new situation. Immediately after COP-2, the Secretariat was absorbed in its long-planned move to Bonn and consequent loss of staff. The EU continued to elaborate its list of policies and measures, and initiated more serious analysis and discussion of emission targets, but failed to reach internal agreement. Japan became increasingly concerned about the implications of legally binding targets, and almost everyone remained unsure about what the United States really had in mind with its references to 'realistic, medium term', and 'binding commitments implemented flexibly ... with instruments like emissions trading'. With hindsight, the novelty of and uncertainty about the US proposals are not easy to recreate, but they may be inferred from the fact that as late as December 1996 Chairman Estrada asked publicly how a binding emission commitment could be flexible.[17]

The United States itself expended much of its international effort – when not occupied with the elections – in clarifying its tentative proposals to the rest of the industrialized world and especially Russia which, as explained later, could be one of the main beneficiaries. Over the period after COP-2, Russia responded by shifting its opposition to binding targets (so much that by January 1997 Russia emerged with its own proposal for emission targets implemented through emissions trading). The victory of Clinton and Gore in the US presidential election of November 1996 set the seal on the new situation.

Perhaps the most significant shift in the aftermath of COP-2 was that of the OPEC countries. The rout of the IPCC's critics and growing momentum for a meaningful agreement left its former position of scientific scepticism and sometimes obstructive tactics increasingly exposed and isolated. Furthermore, with the spectre of carbon taxes receding, and the Saudis' intimate and most powerful ally, the United States, seizing the helm of the negotiations, the costs of obstruction for Saudi Arabia in particular seemed

---

[17] R. Estrada, question to Dirk Forrester of the US DOE, at RIIA conference on 'Controlling Climate Change: International Investment and Trading Initiatives', London, RIIA, 4–5 December 1996.

likely to outweigh any benefits. It is also possible that Saudi Arabia and some other OPEC countries were influenced by research that increasingly questioned whether emission constraints really would be the threat to oil markets and revenues that had been supposed.[18]

In fact, in a consultation meeting at COP-2, the head of the Saudi delegation had already signalled his acceptance that 'no action is not an option'. At the December AGBM negotiating session he went further, indicating that Saudi Arabia had no intention of obstructing the process: all it asked was that its interests and concerns be taken into account in designing the agreement. Chairman Estrada said he was unsure what this could mean in practice and asked the delegate to produce specific proposals. Within a month, various OPEC countries tabled proposals for a 'compensation fund' that would recompense them for any lost revenues.

As well as addressing a long-held concern, the compensation fund proposal was a brilliant tactical move. It at last enabled OPEC to return fully to the G77 fold, and paved the way for a united G77 position to emerge in 1997 for the first time since its collapse in Berlin. The G77 grouped around support for deep emissions cuts by the industrialized world, with action to compensate *all* those adversely affected – whether by climate change itself or by actions to mitigate it. The negotiations witnessed the curious spectacle of OPEC and the small island states working for the same cause. Of course, a compensation fund was unacceptable to the industrialized countries, which reacted very strongly against any such idea (see Chapter 3), and it was unclear what would happen when the negotiations struck this reality. Nevertheless, the proposal eventually brought all the 'special categories' of countries listed in Articles 4.8 and 4.9 of the UN FCCC together under a common banner, a development that now forms an important feature of the climate change regime.

The specific headline commitment around which the G77 could group emerged in the context of the EU–US debate, after the EU came forth with a specific position in March 1997. The Netherlands, which had assumed

---

[18] Peter Kassler and Matthew Paterson, *Energy Exporters and Climate Change*, London: RIIA, 1997. The results of this work were in circulation for review during late 1996, and in November of that year the lead author was invited to Riyadh by the Saudi Crown Prince to discuss the work and climate change more generally.

the EU six-month rotating presidency in January, was determined to forge a coherent EU response to the new situation in the negotiations. At the March meeting of the EU Council of Ministers, an agreement was finally hammered out. The EU collectively supported a position that all industrialized countries should reduce emissions to 15% below 1990 levels by 2010. Just as significant, and mindful of its long-standing failure to implement common abatement policies at the European level, the EU finally responded to the international demand that it clarify internal responsibilities for achieving any target –without which its position would have had little credibility. The Council set forth a 'burden-sharing' agreement that defined emission targets for each member state.

The ramifications of this agreement, and the EU's position overall, are described in Chapter 3. In fact, the individual national targets added up to a 10% reduction – it was unclear how the remaining 5% would be achieved – and the EU's negotiator had to stress that the EU target-sharing agreement was indicative and not binding. Nevertheless, it marked a decisive shift in the EU's position back towards defining responsibilities for each member state; this, and the fact of being the first major player to put forward a specific target, gave the EU renewed profile and initiative in the negotiations.

Meanwhile, various developing countries were engaged in internal reviews (there was not much chance for the G77 to meet and discuss outside the negotiations themselves). Most strikingly, Brazil made a sweeping proposal in June 1997 for global allocations based on historic contributions to observed temperature increases, and a proposal to create a clean development fund financed out of charges for non-compliance by the industrialized countries. However, it was generally recognized that the allocation proposals were far too complex to be considered in the time remaining (and went far beyond the terms of the negotiations). These were shelved for later consideration, but the G77 lined up behind support for a 15% emission reduction by the industrialized countries (albeit different in detail and for widely divergent reasons),[19] and for the fund proposal. They

---

[19] The G77 called for a 15% reduction in each greenhouse gas individually, whereas the EU position was for CO_2 plus nitrous oxide plus methane (see Chapter 3); and the G77 sought to exclude any flexibility in implementation. Reasons probably ranged from the AOSIS

also reaffirmed opposition to the flexible instruments proposed by the United States, regarding these as a way in which that country would use its political and economic might to avoid significant domestic action and somehow transfer responsibility for action onto others (see Chapter 3).

On the topic of emissions trading, the EU had remained silent in its March statements, riven by internal divisions. As described in Chapter 3's analysis of emissions trading proposals, however, in June 1997 the EU managed to turn this to advantage when it indicated that its attitude to emissions trading would be contingent upon the strength of commitments offered by other countries. This further intensified pressure particularly on Japan and the United States to come forward with specific numbers.

In Japan, bitter and long-standing rifts between the Environment Agency and the Ministry of International Trade and Industry (MITI) were finally resolved when Prime Minister Hashimoto intervened, and Japan came forward in October with a complex proposal based on a 5% indicative reduction with derogations according to various criteria (Chapter 3).

In the United States, however, the pressure reopened old divisions. Within the administration, the Environmental Protection Agency and the State Department argued for stronger action while the Departments of Energy and of Commerce took the opposite view; the Department of Defense also joined the fray, worried about the possible impact of emission constraints on its military operations. Opponents seized upon the new divisions, the alarming estimates of abatement costs produced by some economic models, and the lack of participation of developing countries, to renew assault upon the whole US strategy. The Senate was seized with sudden concern about what the administration was up to, and passed a resolution by 95–0 stating that it would not ratify an agreement that did not include commitments by developing countries. This challenge to the Berlin Mandate placed the administration in a very awkward position.

In the summer of 1997, one of the leading architects of the US strategy, Eileen Claussen resigned from the US State Department, and for a while

---

countries' desire for really strong action (albeit considerably weaker than the 20% by 2005 that they had proposed) to the conviction in much of OPEC that the −15% target was wholly unrealistic and unacceptable to Japan, Russia and the United States, and that sufficient pressure for it might even lead to the collapse of the negotiations.

there were rumours that the United States might retreat from its fundamental position of support for legally binding targets. But with both the President and the Vice-President fully committed, the position held. As the final round of pre-Kyoto negotiations opened in October 1997, and with expectations already lowered by the Japanese pronouncement, President Clinton reaffirmed the basic position and announced that the United States supported legally binding commitments, defined flexibly with joint implementation and emissions trading, to return greenhouse gas emissions to 1990 levels during the period 2008–12.

With Clinton's announcement, all the major players had staked out their proposals and positions. The negotiations initiated in 1990, that established the Convention's principles and procedures and led on to the Berlin Mandate's requirement to negotiate more substantive commitments, had reached a decisive moment. The foundations had been laid, and the battle lines drawn. Now it was the details that really mattered.

Chapter 3

Negotiating the Kyoto Protocol

The Kyoto conference was the biggest and most high-profile event on the international environment since the Rio Earth Summit itself, more than five years previously. The United States alone brought almost a hundred people in its official negotiating delegation; they in turn were swamped by the entourage of industry and environmental NGOs seeking to lobby them. Almost 10,000 people attended in all. Ministers from around the world flew in, including the UK Deputy Prime Minister and, more briefly, US Vice-President Al Gore. In contrast, some European NGOs travelled for six weeks in a 'climate train' across Siberia, making the point about the emissions from air travel generated by such a massive gathering. The media swarmed, anxious for every morsel of gossip about the progress of negotiations.

Unlike the Earth Summit, however, the Kyoto conference was fundamentally about something other than raising awareness and setting out principles and broad plans for action. It was a conference to take specific binding decisions about the commitments of nation-states, in the recognition that the exhortations and aims of previous agreements had not proved sufficient. That involved tremendous complexity in how commitments might be defined, distributed, measured, monitored and implemented in ways that were fair enough for all countries to accept.

Kyoto was of course the culmination of a long preparatory negotiating process. It is not the purpose of this chapter (or this book) to explain the full negotiating history, or to detail the various proposals.[1] Rather, this chapter aims to explain the main proposals and debates that underlay the final Protocol. Broadly, these can be divided into the following substantive areas:

---

[1] Authoritative accounts will be given in other books that are in preparation. The full history is a fascinating one which also gives insights into issues that could well resurface in future years, but is beyond the scope of this book. Certain aspects of debates and proposals that did not find their way into the final Protocol, however, are cited where relevant to understanding the basic political history and prospects.

- *Policies and measures.* Following the first leg of the Berlin Mandate, the EU in particular from an early stage sought agreement on lists of specific policies and measures that Parties might adopt. This also became an avenue for expressions of concern, particularly by OPEC, about the potential adverse impacts of response measures, and for possible counter-measures to offset them.
- *Quantified emission limitation and reduction objectives.* The second leg of the Berlin Mandate, on emission targets or QELROs (see Glossary), became the focal point of negotiations after COP-2. Three dimensions emerged to these debates: the *scope* of commitments (the time-scale and coverage of sources and sinks); the *level* of commitments to be adopted, including whether and how they should be differentiated between parties; and the *international mechanisms* by which flexibility might be introduced into national obligations.
- *Developing-country concerns and participation.* Although the Berlin Mandate precluded the adoption of new commitments by developing countries, the requirement to 'continue advancing the existing commitments' of developing countries was but one of several dimensions of debate concerning their role and needs, including issues relating to technology transfer and finance.

This chapter summarizes core elements in these debates and their evolution. In addition, of course, there were important procedural debates, including the whole structure of reporting, review and compliance processes; these are incorporated in Chapter 4, which explains the content of the Protocol as it emerged. We start with a brief outline of the negotiating process itself.

3.1 The negotiating process

International negotiation is a cumbersome process, particularly when the issue is global, the agenda is immense, perspectives are widely divergent, and the stakes are high. Facts, ideas, interests and positions need to be considered domestically, then floated internationally, and (hopefully) reassessed in the light of reactions and statements from other Parties.

Ideas then need to take concrete form, first as oral explanations and justifications, then as written submissions. As the process develops, the elements need to be expressed in a legal form for incorporation into a negotiating text. Finally, the Parties can engage in debates over specific wording and trade-offs between elements that they do and do not like, and that they can and cannot accept.

In the case of the Kyoto Protocol, this process spanned 30 months between the Berlin conference and the last possible date for holding the third annual Conference of Parties, December 1997. It got off to a very slow start. As noted in Chapter 2, uncertainty reigned as to how to start implementing the Berlin Mandate. Many developing countries were unsure of their basic attitudes and the United States was partly absorbed in internal efforts to build consensus on its strategy across the administration. The EU's attempt to kick-start substantive discussions in autumn 1995 with its proposal on policies and measures met with little enthusiasm.

For most of the first year, indeed, the international discussions focused largely on procedural issues. There were extensive wrangles over political representation on the bureau of the negotiations. Long debates on rules of procedure failed to resolve the core issue of voting: the oil-exporting countries maintained that all decisions should be taken unanimously, giving any country a *de jure* power of veto, which others would not accept. Partly because of this, there was not even agreement on whether the negotiations should result in a Protocol, an Amendment or some other legal instrument.[2] As these procedural discussions became bogged down, there were even suggestions to amend the Convention – under the three-quarters majority rule – to define rules for adoption of a Protocol on the same basis, though this carried other complications.

In the aftermath of the second Conference of Parties in July 1996, it was more evident that the negotiations would be taken seriously at the highest political levels; with the new elements injected by the US statement and the Ministerial Declaration, the negotiators' attention began turning towards more substantive issues. Parties submitted various ideas

---

[2] The negotiations throughout referred to 'A Protocol Or Another Legal Instrument'. Ambassador Estrada joked that if the negotiations succeeded he would be tempted to name his god-daughter 'POALI'.

during the autumn, and the Secretariat produced a 'synthesis of proposals' for the December AGBM session. That session agreed that by mid-January 1997, Parties should submit specific textual proposals, couched in language that could be incorporated in a legal agreement. In March 1997 the AGBM met to debate the elements put forward.

Under the terms of the Convention, the 'text of any proposed Protocol' had to be communicated to the Parties by the Secretariat at least six months before the relevant Conference of Parties. After heated debate, notably with China, the chairman gained the authority to combine the submitted proposals as a formal negotiating text. It emerged as a document of more than 100 pages of widely varying and sometimes contradictory proposals and positions.

With this before them, the Parties began to engage in more earnest debate. From the unwieldy morass of proposals and positions, the chairman was authorized to exert his judgment and produce a consolidated negotiating text for the final AGBM session in October 1997.

With an incisive and sometimes ruthless eye to the little time remaining, the Secretariat worked with the chairman to organize and whittle down the proposals into a chairman's proposed text, that emerged with 22 pages of text plus three short Annexes. Proposals that Ambassador Estrada considered hopelessly unwieldy, complete non-starters politically, or outside the terms of the negotiating mandate, were ejected. The 20 pages listing policies and measures proposed by the EU were eliminated; so too were the proposals from OPEC for a fund to compensate them for possible lost revenues, and the US proposals on the evolution of commitments to encompass developing countries. Gone too were the radically new and complex Brazilian proposals, which had been submitted a few days before the June deadline, on ways of assigning responsibilities for emissions. Although he retained some traces of each of these elements, the text that Ambassador Estrada placed before the Parties focused on the most important and politically plausible policy proposals and institutional requirements; these, with one notable exception – the 'Kyoto surprise' – were to form the basis for the final Protocol.

3.2 Policies and measures

The first item in the Berlin Mandate (Chapter 2) was a commitment for the industrialized world to elaborate 'policies and measures' that Parties would implement to mitigate climate change. Since this was a more diffuse and complex area than the proposed targets and timetables for the reduction of greenhouse gas emissions, and given unstinting US opposition, it attracted less attention, particularly at the last stage of the negotiations, in Kyoto.

The term 'policies and measures' encompasses any action which Parties can adopt, either nationally or internationally, to reduce emissions or enhance sinks. In today's increasingly globalized economy, a number of these measures, such as energy or carbon taxation, could potentially affect the international competitiveness of any Party adopting them unilaterally. Even if the country's economy as a whole does not suffer, the competitiveness of particular industries or firms (in the case of energy or carbon taxation, energy-intensive industries such as steel, aluminium or chemicals) could certainly be affected. Some measures, such as the application of minimum standards of energy efficiency to traded products, may not particularly affect competitiveness, but would clearly work more effectively if applied at an international level. And a few measures, such as the taxation of 'bunker' (aviation or marine) fuels, more or less have to be implemented internationally if they are to have any significant impact.

The key debate in the policies and measures area revolved around a fundamental clash of political and governmental cultures between the EU and United States. The EU, familiar and comfortable with internal harmonization and a single market, and concerned about the general lack of action in response to the UN FCCC's stabilization target, argued strongly for a coordinated approach, specifying a wide range of policies and measures, some of which would be mandatory. US negotiators, anti-interventionist by inclination, sensitive to their citizens' and industries' attachment to cheap fuel, and determined to build in to the Protocol as much flexibility in meeting its targets as possible, flatly opposed the EU's position. Their opposition was reinforced by developing countries, particularly oil exporters, alarmed at the implications of some of the measures proposed and seeking to build in various safeguards to protect their economies and exports.

A total of 12 main proposals from different Parties were included in the negotiating text made available at AGBM 7, in August 1997.[3] The EU's text specified that Parties 'shall adopt and implement policies and take measures within national, and, where appropriate, regional programmes … to limit and reduce anthropogenic emissions of greenhouse gases not controlled by the Montreal Protocol from all relevant sectors … and to protect and enhance sinks and reservoirs'.[4] Following the pattern of its initial proposals a year earlier, three lists of measures were appended. Those included in List A were mandatory for Parties listed in Annex X to the Protocol,[5] those in List B were to be given 'high priority' and Parties were 'to work towards their early coordination', and those in List C were to be given 'priority … as appropriate to national circumstances'.

The measures included in the three EU lists together ran to 19 pages of typescript. Lists A and B were not, at this stage, separated; they included minimum fuel excise duties to be applied across all Annex I countries, preparatory work for a common framework of wider environmental taxation, the removal of fossil fuel subsidies, improvements in energy efficiency, including the application of standards and labels, and fuel switching to less carbon-intensive energy sources. List C, which ran to 14 pages, listed a very wide range of measures in the fields of energy, transport, taxation, industry, agriculture, forestry, waste, fluorocarbons and local authority policy; indeed, it is difficult to think of any major greenhouse gas abatement policy that is not included.

Other negotiating proposals were not nearly as comprehensive, most merely suggesting that Parties should actually do something to implement their commitments set out later in the Protocol, and should perhaps set out their intentions in a national plan. Japan – perhaps the closest to the EU in its approach – proposed a menu of policies and measures from which Parties would be required to adopt a certain minimum number. The New Zealand proposal specifically highlighted cooperation through the

---

[3] Report of the Ad-hoc Group on the Berlin Mandate on the work of its sixth session; Addendum: Proposals for a Protocol or Another Legal Instrument (FCCC/AGBM/1997/3/ Add.1, 22 April 1997).

[4] Ibid., p. 22, paragraph 87.

[5] Annex X was the EU's proposal for a slightly expanded list of countries adopting specific commitments (compared to Annex I to the Convention) – see Section 3.8.

International Civil Aviation Organization (ICAO) and International Maritime Organization (IMO) to control emissions from bunker fuels (points which were – along with everything else – included in the EU lists).

The G77 text included a commitment to ensure that any policies and measures adopted would have 'no adverse impacts on socio-economic conditions of developing country Parties',[6] and referred back to two articles of the Convention: 4.8, which says much the same thing at greater length; and 3.5, which refers to the need for 'sustainable economic growth and development' and, borrowing language from the General Agreement on Tariffs and Trade (GATT), seeks to rule out trade-restrictive measures. In a separate proposal, Costa Rica attempted, as usual, to insert a favourable reference to joint implementation (see below).

Finally, a variety of proposals from oil-exporting developing countries – Iran, Venezuela, Saudi Arabia and the United Arab Emirates – sought to protect their own economies and exports. Policies and measures were to be targeted on coal (removal of subsidies) and renewables, technology transfer was to be encouraged, economic development (including of oil-exporting developing countries) was to be protected, commitments were to be fulfilled individually and not through coordinated action (Iran) and carbon, energy or oil taxation was to be explicitly ruled out.

Most of this was tabled in response to the EU's comprehensive listing of measures, and only helped to reinforce opposition to it. The debate was not helped by the fact that from the early days of the process, the EU's proposal was perceived by others as a vehicle for proposals on a coordinated carbon tax across the industrialized world: a thinly veiled criticism of cheap US gasoline (a topic of transatlantic dispute ever since the first oil shock in 1973) and a red rag to OPEC which viewed such proposals as a conspiracy to grab its oil rent revenues.

The EU failed to build any widespread support for its proposed text, because its case was not very strong to start with and it lacked precision. Measures that could plausibly benefit from international coordination were included, along with many for which no such case existed; the innocuous were listed alongside those which some other Parties found

---

[6] AGBM Report, p. 25, para 93(b).

inflammatory; the general ('measures to improve energy efficiency') along-side the specific ('adopt energy efficiency standards for domestic appliances').

As indicated above, the EU's combined lists were so comprehensive that it was difficult to think of anything else that Parties could reasonably want to do if they were genuinely determined to fulfil their commitments, so the point of incorporating them in the Protocol was not immediately apparent. And if Parties did not intend to fulfil their commitments, the fact that the Protocol listed a range of measures which they could adopt seemed hardly likely to change their minds. No enforcement mechanism was suggested for Parties which would not adopt any or all of the suggested measures. And the list was so long that it maximized opposition from Parties concerned about any single aspect of it. Finally, the EU had spent so much time and effort agreeing its text internally that its negotiators were largely unprepared for the degree of opposition they encountered in the wider forum of the AGBM.

The case for listing a smaller number of key policies which would benefit substantially from international coordination was rather stronger – and it was here that the debate did result in a useful outcome – but in general ran counter to the trend of building in flexibility mechanisms (such as emissions trading) which would make specific domestic action less crucial. The EU's inclusion of highly controversial topics like energy or carbon taxation was not realistic in the prevailing political atmosphere.

Article 2 of the chairman's negotiating text was accordingly a much slimmed-down version, including a mere five sub-paragraphs plus an Annex of 'agreed priority areas'. In the end, however, some of the earlier material was reinstated at Kyoto, ending up with an Article 2 of four substantive paragraphs, and incorporating a list of eight indicative areas for action, as described more fully below in Chapter 4.

3.3 Defining emission targets: time scales, gas coverage and sinks

When a legal commitment is being defined, what should 'quantified emission limitation and reduction objectives' actually refer to? The negotiations highlighted several aspects to this seemingly simple question.

3.3.1 Time scales and time horizons

One aspect was the time scale over which commitments would be defined. The discourse in the EU, and in the Convention's non-binding aim, had focused on specific target year commitments. The United States argued that a legally binding commitment had to be defined over a more extended period, to give some flexibility over the exact timing and because emissions could fluctuate significantly from year to year with changes in weather and economic cycles. A single year would be arbitrary and unstable; rather it would make sense to average emissions over a longer 'budget period'. Three years was felt to be too short to even out fluctuations, and four years would lock the commitment cycle into the US political cycle; as this was not necessarily desirable, the United States proposed a five-year averaging period. The EU and Japan accepted the logic of the argument at the final pre-Kyoto negotiating session. The developing countries remained opposed, apparently in part because of semantic misunderstandings about the implications of the term 'budget' and its monetary overtones; in Kyoto they accepted the principle, recast as a 'commitment period'.

A far more contentious issue concerned the future dates upon which any commitment should be centred. The EU, long a proponent of early action, had proposed an initial focus on 2005; the United States believed that far more time would be needed to make the necessary adjustments. This division had already been flagged in the Berlin conference; the Mandate called for commitments to be defined for dates 'such as 2005, 2010 and 2020'.

The argument for an early target was simple: it was feared that commitments that were outside the range of visible electoral cycles or typical industry financial horizons would be taken as an invitation to delay. The converse US concern was that early targets would prove costly to implement – also that they would leave insufficient time for the institutional and political developments required, particularly to get action through the domestic US legislature. US opposition was reinforced by economic studies that purported to show that it would be cheaper to defer abatement action, and do more later (see Appendix 2 for discussion). Though the administration had forcefully rejected this view as a general principle, US industry – and in particular the electricity sector, which was concerned about the possible costs of being forced prematurely to retire its coal-based power stations – lobbied strenuously against early commitments. Desperate to mollify at least some of the domestic opposition, the administration let it be understood that it would not accept any emission restrictions that would be binding before 2010.

The EU itself found defining a nearer-term target more difficult than it had hoped: the March 1997 EU Council reached agreement on a reduction target for the year 2010, but a figure for 2005 had to wait until June, when it was set at half the 2010 reduction and without any agreement on its distribution between member states. Both Japan and the United States simply refused to offer positions on 2005 targets, and 2010 became the focus of debate. The EU was then left somewhat bereft of a strategy to keep 2005 on the table, as negotiations on specific national commitments for that date became untenable. In Kyoto, Ambassador Estrada sought to retain some of the EU's focus on early action by producing a draft with the commitment period brought forward to 2006–10. EU support for this, however, appeared muted, since it would complicate all the existing calculations and would probably imply lower numerical commitments, amplifying domestic criticisms; and the United States objected. The dates were shifted back to 2008–12, but some remnant of the EU's concern was rescued with a requirement that the Parties show 'demonstrable progress' towards their target by 2005.

The United States in its original Protocol submission had proposed a second commitment period to follow the first, with an allowance for banking and borrowing of emission commitments between the two periods. The difficulties in negotiating – and even developing positions – on a single set of commitments were so huge as to make this impractical. The concept of a second commitment period was retained, however, together with the principle that countries that over-achieved their commitments in the first period could 'bank' their unused allowances for use in the subsequent period. The proposal that countries might 'borrow' emissions from subsequent periods gained no support, however: as critics noted, under these circum-stances there would be no point in time at which a country could be assessed as being out of compliance, hence no point at which to apply any enforcement procedures – a strange interpretation of the term 'binding'. The United States recast its borrowing proposal in the form of a penalty for non-compliance (a deduction from allowances in the subsequent period) but it was too late to be reconsidered in that context: specific compliance procedures were deferred for subsequent elaboration (see Chapters 4 and 7).

One topic associated with these various questions about the definition of future commitments was the base to which they should relate. The negotiations never questioned that Annex I commitments should be defined in terms of changes from historic levels: proposals for other indices, such as defining emissions relative to population or GDP, remained confined to academic literature as they involved changes far greater than countries were willing to contemplate. The Convention had used 1990 as the base year for its non-binding aim – a date which had a huge significance as the year in which the world, by endorsing the first IPCC report, formally recognized climate change as a serious issue, and launched the negotiations that led to the Rio Convention.

The 1990 base year had from an early stage caused complications with the EITs, whose centrally planned economies had led to high emissions that had undergone rapid transition around and since 1990. These countries had obtained a 'degree of flexibility' in the Convention and after Berlin they lobbied successfully to extend this to include the possibility of defining a base year other than 1990. COP-2 accepted base years of 1988 or 1989 for various EITs (years in which their emissions were higher than in 1990), and the final Protocol allows other EITs to propose different base years for consideration by future COPs.

Japan and France had also long complained that a 1990 base year did not acknowledge the considerable progress they had made in emission reductions (due to energy efficiency and nuclear power programmes) prior to 1990. In the run-up to Kyoto there was some discussion of moving forward the default base year for all countries to 1995. This would have made life much easier for those, like Japan and the United States, whose emissions had risen since 1990, and it would have allowed a more impressive headline figure to emerge for these countries' commitments. Arguably, it would also put the economies in transition on a more comparable footing. But it would have created a whole new set of problems for handling EIT commitments, and would have put the EU – whose emissions had declined since 1990 – at a considerable disadvantage (emission trends are discussed in Chapter 5). The EU argued forcefully, and successfully, that such a change would simply reward those countries that had done nothing to limit emissions since the Convention process was launched. The EU prevailed, though a 1995 base year was accepted for the trace industrial gases (see below).

The idea that a five-year commitment period should be matched by a five-year averaging of base year emissions over 1988–92 was also rejected, this time on the more pragmatic grounds that a lot of effort had already gone into establishing inventories for the default base year, and credible data were simply not available against which to assess a 1988–92 base, especially (but not exclusively) in the EITs.

3.3.2 Emission sources

The next issue concerned the range of sources to be included and the means by which they might be compared. Table 3.1 shows the main direct greenhouse gases, together with their main sources, and data on trends, lifetimes and relative contributions. Three major greenhouse gases – CO_2, methane and nitrous oxide – had been the focus of attention for a long time. Three other groups of 'trace industrial gases' had gained increasing attention because of their rising emissions and very long atmospheric lifetimes. How would the Protocol address this range?

Way back in the negotiation of the Framework Convention, the United States had advocated the 'comprehensive approach' of including different gases together in a basket, with the atmospheric impact of the gases compared according to their global warming potential (GWP). The United States argued that this would be a more economically efficient approach to attaining a given degree of control, with emissions being limited from those sources that were cheapest to control; the EU perceived it more as a way of avoiding serious action on the core problem of CO_2 emissions. For the Convention, the dispute was managed with the creatively ambiguous wording of the Convention's non-binding aim noted in Chapter 2. For a binding commitment, that path was not open.

Table 3.1: Greenhouse gases in the Kyoto Protocol

Gas	Qualifying sources	Emission trends since the late 1980s	Lifetime (years)	GWP–100	% GHG 1990, Annex I
Carbon dioxide (CO_2)	Fossil fuel burning, cement	EU static, increases other OECD, sharp decline EITs	Variable, with dominant component c. 100 years	1	81.2
Methane (CH_4)	Rice, cattle, biomass burning and decay, fossil fuel production	Decline in most countries (big increase only in Canada, USA, Norway)	12.2±3	21	13.7
Nitrous oxide (N_2O)	Fertilizers, fossil fuel burning, land conversion to agriculture	Varies, small increases in many countries, decline expected before 2000, decline in EITs	120	310	4.0
Hydro fluoro- carbons (HFCs)	Industry, refrigerants	Fast-rising emissions due to substitution for CFCs	1.5–264, HFC 134a (most common) is 14.6	140–11,700; HFC 134a (most com- mon) is 1,300	0.56
Perfluoro- carbons (PFCs)	Industry, aluminium, electronic and electrical industries, fire fighting, solvents	Static	2,600–50,000	Average about 6,770; CF_4 is 6,500; C_2F_6 is 9,200	0.29
Sulphur hexafluoride (SF_6)	Electronic and electrical industries insulation	Increase in most countries, further rise expected	3,200	23,900	0.30

Source: Main data from IPCC, SAR WG I, Table 2.9, p. 121.

At stake were two technical issues, and one major political one. The first issue – how to compare the effects of different greenhouse gases – found an easy if imperfect political resolution in the *de facto* understanding that any comparison would be made using the 100-year direct GWPs that had been estimated by the IPCC in its Second Assessment Report (see Table 3.1).[7] Academic literature had long pointed out that

[7] The GWP compares the radiative forcing of a given mass of different gases, relative to CO_2, integrated over a specified period after emission. The IPCC produced estimates of direct GWPs for 20-, 100- and 500-year time horizons, for the main direct greenhouse

GWPs were both uncertain and logically imperfect, but the only alternative put forward in the negotiations was in the sweeping Brazilian proposal that sought a wholly different and more explicitly science-based approach to defining targets.[8] For other negotiators, the IPCC's 100-year GWPs sufficed.

The other technical difficulty is that the precision with which emissions can be monitored, either for the historical base year or for future years, varies considerably between gases, sources and countries. The difficulty of monitoring emissions of some non-CO_2 gases, notably agricultural methane and N_2O for which uncertainties can be at least 50%, was a principal objection raised against including different gases and sources together. Some major changes in inventories between the First and Second National Communications of Parties reinforced these concerns, though it was countered that this showed that major improvements had been made and it would be difficult for Parties to try to make further substantive alterations in their base year estimates. The problem would then be one of monitoring emissions during the commitment period, and (it was hoped) there would be enough time to resolve this satisfactorily.

Ultimately, in the process of negotiation such technical concerns became eclipsed by the economic and political arguments in favour of including a range of gases together. If significant gases were excluded altogether, it would weaken the scope and impact of the Protocol. If they were included separately it would add yet more tracks of separate negotiations. But most important of all to the politicians, the inclusion of some other gases – especially methane, emissions of which were easier

---

gases. The IPCC concluded that indirect effects arising from chemical interactions and aerosol effects were too uncertain to be quantified. Methane is the shortest-lived of the greenhouse gases; the lifetime of nitrous oxide is comparable to that of the dominant carbon dioxide decay lifetime of around 100 years, while many industrial trace gases have lifetimes of several hundred years.

[8] No single number can compare accurately the relative effects of the different gases, in part because they have different patterns as well as time-scales of decay in the atmosphere. The Brazilian proposal included allocating emission reductions in proportion to observed global average temperature changes which would be traced back to the emissions considered responsible.

to control and in several countries were already declining – made it easier to adopt stronger emission targets. In its March 1997 Council Agreement, the EU defined its targets in terms of a basket of CO_2 + methane + nitrous oxide; the inclusion of methane was essential for enabling the strong numerical targets to be declared.

Alone among the industrialized countries, Japan probably stood to lose from a move to include other greenhouse gases (at least against a 1990 base year), and was concerned that a multi-gas agreement would be neither clear nor easy to monitor.[9] Nevertheless, in its October declaration Japan accepted the trend and moved to the EU basket of three gases. The United States sought to include an additional three industrial trace gases:

- the 'ozone-friendly' hydrofluorocarbon (HFC) replacements for CFCs, that are unfortunately powerful greenhouse gases, use of which had been increasing since the late 1980s;
- perfluorocarbons (PFCs); and
- sulphur hexafluoride (SF_6), principally a by-product of aluminium smelting, and molecule-for-molecule the most potent of all the greenhouse gases.[10]

The EU and Japan opposed including these gases in a 'basket' with the others. An underlying political concern was that the United States had a more fully developed chemical industry in 1990 and had already started its transition towards HFCs by then, so that it stood to gain comparatively; indeed it appeared that chemical industries were lobbying for opposite things on opposite sides of the Atlantic, though the reality was more

---

[9] Non-fossil fuel CO_2 emissions accounted for less than 6% of Japan's 1990 emissions, compared to 15–20% in the United States and most of Europe, and up to 50% in some other OECD countries (see Chapter 5). Including the three industrial trace gases was to Japan's further disadvantage, since these emissions had been growing rapidly with CFC replacements.

[10] Various other gases, such as sulphur dioxide and carbon monoxide, can have an indirect impact on the radiative properties of the atmosphere, but their effects are still more uncertain, and can vary geographically; most are also subject to control for other reasons. The Protocol focuses upon the direct greenhouse gases.

complex.[11] A more respectable problem was that by including these exceptionally long-lived gases alongside others on the basis of a 100-year GWP, the Protocol would not address adequately the distinct, very long-term threat posed by a build-up of gases that would persist in the atmosphere for centuries. In the final days of Kyoto, there was serious talk of a twin-basket approach, deferring separate commitments for the three industrial trace gases to subsequent negotiations. In the end, however, it was agreed to include all six gases together, with the provision that countries could choose a 1995 base year for calculating reference emissions of the industrial trace gases.

3.3.3 Sinks

The possible role of sinks – activities that absorb CO_2 from the atmosphere – formed one of the most technically complex issues in the entire negotiations, which rose to high controversy in the late stages of the Protocol negotiations. Proponents argued that CO_2 absorption should be directly offset against emissions (the 'net' approach) because, from an atmospheric standpoint, absorption is equivalent to reduced emissions. New Zealand, Norway and other Scandinavian countries, and to a lesser extent the United States and Canada, all have large land areas with high actual or potential CO_2 absorption; thus (depending on how future absorption is compared against that in the base year) they stood to gain considerably from the net approach.[12] Opponents feared that this would detract from the pressure to limit emissions; that it might allow countries

---

[11] Greenpeace claimed that an agreement between the Dutch government and its powerful chemical industry was at the root of EU opposition to any inclusion of the trace industrial gases. In the United States, the chemical industry judged that inclusion alongside the mainstream greenhouse gases was the best way of moderating the pressures for control, fearing that they would be subject to relatively more stringent control if they were isolated for separate treatment. Industry in France took a similar view.

[12] The issue also became linked to the debate on joint implementation with non-Annex I countries that could offer forest plantation or protection as a way of 'offsetting' emissions in Annex I countries; see Tim Forsyth, 'Joint Implementation and Technology Transfer: Joint Benefits?', *EEP Climate Change Briefing*, No. 7, October 1997, and the references on joint implementation (note 23 to this chapter).

to claim credit for the massive ongoing naturally occurring absorption; that such sinks were inherently far too difficult to monitor accurately; and that including sinks could give incentives to replace mature, old-growth forests with fast-growing monoculture plantations.

The implications of sinks also varied greatly between countries. Table 3.2 summarizes estimates of total emissions from land use and forestry in 1990 for those countries for which sinks (or emissions) were reported and constituted more than 5% of 1990 emissions. They are certainly significant for all the 'New World' and Scandinavian countries; in all but Australia they represented annual sinks of at least 10% of national gross emissions, and this was *before* any policies directed specifically towards enhancing domestic sinks. In some countries the figures are much larger: New Zealand, Sweden and Latvia had estimated that their total net sinks absorbed more than half their total emissions; this may also have been true for Finland. For Latvia these data result in overall greenhouse gas emissions in 1996, including land-use change and forestry, being only 15% of their base year level.[13] Furthermore, the issue remains plagued by uncertainties: during 1998, one study concluded that US soils were absorbing as much CO_2 as its industries emitted, though these claims were treated with considerable scepticism by other scientists.[14]

Whether to include sinks as 'negative emissions' had been an issue for a very long time; the United States had proposed it as part of the 'comprehensive approach' since before the Convention negotiations themselves were initiated – as indeed it had done for emissions trading.[15] Yet sinks only became the focus of intensive and high-level political debate in the closing stages of the Protocol negotiations, as negotiators realized just how much was at stake. Global carbon dioxide emissions from fossil fuels amount to about 6 billion tonnes annually, to which deforestation adds

---

[13] FCCC/CP/1998/INF.9.

[14] *Science*, vol. 282, 1998, p. 442; a popular account of this and responses are given in *New Scientist*, 24 October 1998.

[15] The US government submitted 'Concept papers' on the comprehensive approach and on emissions trading to the IPCC's Response Strategy Working Group in 1989, and followed this up with a high-level government seminar on these topics for delegates attending the IPCC meetings in Washington in February 1990.

Table 3.2: Land-use change and forestry (LUCF) emissions for Parties reporting with LUCF emissions greater than 5% of 1990 emissions

	GHG emissions excluding LUCF (MtC-equivalent)	Net emissions from LUCF, 1990 (MtC)	(% of total)
Australia	113	+ 16	+ 14.2
Austria	21	– 4	– 17.2
Canada	163	– 12	– 7.4
Estonia	11	– 3	– 27.8
Finland	18	(– 8) to (– 5)	(– 46.5) to (– 29.4)
France	152	– 8	– 5.4
Ireland	16	– 1	– 9.1
Japan	333	– 23	– 6.9
Latvia	10	– 3	– 30.4
Lithuania	14	– 2	– 17.2
New Zealand	20	– 6	– 29.4
Norway	15	– 3	– 17.4
Poland	154	– 9	– 6.2
Russian Federation	829	– 107	– 12.9
Slovakia	20	– 1	– 5.9
Slovenia	5	– 1	– 11.9
Spain	82	– 8	– 9.6
Sweden	18	– 9	– 52.8
Switzerland	15	– 1	– 8.1
Ukraine	247	– 14	– 5.7
United States	1636	– 312	– 19.0

Source: FCCC/CP/1998/INF.9 and FCCC/CP/1998/11/Add.2. Russian data from 2nd National Communication.

another 1–2 billion tonnes. Only about half of this stays in the atmosphere; the rest is quite quickly absorbed in the ocean surface waters and by plant growth.[16] It had long been accepted that Parties should not be able to claim for purely natural sink processes; the Convention refers to 'anthropogenic emissions by sources and removals by sinks' in its general references.[17] Yet if sinks were to be included in specific commitments, what

---

[16] The reality, needless to say, is rather more complex than this indicates, since absorption is driven by concentration trends – and other factors – not annual emissions *per se*.

[17] The convoluted emissions 'aim' in Article 4.2 appears to refer to emissions excluding sinks, though it succeeds in being ambiguous on this matter, as on many others.

actually constitutes an 'anthropogenic sink'? Various definitions were put forward, but 'none of the definitions proposed could guard against the "Amazonian picket fence" problem, i.e. allowing Parties simply to erect a fence around natural forests to be able to take credit for what nature was already doing.'[18] At worst, such a loose definition could negate the entire impact of the quantified commitments under negotiation.

This was another politically divisive issue, partly setting the Old World against the New. Most of Europe and Asia, having high population densities and highly developed land-use patterns, had limited prospects for enhancing sinks on their own territory. To the EU and G77, sinks appeared as another way of weakening the focus on the core problem of curtailing greenhouse gas emissions, and one which, moreover, could carry potentially troublesome overtones concerning sovereignty over land use for developing countries. They thus mostly opposed the inclusion of sinks, fearing it would add new loopholes from which the profligate 'New World' countries would benefit the most (though as countries began to crunch numbers, important divisions also opened within the EU). They argued that anthropogenic sinks could not be adequately defined, delineated, measured and monitored, and also that they were inherently transitional and not part of a long-term resolution of the climate problem.

Some other countries, however, such as New Zealand, realized that they would be unable to agree significant limitations unless sinks were included, and they argued that it made no sense to exclude the area in which they felt they could contribute most. There were also strong industrial pressures to include sinks, particularly in the United States, which conceded that their inclusion could strengthen its stabilization offer by several percentage points. Japan also strongly supported the inclusion of sinks in a form that might allow some credit from forests in its mountainous regions.

Key questions concerning the inclusion of sinks in Annex I commitments included the following:

---

[18] F. Yamin, 'The Kyoto Protocol: Origins, Assessment and Future Challenges', *Review of European Community and International Environmental Law*, Vol. 7, No. 2, 1998.

- should sinks be included only on a monitoring project-by-project basis, or on a basis of overall 'net' anthropogenic emissions?
- how could one realistically distinguish between 'anthropogenic' and 'natural' sinks?
- should qualifying sinks be restricted to visible biomass or should soil carbon be included?
- should other forms of sequestration (e.g. reinjection of CO_2 from gas production) be counted as a sink or as an avoided emission?
- what procedure would best protect against multiple accounting of sinks that were destroyed and grew again?
- should accounting of sinks be symmetrical between the base year and commitment period (the 'net–net' approach), or should it take gross emissions in the base year and account for absorption only up to and/or during the commitment period (different forms of 'gross–net' approach)?

As the EU accepted that sinks would not be kept off the negotiating agenda, the negotiations became more focused, and the answer to a number of these questions as to what sink activities should be allowable crystallized. The EU – responding in part to internal pressures as well – accepted that there would have to be some allowance for netting of emissions against sinks during the commitment period, but they and others succeeded in drawing relatively tight – albeit still ambiguous – boundaries around the scope of qualifying activities. As explained in Chapter 4, the Protocol established a regime on sinks within Annex I that allows only absorption due to 'direct, human-induced' changes, limited initially to forestry-related activities after 1990, while allowing the possibility of additional allowable activities as knowledge and managerial and monitoring capabilities expand and negotiations proceed. The Protocol answered a few, but by no means all, of the outstanding questions, and a number were referred to the IPCC for technical assessment while debate on others simply continued in the aftermath of Kyoto.

3.4 Assigning emission targets: differentiation and the 'EU bubble'

In negotiating emission targets, two fundamental questions emerged: what level should be agreed, and should all the industrialized countries face the same commitment or should they be differentiated according to national circumstances?

The situation facing negotiators was varied in terms of trends already observed during the 1990s, as presented in Table 3.3. Figure 3.1 displays graphically, for most OECD countries, the average annual percentage change of emissions between 1990 and 1995, supplemented with the (more preliminary) data for 1995–6. Within the OECD, emissions had declined significantly only in Germany and the UK; their emissions in 1996 were respectively 10% and 5% below 1990 levels, though both increased slightly in 1995–6. In both cases the reductions are attributable in large part to changes that had nothing to do with climate policy and cannot readily be sustained: the absorption of East Germany, and the 'dash for gas' that accompanied UK electricity privatization.[19] Emissions in France (and Switzerland) had remained roughly stable, and although emissions in all other EU countries increased, these trends were sufficient to keep overall EU emissions in 1996 about 2.5% below 1990 levels.

Elsewhere in the OECD, emissions rose, and more sharply during 1995–6. In the United States, after moderate growth, emissions were said to have jumped by 3.6% in 1996 (though this was later revised downwards) to 9% above 1990 levels; they were more than 10% up in Canada and Japan. The trends are dominated by growth in CO_2 emissions from fossil fuels which, as noted, comprise more than 80% of 1990 emissions from Annex I countries and are the most politically sensitive. Methane emissions generally declined apart from significant growth in the United States, Canada, Norway and Greece. Trends in other greenhouse gases were more varied.

The situation in central and eastern Europe is radically different. There, the economic transition led to a large decline in emissions as economies

---

[19] In fact it seems likely that emissions would still have declined in the absence of these factors, though only very slightly: Nick Eyre, in J. Goldemberg and W. Reid, *Trends and Baselines: Promoting Development whilst Limiting Greenhouse Gas Emissions*, Washington: WRI /UNDP, 1998.

Table 3.3: Total anthropogenic CO_2 emission changes, excluding LUCF, 1990–96

Country	Base year (MtC)	1995 (% from base year)	1996 (% from base year)
EU	**906.7**	**− 3**[a]	**n.a.**
Austria	16.9	+ 2	+ 5
Belgium	31.7	+ 5	+ 11
Denmark	14.3	+ 14	+ 40
Finland	14.7	+ 4	n.a.
France	106.6	+ 1	+ 4
Germany	276.6	− 12	− 10
Greece	23.3	+ 6	+ 8
Ireland	8.4	+ 11	+ 13
Italy	117.9	+ 1	n.a.
Luxembourg	3.5	− 25	n.a.
Netherlands	44.0	+ 10	+ 15
Portugal	12.9	+ 8[c]	n.a.
Spain	61.8	+ 2[c]	n.a.
Sweden	15.1	+ 5	+ 14
UK	159.0	− 7	− 3
USA	**1,348.2**	**+ 5**	**+ 9**
Japan	**306.7**	**+ 9**	**+ 10**
Australia	**75.1**	**+ 8**	**+ 12**
Canada	**125.7**	**+ 7**	**+ 10**
Other OECD			
Iceland	0.6	+ 6	n.a.
New Zealand	6.9	+ 7	+ 15
Norway	9.7	+ 8	+ 16
Switzerland	12.3	− 2	0
EITs			
Bulgaria (1988)[b]	26.4	− 36	n.a.
Croatia[e]	6.4	− 42	− 29
Czech Republic	45.1	− 22	− 20
Estonia	10.3	− 45	n.a.
Hungary (1985–7)[b]	22.8	− 29	n.a.
Latvia	6.8	− 51	− 55
Lithuania	10.8	[d]	n.a.
Poland (1988)[b]	130.0	− 22[c]	n.a.
Romania (1989)[b]	53.1	− 36[c]	n.a.
Russian Federation	647.0	− 30	n.a.
Slovakia	16.4	− 19	− 23
Slovenia[c e]	3.8	[d]	n.a.
Ukraine	190.9	[d]	n.a.

Source: Adjusted from FCCC/CP/1998/INF.9 for countries with 1996 emissions, FCCC/CP/1998/11/Add.2 for other countries. Data for Croatia supplied by Fanny Missfeldt from UN ECE and EKONERG. Data for Romania from Second National Communication.

Notes: [a] 1994 data for Portugal and Spain.   [b] EIT that uses a base year other than 1990.
[c] 1994 data.   [d] no data available after 1990.   [e] base year not yet agreed, data refer to 1990.

Figure 3.1: Trends in emissions of the Kyoto greenhouse gases, 1990–95 and 1995–6

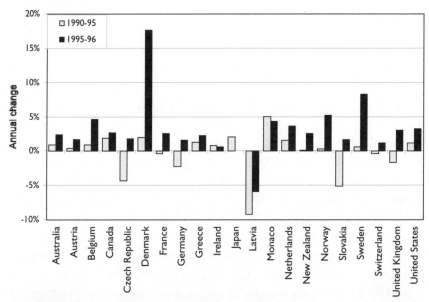

Source: FCCC/CP/1998/INF.9.
Notes: The data show the percentage annual change in the basket of the six greenhouse gases covered by the Kyoto Protocol, excluding land-use change and forestry. Data for 1995–6 must be considered preliminary and influenced by short-term fluctuations.

contracted and old and inefficient plant was scrapped – though much still remains in operation. In the former satellite countries of central Europe, by 1995 emissions had declined to 20–35% below their base year levels, but appeared to have stabilized by the mid-1990s with some reporting a slight increase in 1996. In the former Soviet Union the decline has been more dramatic and extended. The steepest decrease of emissions was experienced in the Baltics and Ukraine, where by 1995 emissions had declined to around half their 1990 levels. Economic change in Russia was slower and collapse in energy demand was cushioned by the maintenance of subsidies or non-payment for its domestic energy production (see Chapter 5, Box 5.2), but even here, CO_2 emissions declined by an estimated 30% over 1990–94. These trends and prospects are discussed further in Chapter 5.

The dominant and almost obsessive focus in the negotiations, however, was on OECD commitments. It was recognized that the EIT commitments would allow growth back towards former levels; the question was how emission cuts in the OECD might be distributed. Flat-rate emission targets appeared attractive because of their simplicity, and have indeed been a feature of the first round of several previous international environmental agreements, which have become subsequently more differentiated over time.[20] To that extent, flat-rate reductions formed a natural initial focal point of discussions. The EU in particular supported flat-rate reductions across all the industrialized countries, relative to 1990 base year emissions. It had the lowest projected increases and judged (rightly) that if targets were differentiated, it would end up with a relatively more stringent commitment than it would with a flat rate; the EU also genuinely feared that differentiation would prove unmanageably complex and derail the negotiations.

Indeed in the central political dialogue between the United States and the EU there was a certain logic to equal percentage cuts from 1990 levels. The United States, with per capita emissions almost twice those of most other OECD countries except Canada and Australia, was vulnerable to accusations that it had a huge potential for reductions and should cut back by more than other countries. At the same time, internal political pressures pointed in the opposite direction: the United States had the greatest difficulty in mustering any domestic support even for stabilizing emissions. Equal reductions between the United States and the EU seemed the only safe solution in such a peculiar political context, and the most obvious way of keeping the US commitment 'in line' with the international community, in some basic psychological sense.

Yet from a wider perspective, flat-rate reductions were neither efficient nor environmentally effective, nor were they feasible as a means of achieving emission reductions. As numerous studies showed, different countries faced very different costs of abatement. There was a danger that agreement could only be reached on a 'lowest common denominator target' which would require very little effort from some countries; or, if the

---

[20] O. Greene, 'Lessons from Other International Environmental Agreements', in M. Patterson and M. Grubb (eds), *Sharing the Effort: Options for Differentiating Commitments on Climate Change*, report of a workshop, London: RIIA, 1996.

pressures for greater reductions were overwhelming, countries that faced insuperable difficulties might simply ensure that the agreement was full of loopholes. And, while no international agreement can plausibly seek fully efficient targets, there were raw political realities behind the data. Some countries, such as New Zealand, which was one of the lowest per capita emitters in the OECD, face huge difficulties in stabilizing emissions. So does Japan, which argued bitterly that flat-rate reductions would penalize it for its massive strides in energy efficiency prior to 1990.

Proponents of flat-rate reductions were forced to accept the principle of differentiation, but, they argued, only for the longer term, because there was no time to solve the immense complexities of working out a rational basis for differentiating commitments before Kyoto. They had a point: every country that supported differentiation had a different idea of how it should be calculated.[21] Many different indicators were proposed, relating to GDP, energy intensity, carbon intensity, historical emissions, trade patterns, etc. Most 'differentiators' argued that low carbon intensity (i.e. low carbon emissions relative to GDP) in 1990 should be a basis for a weaker target. Unfortunately, Australia argued precisely the opposite, claiming that high carbon intensity showed an innate dependence upon fossil fuels that could only be broken at great expense. Almost the only common theme to emerge was that each country proposed indicators that would be most beneficial to itself.

Into this *mêlée* was thrown the question of how the EU would allocate emission commitments among its member countries. Clarification of this was a political imperative if an EU target was to be taken seriously, for the lamentable failure of the EU's harmonized policies since 1992 was apparent for all to see. Other countries were unanimous that the EU must clarify how it would resolve internal responsibilities for implementing any commitments adopted at Kyoto. The breakthrough came in March 1997, with the EU's internal agreement on sharing out a collective 10% emission reduction between its member states; this spanned 25% reductions (for Germany, Denmark, Austria; and even 30% for Luxembourg) to increases for the poorer southern states, ranging up to 40% for Portugal.

---

[21] Ibid.

This 'EU bubble' agreement had political and legal ramifications that reverberated throughout the rest of the negotiations. The EU maintained that its position was justified by its unique character as a supranational institution. However, the original agreement did not specify what would happen if the collective commitment was not achieved. Far more serious was the EU's untenable negotiating position of insisting on its right to have such a huge internal differentiation of commitments among its members, while calling for equal reductions by all other countries. Urging Japan, Norway and New Zealand to agree to a 15% reduction while allowing France, a fully developed country, to stay at 1990 levels and others to rise was inflammatory, and earned the EU bitter condemnation. The United States backed a flat-rate reduction and was happy to accept the EU's internal bubble in principle as 'zero-cost emissions trade', giving it a stranglehold over EU resistance to emissions trading. Nevertheless the United States, struggling to persuade Turkey to sign the Convention and developing countries to engage in constraints, deplored the 40% increase allowed for Portugal.

The bubble thus earned the EU almost universal hostility from the rest of the OECD, and it had many substantive consequences for the Kyoto agreement as set out in Chapter 4. First, it was the precedent for a general provision allowing, in principle, any group of countries to get together and redistribute their commitments. In October 1997 the EU offered the 'Luxembourg compromise', defining liability rules and offering 'bubbling' as a more general principle. This formed the basis of the text in the Protocol, as set out in Chapter 4 below.

Other consequences emerged as a natural result of the refusal to engage in any serious debate about wider differentiation on the grounds that there was not time to develop a rational scheme. The EU made enemies of some countries that were desperate for differentiation and might have otherwise been allies (such as Norway, New Zealand and, possibly, Japan); and in the event differentiation was agreed, in the final days, on a largely irrational basis, with limited analysis. When combined with the international transfer mechanisms this may dog the implementation of Kyoto for years to come.

On the other hand, the EU's stand on strong, flat-rate commitments constituted the only real focal point for negotiations on targets. It provided a point of reference for long enough to ensure that the eventual targets, as set out in Chapter 4, were linked to a significant reduction with relatively few derogations. Since the 'differentiators' could not agree on any coherent alternative, it is hard to imagine what the outcome might have been without such a reference reduction. The EU's stand was from a number of viewpoints inconsistent and unfair, but in the realities of global politics it was ultimately effective in its main aim.

3.5 International flexibility within Annex I: Joint implementation and emissions trading

3.5.1 Introduction: the drive for international flexibility

While the EU's main focus was on the internal policies and measures that its member states might adopt, and the effort to hold other countries to flat-rate emission reductions, the eyes of the United States and many other OECD countries in the JUSSCANNZ group (see Chapter 2, section 2.1) were focused in the opposite direction: on what they might exact from other countries, and thus lessen domestic pressures. All these countries' emissions had risen since 1990 and, for varied reasons, they were uncertain how much they could or would achieve domestically. Against EU insistence on flat-rate emission reductions for all, these countries argued as hard as possible for international flexibility.

The United States had additional reasons for promoting international flexibility. Its economic analysis included many global economic models that underlined that emission reductions could be achieved most cheaply in the economies perceived to be least efficient – the EITs and developing countries. Furthermore, the political debate in the United States had sought – with astonishing effectiveness – to turn the spotlight onto the growth in emissions from developing countries, arguing that if they were not involved, any US action would be swamped by growth in their emissions. In addition, the United States saw that international flexibility could provide opportunities both for assisting EITs and for promoting investment by US industry globally.

But both confusion and concern surrounded international flexibility. With the benefit of hindsight and debate the basic options are clear. As late as December 1996, however, there was much uncertainty about what the United States really meant by commitments that were 'binding but flexible'. Was this not a contradiction in terms?

As indicated in the US statement at Geneva, there were two main specific options: crediting to a sponsor country of emission reductions achieved by investment abroad; and international emissions trading. The first, to eternal confusion, became identified with the term 'joint implementation', which, as noted in Chapter 2, already had a long and tangled political history.[22] The second was more concrete and clear, and to many, more startling.

3.5.2 Project-based joint implementation within Annex I

Project-based 'joint implementation' is a system in which a country subject to a quantified emission commitment, or a company from that country, invests in projects that reduce emissions in another country. The sponsor company and/or country can claim the emission reduction as a 'credit' against its own required reduction in emissions. The host country has an incentive to allow or invite the investment on the basis of its perceived contribution to economic and/or environmental objectives. The incentive for participating host companies is the profit potential of the investment and related stakeholder benefits. This definition is closely linked to the concept of emissions 'crediting' in some US domestic environmental legislation, in which companies subject to an emissions limit can earn credit by reducing emissions from some other source outside the quantified control system.

---

[22] In reality joint implementation is best conceived as a broader term encompassing all forms of international flexibility in implementing emission commitments. Jackson et al. suggest that JI should be understood as a general term for cooperative implementation, of which specific forms are defined according to the openness of the system ('JI' with countries that do not have constraints being at one extreme of this), the level of actors (governments or companies at different levels in the supply chain), and the degree of crediting. (T. Jackson, K. Begg, and S. Parkinson, 'The Language of Flexibility and the Flexibility of Language', *International Journal of Environment and Pollution*, Vol. 10, Nos. 3/4, 1998, special issue on 'EU Climate Policy: The European Commission Policy/ Research Interface for Kyoto and Beyond', pp. 462–75.)

In the eyes of its proponents, project-based international crediting was an obvious tool for promoting efficiency, a 'win–win' option for all concerned. As the debates around the Berlin decision on 'activities implemented jointly' had shown, however, it was in fact highly contentious, particularly in a North–South context, for reasons sketched below. Throughout most of the negotiations, the G77 and China maintained their basic opposition to joint implementation, despite sustained pleas and pressure from the United States and others.

There had been various experiences with 'activities implemented jointly' under the terms of the AIJ decision taken at Berlin, as summarized below, and this had helped east European enthusiasm for the concept. In practice, the impasse on joint implementation was confined to the North–South context. Far less concern was raised about its operation between industrialized countries, and the structure of the negotiating text treated this separately. Concerns were raised about adequate monitoring and oversight, but it was not felt that any fundamental principles were at stake in allowing project investments between two industrialized countries that were bound by specific commitments to generate a transfer of credits. East European countries almost unanimously supported the idea as a possible way of attracting additional investment to refurbish and replace their ageing infrastructure, and the concept of joint implementation in this context entered the final Protocol with relatively little dispute.

3.5.3 International emissions trading

The same could hardly be said of emissions trading. Emissions trading is in principle a simple idea. Each participant in the system is subject to an allowed level of emissions, such as total allowable greenhouse gas emissions over a given period. However, they are also allowed to trade these allowances: one may emit more, if another agrees to emit less and sells its corresponding allowance. Economically, emitters with low abatement costs will reduce emissions and sell their excess allowance, whereas polluters with high abatement costs will prefer to buy. If necessary, the overall allowable emissions total can be reduced over successive time periods.

Such a system has economic benefits, because it creates an incentive for participants to reduce emissions in ways that are most cost-effective. A given level of emissions reduction is therefore achieved at least cost. The administrative requirements may be modest: as with any binding constraint, emissions need to be monitored effectively for each participant, and trades need to be registered and tracked, but this is not an onerous task.

Interest in emissions trading for controlling greenhouse gases had grown during the 1990s, particularly in the United States, for several reasons.

First, the US experience with its tradable permit system to control sulphur dioxide (and ozone-depleting substances) was widely seen to have been a success: emissions have been reduced at much lower cost than originally predicted, because of the flexibility of the permit systems and the incentives they give for innovation.

Second, emissions trading is more attractive to industry than other economic instruments (such as carbon taxes), because the system creates a tradable asset: the permit or allowance. Even if a company has to buy permits to cover all its emissions, it still acquires the value of those permits as an asset which can in principle be sold in the future if emissions can be reduced. This contrasts with a tax, which extracts revenue from the firm without adding any compensating value. In addition, governments may give out some or all of the permits initially, reducing the cost to industry. In many countries taxes have proved extremely controversial and difficult to implement, primarily because of industrial opposition, so that other market-based approaches are attracting growing interest.

In the context of climate change there were additional reasons. Proponents argued that emissions trading increased the flexibility and efficiency of commitments, making it more plausible for governments to agree to greater emissions control, at lower cost. International emissions trading was seen as the most efficient and direct route to the international flexibility sought by the United States and other JUSSCANNZ countries – and was viewed with suspicion by most of the rest of the world. It was to become the crux on which, from some perspectives, Kyoto stood – and nearly fell.

The first reaction to the US proposal at Geneva was one of uncertainty, backed with a lot of suspicion. For most negotiators, the concept was entirely new. Indeed the United States itself was at the time unclear as to how it would develop the proposal in the context of an international agreement, and it mixed references to governmental commitments with examples of its sulphur trading programme, citing the need to let industry invest in emissions reductions wherever it was cheapest to do so. The import of the US proposals only really became clear after a group of US officials retreated over Christmas 1996 to hammer out specific textual proposals. Their subsequent Protocol proposal stated that countries – Parties to the Protocol – should agree to a specific commitment of 'carbon equivalent emissions allowed for a budget period', and that then: 'A Party ... may ... transfer to, or receive from, any other [participating] Party, any of its tonnes of carbon equivalent emissions allowed for a budget period, for the purpose of meeting its obligations.' The proposed basis on which governments could exchange their emission commitments was not specified and did not need to be: all that matters is that combined emission allowances after the 'trade' do not exceed the initial combined total allocation. The responsibility rests with the governments, but the way in which countries limit emissions internally would remain entirely a sovereign issue, as would the terms of trading.

The US draft language then stated that: 'A Party may authorize any domestic entity ... to participate in actions leading to transfer and receipt ... of tonnes of carbon equivalent emissions allowed.'

This could allow a government to create an internal system of tradable emission permits along the lines of the system used for controlling sulphur in the United States, and then to engage in international trading. Each participating industry would be required to obtain permits equivalent to its emissions. Direct participation in such a system could be limited to major industries (such as power generators). The new feature would be that industries could trade internationally with companies in other countries that adopted a similar system.

Although many details remained to be resolved (see Chapters 4 and 6), the essence of the US proposal on emissions trading was now clear; and the United States embarked upon strenuous diplomatic efforts to explain and promote the concept to other countries.

The proposals found ready favour with most of the JUSSCANNZ

group, which were desperate for ways of promoting international flexibility and easing the domestic costs and difficulties that would be caused by inflexible flat-rate emission reductions. Elsewhere, however, scepticism was harder to overcome.

Initial reactions included concern that emissions trading would somehow confer a morally objectionable right to pollute. In the OECD, this concern did not go far beyond bar-room rhetoric and denunciations in the press. Every country was emitting and would continue to emit greenhouse gases at some level; indeed the very concept of emission targets could be similarly construed as condoning this. It was hard to argue that making targets tradable necessarily makes them immoral.

A more practical objection raised in Europe was that such a novel and complex idea in the international scene could not possibly be negotiated in time for Kyoto. More Machiavellan commentators suspected the United States of introducing emissions trading as a way of deliberately confounding the negotiations. For Kyoto to establish industry-level trading – in which industries hold the permits and can trade them internationally – would indeed have been an impossibly complex task in the time available. But the US submission of its negotiating text made it plain that the goal for Kyoto was altogether more practical, focusing upon a structure of intergovernmental trading that could allow the subsequent evolution of industry-level trading by those countries desiring to participate. It sounds complex but is in fact astonishingly simple: the text would just establish that governments can exchange their 'tonnes of CO_2' allowed under the Protocol, and that they are entitled to pass this authority on to subnational entities (e.g. industries) providing that the governments ensure the integrity of any resulting trades.

Thus concerns that such a system was not practical, that there was not enough time to set it up, or that administrative requirements would be too onerous, proved hard to sustain. As these objections foundered, two core concerns emerged.

The first was the general concern that international flexibility might provide a way for the leading emitters to avoid serious domestic action. As the Kyoto debate proceeded, this began to take a very specific and startling form, arising from the peculiarities of the situation in the EITs. In most of the OECD, as noted, emissions had continued rising. But in the EITs, and most notably Russia, emissions fell dramatically with the economic transition. If emissions there stayed quite low, a flat-rate emissions target with unconstrained trading could imply a huge transfer of allowances from most of the transition economies, which would have a surplus, to the OECD countries, above all the United States and Japan. The United States understood this, and over successive months teams of US officials went to the East to explain the windfall that could be waiting. No specific deal was concluded but the Russians got the point, and in March they submitted their own negotiating proposals in favour of emissions trading with flat-rate initial allocations from 1990 levels.

There are two ways of looking at this. The primary incentive for the United States is that trading provides perhaps the only means by which it can agree to substantial flat-rate emission reductions. It was obvious that the US Congress would never agree to bind the country to significant cuts in its domestic emissions, and it was equally obvious that the EU would place intense pressure for such cuts on the United States and the rest of the OECD. The only way of squaring the circle was to have emissions trading, preferably with a big seller in the market. Russia, with its probable surplus allowances, was the only player big enough.

A more charitable interpretation was possible. To help sustain its old foe in its conversion from communism, the United States was proposing to use the Climate Change Convention as a way of endowing Russia with desperately needed resources. Additionally, Russia would be given an incentive to restructure its staggeringly inefficient and polluting economy in more efficient and sustainable ways, so as to enhance the allowances it could sell. The US proposal was not merely revolutionary in its structure, but audacious in its implications.

The problem was that it still looked like cheating on the basic commitment. The EU concern that emissions trading would undermine any success in imposing more stringent US emission targets was reinforced by the east European situation. The United States might not even have to take significant action elsewhere through joint implementation, or ensure that other countries took compensating action: it might simply buy a surplus of what the EU termed 'hot air' from Russia (this concern in fact was aggravated, not alleviated, by the final outcome at Kyoto and is explored more fully in Part II of this book).

Against this background the European Council of Ministers met in June, just before the Denver Summit, and under the guidance of the Dutch presidency crafted a compromise between the majority of governments that remained hostile to emissions trading and the small band of supporters. 'The Council considers', they concluded, 'that mechanisms such as emissions trading are supplementary to domestic action and common coordinated policies and measures, and that the inclusion of any trading system in the Protocol and the level of the targets to be achieved are interdependent. It therefore calls upon all industrialized countries to indicate the targets they envisage for 2005 and 2010.' In other words, the EU was prepared to accept the logic of emissions trading, but only if it resulted in practice in the benefits which its proponents claimed, with greater efficiency enabling a stronger overall outcome. The EU had worked out how to catch the ball tossed by the United States, and thrown it down as a gauntlet.

The Japanese focused on a different concern. They were worried that the United States would use its enormous political leverage over Russia to monopolize that country's surplus, and so – along with several other countries – Japan demanded conditions that any trading should be transparent, competitive and open. To this were added other factors. As the EU recognized that the overall level of commitments would indeed be much weaker than its proposed 15% reduction, it pressed for constraints to ensure that any international action would be supplementary to domestic action – a 'supplementary cap' on international trading (see Chapter 6, Section 6.4). The French, in particular, also pressed for some common rules about how any domestic allocations of emission permits might be made. Against this complex background, in Kyoto the OECD countries edged towards an agreed text on emissions trading (Chapter 4); for whatever doubts the EU had about emissions trading, it knew full well by now that an agreement was impossible without it.

But the OECD was overlooking a deeper challenge. Developing countries had been almost ignored in the internecine OECD debate on emissions trading. This fed a resentment that grew as they began to understand the game: not only might such trading allow the United States to avoid serious domestic action, but Russia, an industrialized country, could be the major beneficiary. Behind this was a more principled objection. Although the United States was at pains to emphasize that the allowances agreed at Kyoto would not constitute any more basic or long-term 'right' to emit, there was a deep fear that the whole question of long-term emission entitlements was being pre-empted. African countries had already started supporting a principle that emission entitlements should be allocated on a per capita basis, or at least that they should converge towards this. China and, most stridently, India backed similar proposals, arguing that fundamental issues of principle were at stake on which they had not been adequately consulted and on which they could not retreat.

The OECD, having spent much of its energies on internal debate, hit a brick wall when it then tried to place its delicately crafted (and still not finalized) proposed text into the final negotiating texts at the climax to the Kyoto conference. The G77 adamantly refused to accept an article on emissions trading. Chairman Estrada rejected the OECD's complex text on the grounds that it was submitted too late; in reality, he knew that every sentence would be opposed as a matter of principle and that such complexity at the final hour would destroy the negotiations. The proposed article was omitted and replaced by a simple paragraph allowing countries to exchange parts of their assigned amounts.

But developing-country opposition to emissions trading was rooted in principles, fed by anger that it might enable the United States to avoid significant domestic action, and magnified by resentment about the prospective Russian windfall. They were not going to budge. In an almost comical touch, in the evening of 10 December at Kyoto senior OECD officials were desperately searching for anyone who might know how to get in immediate contact with the Indian Prime Minister, hoping that he could be persuaded to override his negotiators.

At about 2.00 a.m. on 11 December, with the conference already long past its official deadline, the negotiations reached the offending paragraphs. On behalf of the EU, the UK proposed an amendment that trading could not start until rules were agreed at the subsequent Conference of Parties, a condition that the United States resisted, fearing it would be used to block emissions trading altogether. Then China said it could not accept the paragraph, amended or not, as there had not been sufficient time to consider the implications. India followed with its fundamental objections to the ad-hoc creation of tradable entitlements without debate about the global principles involved. A dozen developing countries raised their objections.

At 4.00 a.m. Ambassador Estrada called a halt to the negotiations, for short consultations. A pleasant (alas, unconfirmed) rumour holds that instead of yet more frantic consultations, he went back to his hotel room for a short rest, a shower and a change of clothes. In reality he remained locked in consultations, testing out compromises on emissions trading repeatedly, also in relation to other parts of the package. Nevertheless, he returned with renewed vigour. He reconvened the conference and announced that he proposed to delete the offending paragraphs, and instead insert a new article within the section on implementation procedures. The new article would consist of three short sentences stating basically that the Parties would subsequently negotiate principles, rules, etc. for emissions trading; that they may trade emissions; and that any trading would be supplemental to domestic action (see Chapter 4 for the text and interpretations). He read the sentences twice, very slowly, and paused. He asked if there were any objections and brought down his gavel as India – some say joined by both China and the EU – raised flags to object. He ignored the flags and stormed ahead to the next paragraph on commit-ments. Any country that openly challenged his authority would almost certainly have been held responsible for destroying the Kyoto negoti-ations. None did. Thus in the defining moment of the Kyoto conference, the objections in principle of major developing countries to emissions trading – and countries altogether probably representing almost half the world's population – were overridden. In another area of the agreement, however, they were to get their *quid pro quo*. First, however, the Kyoto negotiations had another surprise in store in the area of international flexibility.

3.6 International flexibility with developing countries: objections to JI and the emergence of the CDM[23]

3.6.1 Objections to North–South 'joint implementation'

The concept of joint implementation – generating emission credits by investing in projects outside the industrialized world – faced far more severe opposition in relation to developing countries than when applied between industrialized countries. This sprang from a number of long-standing concerns.

Locus of responsibilities and 'cream-skimming' Perhaps the most fundamental and principled criticism of JI was that it could allow developed countries – those with the highest emissions and an acknowledged responsibility to lead – to achieve their emission targets without taking adequate action at home. JI was seen as enabling developed countries to escape their prime responsibility to put their own house in order.

Related to this was the fear that JI projects would focus on and 'use up' the cheapest reduction options in developing countries, so that if and when developing countries came to adopt emission commitments, they would only have more expensive options left. While, strictly speaking, this is an issue that could be simply resolved in terms of allocations, the underlying fear was a potent one. Many developing countries consequently suggested that current JI proposals should be re-evaluated in order to reflect the total economic cost of carbon abatement in the short and long term, rather than

---

[23] Drafted with Tim Forsyth. An enormous literature has emerged on joint implementation. Relevant pre-Kyoto publications include R.S. Maya and J. Gupta (eds), *Joint Implementation: Carbon Colonies or Business Opportunities? Weighing the Odds in an Information Vacuum*, Harare: Southern Centre for Energy and Environment, 1996; D. Anderson and M. Grubb (eds), *Controlling Carbon and Sulphur: Joint Implementation and Trading Initiatives*, London: RIIA, 1997; L.D. Harvey, Danny and Elizabeth J. Bush, 'Joint Implementation: an Effective Strategy for Combating Global Warming?', *Environment*, October 1997, pp. 14–43; IEA (International Energy Agency)/OECD (Organization for Economic Cooperation and Development), *Activities Implemented Jointly: Partnerships for Climate and Development*, Energy and Environment Policy Series, Paris: OECD, 1997; P. Cullet and P. Kameri-Mbote, 'Joint Implementation and Forestry Projects: Conceptual and Operational Fallacies', *International Affairs*, Vol. 74, No. 2, 1998, pp. 393–408. The journal *Joint Implementation Quarterly* publishes regular details of projects and debates.

on the basis of short-term single projects undertaken by individual companies, but this was anathema to the idea of a market-based process promoting
private investment.

Baselines for measuring 'additional' emission reductions There are serious
methodological difficulties in estimating the emissions saving resulting
from a JI project, since this hinges upon an estimate of how much
higher emissions would have been without it. It is a 'counterfactual'
problem of measuring emissions avoided rather than simply the emissions
themselves. Would certain alternative projects really have gone ahead?
Whose assessment of the alternative favoured and feasible options should
be believed? Developing countries have feared that investing countries
would get unfairly high credits for JI projects. One solution is to
incorporate wide margins of error into the predicted savings achieved by
JI projects. However, this would reduce the scope and profitability of
projects. Issues of baselines and additionality are discussed further in
Chapter 7.

In addition, an often-questioned aspect of JI between North and South
is the system of measurement used for GHG monitoring. The IPCC-
agreed approach is to evaluate GHG emissions on the basis of a total for
each country. However, many developing countries see this as inappropriate for determining either international responsibility or the impacts of
policies adopted. For example, some argued that inventories should focus
on the location of economic demand for GHG-emitting practices rather
than on the site of production. Under such a scheme, the carbon released
from tropical deforestation for traded timber would be located in
consuming countries such as Japan and the United States rather than in
producing countries such as Malaysia and Brazil.

JI for carbon sequestration or technology transfer? The difficulties were
aggravated by the dispute about the role of sinks. To many in the North,
forest management and reforestation offered an obvious and cheap way to
offset emissions, a view supported by some organizations seeking to protect wildlife and biodiversity in the developing world, and also by some
forestry organizations in developing countries that welcomed the oppor-

tunity to attract more resources. Costa Rica led the field with extensive offers of forest protection under the aegis of JI.

However, many other developing countries opposed using their land to generate credits that would allow industrialized countries to continue emitting GHGs. Furthermore, many reforestation or land-management schemes in developing countries risk being ineffective or politically unsettling if they take land away from agriculture or otherwise cut across already disputed aspects of land rights.[24] Concentrating on carbon sequestration projects in the developing world provoked the accusation of imposing developed-world environmental values onto the poor and weak – at worst, cast as 'carbon colonialism' – as well as being difficult to measure and enforce.[25]

Selection of AIJ/JI projects The selection of AIJ/JI projects entails identification of which type of project to approve and in which country. Both have proved controversial. The usual approach adopted by government authorities in charge of JI (such as the United States Initiative on Joint Implementation, see below) is to ensure that local governments have agreed to proposed AIJ/JI projects before finally approving them. However, not all projects, such as reforestation plans or hydro dams, necessarily have the support of the host population. Not infrequently there could be clashes of priorities between donor and host governments, and/or between the host government and the local population.

Similarly, it is often assumed that AIJ/JI may proceed on an equal footing anywhere in the world. In reality, however, many developed countries invest only with traditional trading partners, and developing countries are

---

[24] For example, in 1993 reforestation schemes in northeastern Thailand caused public demonstrations and marches on parliament because they were seen to be unfair to poor farmers. Pasuk Phongpaichit and C. Baker, *Thailand: Economy and Politics*, Oxford: OUP, 1995, p. 390.

[25] The depth of feelings – and the complexities – became visible from an early stage. The most striking example the author witnessed was at a meeting in the early 1990s. An economist from a US environmental NGO (one long associated with promoting market instruments) had expounded the virtues of JI and explained how much cheaper it could be to absorb CO_2 in Africa than to limit emissions in the United States. Shaking with anger, an African present rose and asked 'why should African governments let their land be used as a toilet for absorbing emissions from Americans' second cars?' Needless to say, the ensuing debate was not a very productive one.

not always in a position to reject foreign countries' inappropriate proposals. The result may be a weak negotiating stance for many developing countries, and the emergence of competition between them to attract investment. This in turn may imply that most financial benefits of JI accrue to investing countries rather than host countries.

3.6.2 Interest and experience with AIJ/JI projects

While the academic and political debates raged, various projects were proceeding, initially just on the basis of corporate and national initiatives and later under the formal terms of the Berlin decision on activities implemented jointly. A few companies, principally from North America, proceeded with carbon-saving investments (particularly in Latin American forestry), both for public relations purposes and as a hedge against possible future regulation. In addition, the Global Environment Facility became involved in sponsoring AIJ projects, and the World Bank created a prototype Carbon Investment Fund for governments and companies wishing to invest in CO_2-reducing projects in developing countries.

Attitudes in the developing world varied. In Central America, Costa Rica welcomed 'carbon offset initiatives' from an early stage. In 1994, the Costa Rican government established an office for JI, and a Carbon Fund of Costa Rica aiming to create a portfolio of carbon-offset projects. At the opposite extreme, the Indian government refused to host projects even after the Berlin decision formally established the AIJ pilot phase.

Progress with projects was thus slow and uneven, but gradually gathered pace. Latin America attracted the bulk of projects, much of Asia – especially India – remained resistant to repeated requests and much of Africa was perceived to be unattractive to potential investors anyway. But the activities fed a growing perception that JI could be a way of attracting investment to developing countries. An invisible competition began to emerge and the constituency of interests potentially more favourable to JI slowly expanded. However, most of the objections remained and the G77 as a bloc stayed firm and resolute in its opposition to North–South JI. Though the US proposals were included in the chairman's negotiating text, there seemed little chance that they would be agreed.

Box 3.1: Progress on AIJ/JI projects before the Kyoto Protocol

Some developed countries, notably the United States, Norway and the Netherlands, were quick to investigate the potential of JI by establishing a variety of projects after the 1992 Earth Summit, and activities accelerated after the Berlin decision establishing the pilot base of AIJ. The United States Initiative on Joint Implementation (USIJI) was established in October 1993, and had approved 25 AIJ projects by July 1997. These represented total investment of $450 million, at a cost of between $1 to $7 per tC. In the same time period, the Netherlands approved eight projects, Norway four, Germany three, France two, and Japan (which was repeatedly rebuffed by India) one.

Compared to the problem this is a relatively low level of activity, due both to the lack of incentives in the absence of crediting, and resistance in some developing countries even to AIJ projects. India gained a reputation for being hostile to foreign-invested climate mitigation projects.

Many projects depended upon private-sector funding, with an uneven approach to underwriting risk by international organizations. French, German and US AIJ programmes offer funds for feasibility studies and monitoring of projects. Norway and the Netherlands also offer funds for project implementation. The World Bank's prototype Global Carbon Fund emerged to provide capital to cover the initial costs of mitigation projects involving foreign investment. There was relatively little AIJ activity in areas of the world considered unattractive for foreign investors. Up to 1998, for example, Africa had only attracted one Dutch forest protection project in Uganda, a Norwegian fuelwood project in Burkina Faso, and a French hydroelectric project in Zimbabwe (and only some of these were official, that is approved, AIJ projects). The USIJI had yet to accept a project in Africa.

In the USIJI programme, by December 1996 there were 12 projects featuring reforestation, forest conservation or managing land use, and 13 featuring renewable energy. To date, 19 projects have been in Latin America (17 in Central America), five in eastern Europe (three in Russia), and just one in Asia (a carbon sequestration project through reduced logging in Indonesia).

The Global Environment Facility (GEF) had by June 1997 financed 14 AIJ projects. All of these have focused on renewable energy or sustainable industrialization projects. Seven of these projects are in Asia (including two in China), two each are in eastern Europe and Latin America, and Africa and the Middle East have one project each (in Senegal and Jordan, both for renewable energy); the 14th is global.

Programmes continued to grow after Kyoto, with over 100 AIJ projects approved by the end of 1998.

3.6.3 The 'Kyoto surprise'

It was against this background that the single most remarkable development of the entire Kyoto negotiations occurred. It arose from a most unexpected quarter. One element of the sweeping proposals that Brazil had put forward in June 1997 was that Annex I Parties should be subject to a financial penalty if they did not comply with their quantified commitments under the Protocol, with the fine being levied in proportion to the degree of non-compliance. The money would be paid into a Clean Development Fund that would be used to support appropriate projects in developing countries, for limiting emissions and potentially for adaptation. The proposal included ways of apportioning the proceeds between developing countries, and suggested a level of $10 per tonne of excess carbon-equivalent emitted.

The idea that industrialized countries would agree to being subject to assessed financial penalties as a compliance measure seemed far-fetched to any seasoned politician – the debate about enforcement mechanisms had barely begun – and Ambassador Estrada omitted it (along with the rest of the complex Brazilian proposal) from the chairman's negotiating text for the final pre-Kyoto session in October. But the G77, while uneasy about several aspects of the Brazilian proposal, could unite around the suggestion for financial penalties channelled into a Clean Development Fund, and they insisted that it be reinserted into the text.

The Annex I countries pronounced their opposition and it seemed a fruitless debate. But by the end of the final pre-Kyoto negotiating session, a remarkable twist on the proposal had occurred to one or two key people.[26] If the penalty were levied at a sufficient rate to fund carbon-saving projects in the developing countries that would save emissions equivalent to the excess emissions from Annex I, the practical consequence would be almost identical to JI, though the legal and institutional framework would be completely different. From this perspective, one of the most apparently aggressive proposals from the G77 could be considered as consonant with the US proposals for JI that they had been fiercely rejecting for five years.

The idea was first floated in the negotiations at the small ministerial discussions that Japan hosted early in November 1997. Fired with the

---

[26] The idea is generally believed to have emerged out of discussions at the subsequent ministerial consultation in Tokyo. However, on the final day of the October AGBM, the head of the Brazilian delegation took the author aside and asked him to think about the implications of levying the proposed penalty at a rate set to fund projects that would save emissions equivalent to the degree of non-compliance. In outline, it did not need much thought to realize that the result might be something very much like joint implementation.

possibilities, a US team dashed down to Rio to explore the options. The United States managed to shift the 'line of compliance' to encompass such investments as *contributing to* compliance, rather than being a *penalty for not* complying. The idea of a penalty on governments was transformed into a mechanism for investment by companies. The multilateral character of the framework was retained, but it became a 'mechanism' rather than a 'fund'; and in the final days of Kyoto, the clean development mechanism, described in Chapter 4, was born.[27]

3.7 Developing-country concerns and participation

3.7.1 Compensation and the minimization of adverse impacts

As noted at the end of Chapter 2, one of the key political developments in the negotiations was the change in the stance of OPEC countries towards accepting action to limit greenhouse gas emissions, providing that adequate efforts were made to recognize and protect their interests. Specifically, they had proposed that they should be compensated for any lost revenues, and they linked this directly to the question of the responsibility of industrialized countries, under Articles 4.8 and 4.9 of the Convention, to assist vulnerable countries which would be adversely affected by climate change or by response measures

The whole idea of compensation was anathema to the industrialized countries for obvious reasons, and political responses were easy enough: there were mutterings about making OPEC compensate the rest of the world for the oil price shocks of the 1970s, and other arguments that led *ad absurdum* if each country had to compensate others for actions that

---

[27] Far more could be written about the rapid evolution of the CDM and attitudes towards it, including agonizing divisions within both the G77 and the OECD. One almost comical but heart-stopping moment occurred in the final plenary discussion at Kyoto. As the final text on the CDM was considered for approval at about 7.00 am, a Canadian delegate welcomed it 'because they had always supported joint implementation'. A palpable sense of fear rippled through the conference at the mention of the dreaded words, lest it shatter the fragile G77 decision to support the CDM. Chairman Estrada, who had always disliked JI, and the CDM proposal once its nature became clear, sardonically welcomed the delegate's honesty; but as he knew full well, the deal had already been sealed during the night's discussions.

might have an adverse economic impact. OPEC argued that action adopted under an international treaty was different, and certainly that any action taken should minimize adverse impacts.

The G77 recognized the impossibility of establishing a specific general compensation mechanism, though they did achieve the requirement that some of the proceeds from the CDM should be used to help countries adapt to climate change. The broader idea that industrialized countries should minimize adverse impacts – both of climate change and of the specific responses under the Protocol – gained the full backing of the G77 and this was absolutely central to gaining OPEC's acquiescence in the final agreement (which technically it could have easily blocked). It sounded innocuous enough, but as explained in Chapter 4, it may have quite far-reaching implications.

3.7.2 Technology transfer and funding[28]

'Technology transfer' in the context of international negotiations can be defined as the process of developing and extending the use of technologies among new user groups and in different countries, particularly developing countries. In the context of the climate change negotiations, the transfer of environmentally sound technology (EST) has long been considered essential to help developing or transitional countries industrialize while limiting their emissions. Developing countries have made signing international agreements such as the Montreal Protocol and UN FCCC conditional on agreement on technology transfer.

As noted in Chapter 2, the UN FCCC had extensive references to the promotion of technology transfer. The urgency had also been echoed in Chapter 34 of Rio's Agenda 21.[29] However, the success of 'technology

---

[28] Section drafted by Tim Forsyth.

[29] Using language similar to that of the UN Climate Convention, Agenda 21 stated that access to and transfer of environmentally sound technology (EST) should be promoted 'on favourable terms, including on concessional and preferential terms, as mutually agreed, taking into account the need to protect intellectual rights as well as the special needs of developing countries for the implementation of Agenda 21'. In order to achieve this technology transfer, Chapter 34 of Agenda 21 also urged the adoption of 'a collaborative network of research centres' and 'programmes of cooperation and assistance'.

transfer' since the Earth Summit was far from clear. Many developing countries saw the statement that technology transfer should be conducted 'on concessional and preferential terms' as implying that technology should be provided quickly and cheaply by OECD governments at well below market prices. During the discussions, China came forward with a list of specific technologies which it wanted in order to contribute towards CO_2 limitation.

Business interests and many governments in the developed world, however, saw this interpretation as naive and against commercial sense. Furthermore, 'technology' comprises a variety of 'hard' (equipment-related) and 'soft' (personnel- or training-related) technologies, which may have to mature over a number of years in order to develop effective EST.

Official actions to promote technology transfer under the Convention were also largely unsuccessful. After the Convention entered into force, there were extensive discussions aimed at setting up Technology Assessment Panels (TAPs), which were intended to identify the technology needs and aspirations of developing countries in collaboration with investors. In the end, however, these initiatives foundered upon disagreements about who should participate and about transferring technology at sub-market rates.

Another approach was the Climate Technology Initiative (CTI). This emerged largely through the action of Japan operating through the International Energy Agency (IEA). The CTI is still in evolution, but its activities include a variety of voluntary actions by IEA member states, national advice and technological development plans, offering prizes for technological development, enhancing markets for emerging technologies, and collaboration between states on technology research and development. In addition the CTI has also established regional seminars on climate change-related technology, and has worked in collaboration with the UN FCCC Subsidiary Body for Scientific and Technological Advice (SBSTA).

In 1997, the IEA announced that it would increase its work with the CTI and seek new collaboration with existing bodies like the International Standards Organization (ISO). A new Global Remedy for the Environment and Energy Use initiative was also established, aiming to enhance the use of climate change-mitigating technology in ODA and private investment.

However, the impacts of these official attempts to generate technology transfer are as yet unclear, and also relatively small compared with private investment flows. As an illustration of this, funding from the Global Environment Facility between 1990 and 1997 amounted to $5.25 billion. ODA in 1997 was a total of $40–50 billion. Private-sector transfers from FDI reached $240–250 billion. Initiatives to increase technology transfer must therefore target private investment, and this consequently implied rethinking the ways in which the UN FCCC has approached technology transfer. The Kyoto Protocol reaffirmed the importance of technology transfer and moved somewhat in this direction (Chapter 4), but this was one area in which a bigger advance occurred at the subsequent Buenos Aires follow-up (Chapter 8).

3.7.3 Enhancing the implementation of existing developing-country commitments

The topic of 'advancing the implementation' of the existing Article 4.1 commitments that covered all countries had considerable political significance. Developed countries sought greater developing-country commitment to effective policies and measures and stronger reporting; developing countries sought stronger assurances concerning financial assistance and technology transfer.

Specific new proposals made little headway. The United States proposed that countries should be required to implement 'no regret' measures (measures without net economic cost) identified by international reviews conducted by the Secretariat. This would indeed have had interesting consequences given the continued scope for such policies in OECD countries themselves, such as the removal of coal subsidies in Germany, and the tax treatment for some kinds of fossil fuel production, scope for stronger building codes or certain vehicle standards, etc. in the United States and many other OECD countries. As in more general economic debates, therefore, the OECD sought to use the Article 4.1 negotiations to press developing countries to adopt 'good practice' that they themselves were frequently unable to implement fully. The EU, more modestly, sought to extend the existing in-depth review process to developing countries, but this too was rebuffed.

The developing countries, for their part, sought to give the Convention obligations concerning finance and technology transfer from Annex II Parties (principally the OECD) a more concrete form. This in turn was rebuffed by the OECD.

Overall, the advances represented in the final text, as described in Chapter 4, were modest, a disappointing outcome from such extended negotiations. This result can be traced mostly to the inherent reluctance of countries (especially but not exclusively developing countries) to open themselves to international intrusion in their domestic policies and priorities, and to the fundamental objection of most developing countries to being pushed into action on climate change when so little had yet been implemented in the industrialized countries. In the end the United States had to accept this reality if there was to be any Protocol at all.

The struggle on 'advancing the existing commitments' of all Parties was thus ultimately partly symbolic, but it lasted to the end. The Swedish Ambassador Bo Kjellen, co-chairman of the negotiating group on this topic, struggled to formulate an acceptable compromise, but the G77 still refused to accept his final effort. Faced with the impasse, Ambassador Estrada set it aside for reconsideration as almost the last item on the agenda of the final session at Kyoto. When he returned to it, the G77 reiterated its preference for its own text on the topic, which the OECD still considered too limp to be acceptable. The biggest battles over, Estrada asked what the delegate representing the G77 expected him to do. The delegate suggested he use his gavel as he had throughout the night. Estrada shrugged his shoulders, declared in favour of Bo Kjellen's compromise text, and banged his gavel. For the G77 had just won a bigger battle concerning potentially far bigger commitments.

3.8 Extending commitments to new countries: evolution and voluntary accession

The most divisive North–South issue of all in the Kyoto negotiations concerned the desire of most OECD countries – but especially the United States – to draw developing countries into specific quantified emission limitation commitments, or at least into a process that might visibly lead to this conclusion.

The reasons why this was (and remains) such a divisive issue are easy to see. From one standpoint, it is quite unreasonable to expect the developing countries to commit themselves to adopting binding emission constraints at present. Most of them have contributed hardly anything to the climate change problem and their per capita emissions are still but a small fraction of those in the industrialized world (Chapter 2, Section 2.1). Why should they devote resources towards a diffuse and long-term problem, largely caused by others, when they have more basic and pressing priorities ranging from simple health and sanitation to basic infrastructure developments?

From the other perspective, it is equally apparent that the climate change problem cannot ultimately be solved without action by the developing countries. Their emissions are growing rapidly, especially in percentage terms (though less rapidly since the Asian financial crisis erupted). Since they comprise more than three-quarters of the world's population, their long-term potential emissions growth could ultimately swamp any restraint by the current industrialized world. Furthermore, many of the cheapest options for limiting emissions may lie in developing countries, whose economies are inevitably less efficient (in carbon per unit of GDP) than most OECD economies. Altering the trajectory of emissions growth in developing countries is probably the biggest low-cost long-term opportunity for limiting global emissions that exists. From this perspective, excluding them makes economic nonsense.

If the divide itself derives from a classic clash of perspectives, and potentially a clash of equity vs. efficiency (though it need not take this form), it became poisoned during the negotiations as a result of the long history of North–South politics, the internal politics of the G77 group (which encompasses hugely diverse interests) and the specific legacy of the debate on developing-country involvement in the climate change regime.

The Convention had little choice but to enshrine the basic structural division between the Annex I countries and the developing countries. Already by the time of the Berlin conference, the pressures in the United States to include developing countries in negotiations on any new commitments were very strong. However, since the industrialized countries had done so little towards establishing leadership through their domestic actions, and most of the OECD was far from ready to achieve the indicative aim of returning emissions to 1990 levels by 2000, there was never any chance of the Mandate including anything that might lead to new commitments by developing countries.

In fact it was the EU which first formally proposed extending the net of commitments, recommending that they be applied in all countries listed in an Annex X, which was clearly intended to go beyond the existing Annex I. The EU belatedly indicated that this should include at least South Korea and Mexico, and also Turkey which was trying to withdraw from Annex I. It was also implied that this same list should undertake quantified emission limitations. Since the core EU proposal on quantified commitments focused on flat-rate reductions from 1990 levels, this seemed a clear provocation to countries whose emissions in 1990 were relatively low but had already grown by as much as 25–40%. Quite apart from being outside the terms of the Berlin Mandate, the resultant perception that any talk of extending commitments to some new countries might require them to reduce below 1990 levels – clearly impossible and inequitable for most developing countries – was a fear that was very slow to dissipate.

The pressures in the United States were for far broader participation, and they took on concrete form with the specific US protocol proposal of January 1997. This included a section on the 'evolution of commitments' which stated that all countries should have binding quantified commitments by a certain date – 2005 was suggested as a negotiating basis in the US proposal. This was clearly outside the terms of the Berlin Mandate and aroused predictable anger across most of the developing world. 'Evolution of commitments' were dirty words in the negotiations thereafter.

Attempts to extend the negotiations to include new commitments for developing countries continued right up to Kyoto itself, where New Zealand, supported by the other members of JUSSCANNZ and by Poland and Slovenia, proposed that conditional upon industrialized countries having fulfilled their commitments in the first period, developing countries should agree to binding commitments applicable in a subsequent period.[30] This at least attempted to acknowledge the agreed principle of industrialized-country leadership, and universal participation was not proposed (they called for 'progressive engagement' according to levels of development, and explicitly excluded the least developed), but it was still clearly outside the terms of the Berlin Mandate. The proposal provoked several hours of angry rejection from the developing countries, which saw such linkage as a betrayal of trust about the whole purpose of the Kyoto negotiations.

The only creative approach towards developing-country commitments that could plausibly be argued to fall within the terms of the Berlin Mandate was the proposal crafted by Chairman Raúl Estrada-Oyuela to include an article explicitly providing a path for voluntary adoption of quantified commitments – Article 10 (later 9) in the draft Protocol. This appeared innocuous, but many developing countries saw it as a Trojan horse which would leave them vulnerable to being subjected, perhaps one by one, to pressure to adopt commitments. The G77 always felt that its only protection against the might of the OECD was collective strength. Though at least 35 developing countries registered their support for the proposed article in the final night at Kyoto, core countries, which felt they had been slighted in some of the earlier debates, remained adamantly opposed to it. With China, India and Brazil lining up with the OPEC countries and others, forcing the article through could have alienated half the world from the Kyoto regime for years to come. In what must have been the most difficult decision of that long night, Ambassador Estrada declared that the article on voluntary commitments – his own creation and the only credible attempt to resolve the impossible division – had failed to gain sufficient support and had to be removed.

---

[30] The proposal '*considered* that Annex I commitments beyond [the first commitment period] should comprise the widest participation in binding action; *recognized* the dependence of inception of non-Annex I Parties' legally binding emissions limitations commitments on Annex I Parties' implementation particularly of Kyoto Protocol QELROs; *agreed* that there should be further QELROs for Annex I Parties and "quantified emission limitation objectives" for other Parties, except least-developed countries; and *established* a process to set the commitments, to be concluded by 2002' (cited with discussion and reactions in Tata Energy Research Group, *Climate Change: Post-Kyoto Perspectives from the South*, New Delhi, 1998).

The deletion of the proposed article on voluntary accession was the other major plenary decision in the long night at Kyoto, a *quid pro quo* for riding roughshod over the objections of China and India to emissions trading; given the weight that the United States had placed upon developing-country commitments, some waited to see if its delegation would walk out. But the clean development mechanism and slightly strengthened references to enhancing the existing commitments of developing countries (shortly to be confirmed) were enough for the US delegation to stay. No one else walked out; half of the delegates anyway were asleep from sheer exhaustion.

The meeting proceeded to more detailed wrangles over aspects of the CDM and confirmation of the procedural articles (including the terms of entry into force, considered in Chapter 8 of this book), and the struggle over improving the implementation of existing commitments, noted above. With the full terms of the instruments and the political package in place, the final points of the specific numerical commitments were agreed and brought to the plenary hall. Given the delicate balance between the varied concerns finally reached after two and a half years of negotiations, not a single Party wanted to risk the opprobrium of objecting. Ambassador Estrada brought down his gavel, and declared that the Kyoto Protocol had been agreed unanimously and could now be forwarded for formal adoption.

3.9 Conclusions

What general conclusions can be drawn from the negotiations of the Kyoto Protocol? It was an extraordinary process, grappling with an unprecedented problem. Trying to get more than 150 countries with hugely divergent interests and perceptions to agree was bound to be extremely difficult. That it succeeded at all was a considerable achievement, due in no small measure to the generosity of Japan in funding a host of additional facilitating activities, and the skill and authority ultimately wielded by the chairman of the negotiations, Ambassador Estrada-Oyuela.

One broad observation is that, for all the academic speculation about the decline of the nation-state in the era of economic globalization, the Kyoto Protocol is very much an agreement struck by governments. Environmental

NGOs not only wanted much stronger commitments but opposed almost all the transfer mechanisms. The vast weight of US industry was opposed to the whole process and threw probably up to $100 million into fighting it. But governments wanted a deal, and did what they considered was possible to protect the atmosphere while also protecting their own interests.

Within that panoply, US dominance is striking. The United States got virtually everything it wanted in respect of flexibility for Annex I commitments with the sole exception of 'borrowing'. The EU, with greater population and GDP, did score important successes: through its efforts, the United States (and other OECD countries) made a stronger commitment than they otherwise would have done; the EU headed off many of the most potentially dangerous proposals on sinks, and at least kept the door open on aspects of its policies and measures proposals. But to discover the source of most of the ideas in the Protocol, one only needs to read the US proposal of January 1997. The coherence of the US administration contrasted with some disarray in Japan and the unwieldy (and introspective) morass of EU decision making during the negotiating process. This reflects the EU's broader foreign policy difficulties, and should provoke a lot of thought among member states. Only in respect of developing countries did the United States not get what it wanted; indeed US pressure on developing countries frequently served only to inflame their resistance.

The main policy objective of US strategy was to establish flexibility in all dimensions. This was a result of the country's confluence of political interest and economic ideology. Politically (and with good reason), the administration lacked confidence about what measures on CO_2 emissions could be ratified or implemented domestically, and it regarded the ability to meet any commitments through action on other gases, sinks and international mechanisms as a political imperative. Economically, US thinking was dominated by general equilibrium concepts that focus upon economic efficiency and imply that flexibility achieves the same environmental benefits at lower costs: hence, the more flexibility the better. That attitude, combined with US political dominance and the relative paucity of counter-arguments, largely determined the outcome of most of the key policy debates.

Another feature was the internal difficulties experienced by the developing countries in formulating and agreeing positive contributions. The EITs defended their interests sufficiently, and aligned themselves with either the EU or US trading objectives. The developing countries – representing more than two-thirds of the participating countries and three-quarters of the world's population – produced some initial impetus on targets (originating from AOSIS) and potential adverse impacts (originating from AOSIS for climate impacts, and OPEC for response impacts). Otherwise, the most visible positive contributions from developing countries were some involvement in drafting the chairman's proposal on voluntary accession, Costa Rican efforts on JI that never gained G77 support, and the sweeping Brazilian proposal of which one portion was supported and radically rewritten into the CDM. For the most part, however, the only issues on which such a diverse group as the G77 could agree were defensive ones.

To many in the OECD, the greatest failure of the negotiating process was the failure to include developing countries in quantified emission limits. This is simplistic, and not just because of the principles of industrialized-country leadership that have been a feature of every international environmental agreement to date and have been agreed from the onset of the climate change regime. It would also have been of questionable value to force commitments from them in the absence of knowledge on their actual prospects for emissions or their institutional capacity to control them. These questions are difficult enough for certain Annex I countries.

The underlying problem is deeper and more subtle. The Protocol rests upon global institutional structures that divide the world into 'North' and 'South'. The inclusion of countries like Ukraine, which had not even submitted its first National Communication, in commitments and processes designed for the developed economies, gives rise to one set of problems – problems which, as will be seen later, are potent ones in the Protocol as agreed. Grouping together the enormous range of 'developing' countries, from new OECD entrants to the poorest of the poor, and from the most populous countries on earth to the tiny island states, was never likely to yield sensible results. In this respect, the Kyoto Protocol is a victim of the UN system itself. Its structural failures are no greater than those of the UN system, and the task of moving forward no less awesome in its complexity. But first, one needs to understand the specific provisions of the Protocol itself, as it finally emerged from the cauldron of Kyoto. That is the task of the next chapter.

Chapter 4

The Kyoto Protocol

Chapter 3 described the various policy debates behind the Kyoto Protocol and their political resolution. This chapter describes and interprets the key provisions of the Protocol itself, the full text of which is reproduced in Appendix 1.

This chapter focuses mainly upon the Protocol's specific commitments, rather than its institutions and legal procedures. In many areas the Protocol borrows the latter from the Convention itself (see Chapter 2). They include the institutions of the Convention's Secretariat, Subsidiary Bodies and Financial Mechanism, while the procedures for settlement of disputes and amendment are taken directly from the Convention. Disputes about whether the Protocol should have a separate annual conference to oversee its implementation were resolved in an agreement that legally distinct meetings of the Parties to the Protocol would be convened as part of the Convention's annual conference, deemed 'the Conference of Parties serving as the meeting of Parties to this Protocol'. This will be the supreme body overseeing implementation of the Protocol, and in this book the phrase 'Conference/meeting of the Parties' is used in preference to the more precise legal acronym of COP/MOP. Provisions for entry into force are discussed in Chapter 8, in a consideration of the prospects for the Kyoto regime.

4.1 Definition of Annex I commitments

4.1.1 The headline commitments

In its central provisions, the Kyoto Protocol defines allowable greenhouse gas emissions for each industrialized country Party in terms of *assigned amounts* for the *commitment period* 2008–12. The commitments apply to the industrialized countries in Annex I of the Convention, and the numerical commitments are listed in Annex B of the Protocol. Specifically, the latter covers those Annex I countries that had ratified the Convention by the time of Kyoto (thus excluding Turkey and Belarus), plus those whose application to Annex I was accepted during the Kyoto Conference.[1] The commitments add up to a reduction of 5.2% below 1990 levels, and the overall commitment is summarized in the first paragraph of Article 3:

> The Parties included in Annex I shall, individually or jointly, ensure that their aggregate anthropogenic carbon dioxide equivalent emissions of the greenhouse gases listed in Annex A do not exceed their assigned amounts, calculated pursuant to their quantified emission limitation and reduction commitments inscribed in Annex B and in accordance with the provisions of this Article, with a view to reducing their overall emissions of such gases by at least 5 per cent below 1990 levels in the commitment period 2008 to 2012.

The specific commitments, defined as percentages relative to base year emissions, are set out in Table 4.1. The 15 countries of the EU are each listed with an 8% reduction from 1990 levels, commitments that were redistributed under the bubble provision of Article 4 (see below). The base year is generally 1990, which as noted (Chapter 3, Section 3) is in most respects taken as the base year for the international climate change negotiations. A different base year is allowed for some of the transition economies of central Europe, and for industrial trace gases, as outlined below. The numbers can only be understood as the outcome of a highly political process arising from the clash between competing numerical aims, structural visions, and root conceptions of political imperative – all combined with the personal and political dynamics of the final days at Kyoto.[2]

---

[1] Croatia, Liechtenstein, Monaco and Slovenia, together with the Czech Republic and Slovakia replacing Czechoslovakia listed in the Convention's Annex I. The original Annex I comprised all OECD countries except the recent entrants – Mexico, South Korea, the central European countries, Russia, Ukraine and the Baltic states of the former Soviet Union. At COP-4, Kazakhstan announced its desire to join Annex I and take on commitments under Annex B of the Protocol (see Chapter 8).

[2] The central clash was between the EU's aim of flat-rate reductions for all in the range 10-15% below 1990 levels, and US and Japanese support for reductions of 0-5%, with varied ideas about differentiation and flexibility, combined with Russian sensitivities and the special circumstances of some of the smaller countries. The United States offered stronger commitments

The commitments for 2008–12 serve as the engine for almost everything else in the Protocol. There is no earlier commitment period, though Article 3.2 states that 'Each Party shall, by 2005, have made demonstrable progress in achieving its commitments under this Protocol.'

The Protocol makes reference to a second commitment period to succeed the first, and stipulates that negotiations to define such commitments shall start no later than 2005. Emission reductions obtained over and above commitments in the first period may be 'banked' against commitments in the subsequent period:

> 3.13 If the emissions of a Party included in Annex I in a commitment period are less than its assigned amount under this Article, this difference shall, on request of that Party, be added to the assigned amount for that Party for subsequent commitment periods.

The US proposals that excess emissions in the first period should be deducted, or 'borrowed', from subsequent periods – which could be interpreted either as a mechanism for flexibility over time or as a penalty for non-compliance – were not accepted.

4.1.2 Gas and source coverage

The commitments in the Kyoto Protocol cover emissions of six greenhouse gases from identified sources that together account for almost all

---

in return for increases in the degree of flexibility (e.g. inclusion of sinks enabled it to add three percentage points; after Kyoto, the United States argued domestically that in reality it had only had to concede an additional two percentage points from its original zero, the rest being directly tied to increased flexibilities). Japan, the third party in the internecine OECD debates, was dragged reluctantly along to commitments higher than it considered feasible. Russia started with zero and – annoyed by the EU's opening Ministerial reference to the importance of keeping the 'three major Parties' at the same level – refused to budge. All this was overlaid by root political objectives and perceptions that pegged some countries' numbers to those of others. EIT countries aspiring to membership of the EU or OECD wanted to align themselves with the EU's standard-setting commitment, though Poland and Hungary retreated back to 6% in protest at the Russian and Ukrainian figures. Canada honoured its status as a G7 member by staying within the range of the leading OECD countries; Australia, feeling no such constraint and having long resisted the thrust of the Kyoto negotiations, simply insisted on being allowed a big increase.

Table 4.1: Emissions and commitments in the Kyoto Protocol (from base year)

Country	Commitment (% from base year)	1995 emissions (% from base year)	Other GHGs[d] (% GWP in base year)
OECD			
Australia	+ 8	+4.8	33.8
Canada	− 6	+9.3	22.9
European Union[a]	− 8	−3.4	21.6
Iceland	+ 10	+4.3	25.7
Japan	− 6	+10.7	8.0
Liechtenstein	− 8	n.a.	n.a.
Monaco	− 8	+25.2	2.7
New Zealand	+ 1	+ 0.5	65.1
Norway	0	+ 1.7	35.6
Switzerland	− 8	− 1.8	16.1
United States	− 7	+ 5.8	17.6
EITs			
Bulgaria (1988)[b]	− 8	− 35.7	28.8
Croatia[c]	− 5	n.a.	n.a.
Czech Republic	− 8	− 21.5	13.9
Estonia	− 8	− 44.4	7.2
Hungary (1985–7)[b]	− 6	− 24.0	17.7
Latvia	− 8	− 46.2	30.6
Lithuania	− 8	n.a.	23.3
Poland (1988)[b]	− 6	− 22.2	15.5
Romania (1989)[b]	− 8	− 38.1	28.8
Russian Federation	0	− 29.3	22.0
Slovak Republic	− 8	− 25.2	17.2
Slovenia[c]	− 8	n.a.	27.5
Ukraine[c]	0	n.a.	22.7

Source: FCCC/CP/1998/INF.9 and FCCC/CP/1998/11.Add.2. Data for Romania from Second National Communication.
Notes: [a] The fifteen countries of the EU are listed as each having a target of − 8. These targets were subsequently redistributed under the 'bubbling' provisions of Article 4; see Table 4.2.
[b] EIT that uses a base year other than 1990.
[c] Base year not yet agreed, data refer to 1990.
[d] Share of greenhouse gases other than industrial CO_2, but excluding LUCF, in the base year emissions, these account for a differing trend by 1995 compared with CO_2 only (Table 3.1).

anthropogenic greenhouse gas emissions in the industrialized world (Table 3.1). The gases are taken together as a 'basket' compared on the basis of the 100-year 'global warming potentials' (GWP) estimated in the IPCC's

Second Assessment Report for the first commitment period; the GWPs may be revised for any subsequent commitment periods.[3]

On this basis carbon dioxide, principally from fossil fuels, accounted for over 80% of greenhouse gas emissions from the industrialized world in 1990. Emissions of methane and nitrous oxide in many industrialized countries have declined during the 1990s; thus if they are included, a given numerical target is easier to achieve. In most countries the reverse is true for at least some of the three industrial trace gases, emissions of which are increasing rapidly in some cases. As noted, efforts to make 1995 the base year for all greenhouse gas emissions were rejected on the grounds that this would reward countries that had done nothing to constrain CO_2 emissions since 1990, but countries may take a 1995 base year for the three industrial trace gases.

The qualifying gases and sources are listed in Annex A of the Protocol, which is taken directly from the IPCC Guidelines on preparing inventories of greenhouse gas emissions; they correspond closely to those in Table 3.1. Land-use change and forestry (LUCF) are treated separately (discussed below).

Table 4.1 also shows basic statistics concerning the mix of gases in different countries. In general, the New World countries, and the high-latitude countries that are characterized by low population density, have a much higher percentage of non-CO_2 gases in their base year emissions, so these countries tend to benefit more from the move to comprehensive gas coverage. However, the EU has a higher proportion than the United States, due mainly to the combination of relatively low CO_2 emissions from its energy sector and high non-CO_2 emissions from the Scandinavian countries, compared to high US CO_2 emissions. Japan has a very small inventory of non-CO_2 emissions, and stands out as the only country for which 'greenhouse gases' and 'industrial CO_2' are virtually synonymous.

To try to minimize the problems arising from uncertainties, countries must submit national reports on emission inventories for technical review and have them accepted by the technical bodies of the Convention in order for them to come into compliance. Thus, countries have to gain technical and political acceptance for their emission estimates in order to participate, for example, in emissions trading.

---

[3] Article 5, and Decision 2/CP-3.

4.1.3 Sinks, land-use change and forestry (LUCF)

As noted in Chapter 3, the question of 'sinks' emerged as important and highly controversial relatively late in the negotiations leading up to Kyoto. This question was inevitably broadened into debate about the whole approach to emissions and absorption from land-use activities, which occupied an important part of the first week at Kyoto since progress in some other areas was contingent upon it. Eventually, both emissions and absorption from land-use change other than agriculture itself were dropped from the list of qualifying emissions in Annex A and addressed in separate clauses on land-use change and forestry (LUCF). The compromise fought out at Kyoto states that:

> The net changes in greenhouse gas emissions by sources and removals by sinks resulting from direct human-induced land-use change and forestry activities, limited to afforestation, reforestation and deforestation since 1990, measured as verifiable changes in carbon stocks in each commitment period shall be used to meet the commitments under this Article ... [they] shall be reported in a transparent and verifiable manner and reviewed in accordance with Articles 7 and 8.

The subsequent Subsidiary Body meeting in June 1998 clarified this clause as meaning that Parties' assigned amounts should be adjusted by 'verifiable changes in carbon stocks during the period 2008 to 2012 resulting from direct human-induced activities of afforestation, reforestation and deforestation since 1 January 1990'.[4]

Compared with some of the possibilities, this is quite a closely circumscribed definition of sinks, though debate after Kyoto revealed the ambiguities inherent in it, arising partly because there is no agreed definition of

---

[4] 'The SBSTA understands the meaning of Article 3.3 of the Kyoto Protocol to be as follows: The adjustment to a Party's assigned amount shall be equal to verifiable changes in carbon stocks during the period 2008 to 2012 resulting from direct human-induced activities of afforestation, reforestation and deforestation since 1 January 1990. Where the result of this calculation is a net sink, this value shall be added to the Party's assigned amount. Where the result of this calculation is a net emission, this value shall be subtracted from the Party's assigned amount': Report of the Subsidiary Body on Scientific and Technological Advice, June 1998.

reforestation.[5] It is also important in establishing the dividing line not between sources and sinks, but between the Annex A emission sources on the one hand, and net emissions arising from prescribed categories of LUCF changes on the other: Annex A emission sources constitute measured emissions, while net emissions from LUCF changes adjust the assigned amounts that countries are allowed to emit. This is predominantly for reasons of definitional clarity and monitoring feasibility with respect to the cycles of growth and decay in vegetation.

The next paragraph reflected the complexity of these negotiations and the concerns of those who had sought to include additional categories. It states that a subsequent Conference of Parties should review the qualifying categories of LUCF – and of agricultural soils – in the light of uncertainties and further methodological debate including studies by the IPCC. On this basis, future COPs should 'decide upon modalities, rules and guidelines as to how, and which, additional human-induced activities related to changes in greenhouse gas emissions by sources and removals by sinks in the agricultural soils and the land-use change and forestry categories' might subsequently be included. Such additions would be automatically applied to subsequent commitment periods, and countries could choose whether or not to include them in the first period.

One other issue concerned what became known as the 'Australia Clause'. Unlike most countries for which LUCF was a net sink in 1990, Australian land clearance resulted in considerable net emissions, estimated at the time of Kyoto as accounting for around 20% of Australian 1990 CO_2 emissions. In the closing hours of the Kyoto negotiations, Australia successfully inserted a sentence enabling countries for which LUCF emissions were positive in 1990 to include net land-use emissions in their 1990 base inventory.[6]

---

[5] Bernhard Schlamadinger and Gregg Marland, 'The Kyoto Protocol: Provisions and Unresolved Issues Relevant to Land-Use Change and Forestry', *Environmental Science and Policy*, 1998.

[6] Article 3.7 (2nd part): 'Those Parties included in Annex I for whom land use change and forestry constituted a net source of greenhouse gas emissions in 1990 shall include in their 1990 emissions base year or period the aggregate anthropogenic carbon dioxide equivalent emissions minus removals in 1990 from land use change for the purpose of calculating their assigned amount.' The omission of forestry from the qualifying amount was apparently

The first item of the work programme appended to the Kyoto Protocol calls upon the Convention's Subsidiary Bodies to take forward analysis of LUCF issues, and over the following year Parties made submissions totalling hundreds of pages. In effect, therefore, while the Protocol made an important structural decision regarding sinks and LUCF, Kyoto was just a staging post in long-running and convoluted debates about the details of qualifying activities.

4.2 'Bubbling' and the EU redistribution of emission commitments

The 'bubbling' provision (Article 4) enables a group of countries, when they ratify the Protocol, to redistribute their emission commitments in ways that preserve the collective total; it establishes legal responsibilities in the event of the collective commitment not being achieved. Specifically:

> 4.1 Any Parties ... that have reached agreement to fulfil their commitments under Article 3 jointly ... shall be deemed to have met those commitments ... provided that their total combined aggregate ... emissions ... do not exceed their assigned amounts ... calculated in accordance with Article 3. The respective emission levels allocated to each of the Parties ... shall be set out in that agreement.

The Article goes on to establish that (Article 4.3) 'any such an agreement shall remain in operation for the duration of the commitment period': it cannot be modified after submission with the instrument of ratification. Furthermore, for a group of countries acting 'in the framework of, and together with, a regional economic integration organization', the commitments shall then apply irrespective of any changes in the future composition of that organization (Article 4.4). This means that the EU cannot meet its commitment through expansion to include countries of central/ east Europe whose emissions had declined substantially through economic transition.

---

a deliberate sleight of hand overlooked by most participants at Kyoto; absorption from forestry change in Australia had helped to offset some of that country's other emissions from land-use change in 1990.

Table 4.2: The internal distribution of the EU 'bubble'

Country	Internal commitment (% change from 1990 levels)
Austria	− 13.0
Belgium	− 7.5
Denmark	− 21.0
Finland	0
France	0
Germany	− 21.0
Greece	+ 25.0
Ireland	+ 13.0
Italy	− 6.5
Luxembourg	− 28.0
Netherlands	− 6.0
Portugal	+27.0
Spain	+15.0
Sweden	+ 4.0
United Kingdom	− 12.5

Finally, the Article sets out the responsibilities in the event of failure to achieve the collective commitment. In this event, each country 'is responsible for its level of emissions set out in the agreement'; and for the EU, the European Community (which is the legal entity representing the European Union), as a Party to the Protocol, would share responsibility with its member states, in a situation of 'joint and several' liability.

In the aftermath of Kyoto, the EU moved rapidly to negotiate a formal, binding redistribution of its member states' commitments under Article 4. In June 1998 the EU Council reached agreement, guided by the previous non-binding agreement of March 1997, redefining the emission commitments of its member states, as set out in Table 4.2. This will now form the basis of EU ratification and the numbers in Table 4.2 will become the 'quantified emission limitation and reduction commitments' for each EU member state under the Protocol.

In many respects, offering the possibility of 'bubbling' to any group of countries represents a capitulation of the EU position that it should be entitled to unique treatment. The EU offered the provision in the belief that in practice it would be the only group of countries to utilize it. It is, however, a general provision, and a brief storm arose at Kyoto with the rumour that the 'JUSSCANNZ' countries – most of the OECD outside the EU – might form a bubble with Russia and Ukraine. This was probably a negotiating ploy to counter EU insistence on flat-rate reductions for all other countries, but the theoretical possibility of redistributing targets remains as countries approach ratification.[7]

4.3 Policies and measures[8]

Article 2 of the Kyoto Protocol fulfils the first item of the Berlin Mandate, which was itself developed from parts of Article 4 of the FCCC.[9] It sets out a number of activities – 'policies and measures' – which parties are encouraged to adopt in order to achieve their commitments under Article 3 of the Protocol. As described above in Chapter 3, the final text represents a compromise between widely divergent stances. The EU argued for a long and detailed menu of policy options from which parties would choose; some were to be mandatory, though the EU did not specify which.

The United States adopted a far more *laissez-faire* position, arguing against any particular direction in policy choice. Developing countries also joined the debate, seeking to ensure that any measures that were included in the final agreement would not harm their own economies and exports, a matter of particular concern to the oil exporters.

The resulting Article 2 consists of four main paragraphs, summarized in Box 4.1, that reflect a measure of compromise between these various positions. However, to a large degree, the details represent a victory for the US approach. The permissive wording of Article 2.1(a), requesting parties to

---

[7] In the months after Kyoto, EU alarm was further heightened when the United States convened a meeting of the 'Umbrella group' countries – JUSSCANNZ plus Russia and Ukraine – in Washington. But the meeting probably served to underline the political and practical difficulties of more widespread 'bubbling', which would require some countries to accept stronger targets to enable others to do less, and then to coordinate domestic ratification processes on this basis. However, the possibility is likely to be held open to forestall any attempts to block the implementation of emissions trading.

[8] Based on a draft by Duncan Brack.

[9] In particular, Article 4, para 2(a): 'Each of these [the developed-country] Parties shall adopt national policies and take corresponding measures on the mitigation of climate change … These Parties may implement such policies and measures jointly with other Parties … '.

Box 4.1: Elements of policies and measures in the Kyoto Protocol

2.1 Each Party included in Annex I ... shall:

(a) Implement and/or further elaborate policies and measures in accordance with its national circumstances, such as:

- enhancement of *energy efficiency* in relevant sectors;
- protection and enhancement of *sinks and reservoirs*;
- promotion of sustainable forms of *agriculture* in the light of climate change considerations;
- promotion, research, development and increased use of *new and renewable forms of energy*, of carbon dioxide *sequestration* technologies and of advanced and innovative environmentally sound technologies;
- progressive reduction or phasing out of *market imperfections* ... that run counter to the objective of the Convention, and apply *market instruments*;
- measures to limit and/or reduce emissions ... in the *transport* sector;
- limitation and reduction of *methane* ... through recovery and use in waste management ... and [provision of] energy.

There is also one catch-all sub-paragraph encouraging 'appropriate reforms in relevant sectors'.

(b) Cooperate with other such Parties to enhance the individual and combined effectiveness of their policies and measures adopted under this Article ... [the Conference/meeting of the Parties] ... shall consider ways to facilitate such cooperation.

2.2 [Annex I Parties shall] pursue limitation or reduction of emissions ... from *aviation and marine bunker fuels*, working through the International Civil Aviation Organization and the International Maritime Organization, respectively.

2.3 [Annex I Parties shall] strive to implement policies and measures ... in such a way as to minimize adverse effects ... on other Parties ... [the Conference/meeting of the Parties] may take further action, as appropriate.

2.4 [The Conference/meeting of the Parties], if it decides that it would beneficial to coordinate any of the policies and measures in paragraph 1(a) above ... shall consider ways and means to elaborate the coordination of such policies and measures.

adopt rather general measures 'such as', replaces the EU wording which, instead, used the words 'in particular'.

An important exception to the *laissez-faire* approach comes in the paragraph 2 requirement that Parties *shall* (emphasis added) pursue limitation or reduction of emissions from aviation and marine bunker fuels, working through the ICAO and IMO, respectively. This is much tougher wording than in paragraph 1, much more in line with the EU stance. International transport is the area in which coordinated international action is most obviously necessary, and in which other countries joined the EU in arguing for it.

Cooperation with other Parties 'to enhance the individual and combined effectiveness of their policies and measures' is maintained as an aim in paragraphs 2.1(b) and 2.4. Referring back to the FCCC's requirement to 'coordinate as appropriate', the Protocol text encourages experience and information exchange (paragraph 1(b)) and the consideration of coordination of policies and measures (paragraph 4). In each case, Conferences/meetings of the Parties to the Protocol are to consider ways to facilitate such cooperation and coordination.

Finally, paragraph 3 restates the principle of protection of countries from any adverse effects of any of the policies and measures that may be adopted, 'including the adverse effects of climate change, effects on international trade, and social, environmental and economic impacts on other parties, especially developing country parties'. Reference is made to Articles 4.8 and 4.9 of the FCCC, which list categories of developing countries particularly at risk, including obvious ones such as small island countries or those with areas prone to natural disasters, but also including 'countries whose economies are highly dependent on income generated from the production, processing and export, and/or consumption of fossil fuels and associated energy-intensive products'. Reference is also made to Article 3 of the FCCC, which lists general principles, including, *inter alia,* cooperation to promote 'a supportive and open international economic system that would lead to sustainable economic growth and development'. What none of these pieces of text do – of course – is to lay down what principles should be followed when the pursuit of some objectives (such as the promotion of energy efficiency) conflicts with that of others (e.g. avoiding a reduction in fossil fuel imports).

Although the outcome largely represented a defeat for the EU, the more *laissez-faire* approach favoured by the United States was not wholly predominant. Main headings for policies and measures (including energy efficiency, sustainable agriculture, transport, removal of energy subsidies) are listed, though only as examples of areas for action. The commitment to pursue the reduction of emissions from bunker fuels helped to reconcile the EU to the final compromise.

The general intention to protect developing countries' economies is present, as noted, in the requirement to minimize adverse impacts, but – hardly surprisingly – without special treatment for oil exporters.

The principle of cooperation 'to enhance the individual and combined effectiveness of their policies and measures'[10] is included. Paragraph 4 opens up a possible future route to international coordination, listing it as an item for a Conference/meeting of the Parties to discuss. This commitment, tentative though it is, represented a significant compromise by the United States. Canada was important in brokering a deal here, and suggesting the final Protocol text for this element. The efforts expended by the EU, therefore, though perhaps at least somewhat misdirected, were not entirely without effect, and do open the way to further negotiations on the future coordination of appropriate policies and measures to mitigate climate change.

Overall, Article 2 is the outcome of the unsatisfactory and incomplete debate which took place during the negotiations; as described above, attention at Kyoto had to be focused on the targets and timetables debate rather than policies and measures. Nevertheless, the Protocol commitments require domestic action to limit emissions, and Article 2 provides the ground for a future assault on Parties showing reluctance to take such action. More positively, it also provides the framework, particularly in paragraph 2 on bunker fuels, for a revived attempt to coordinate policies and measures where it would make sense to do so. Whether and how rapidly this happens is probably dependent on the willingness of the EU to do battle on policies and measures once more, this time with greater sophistication regarding what is and is not desirable and feasible.

---

[10] Kyoto Protocol, Article 2, para 1(b).

4.4 Mechanisms for international transfer

To complement the 'internal' flexibilities associated with multiple gases and sinks, the Kyoto Protocol includes three dimensions of international flexibility in addition to the 'bubbling' provision (which is hardly a 'flexible' mechanism, given the requirement for any such redistribution to be deposited as part of ratification and fixed thereafter). The three mechanisms share a common basis in the basic article on Commitments:

> 3.10 Any emission reduction units, or any part of an assigned amount, which a Party acquires from another Party in accordance with the provisions of Article 6 [Project JI] or of Article 17 [Trading] shall be added to the assigned amount for the acquiring Party.

> 3.11 Any emission reduction units, or any part of an assigned amount, which a Party transfers from another Party in accordance with the provisions of Article 6 [Project JI] or of Article 17 [Trading] shall be subtracted from the assigned amount for the acquiring Party.

> 3.12 Any certified emission reductions which a Party acquires from another Party in accordance with the provisions of Article 12 [Clean Development Mechanism] shall be added to the assigned amount for the acquiring Party.

The basic mechanism is thus quite simple. The devil lies in the detail – or lack of it.

4.4.1 Emissions trading

Emissions trading – the ability for two entities that are subject to emissions control to exchange part of their emission allowances – has evolved principally in a domestic context as a means for controlling domestic emissions (Chapter 3). In the Kyoto Protocol, it enables any two Parties to the Protocol to exchange part of their emission commitment, in effect redistributing the division of allowed emissions between them, at any time.

As noted, this proved to be one of the most controversial areas of the negotiations, though for different reasons in different quarters. Among the industrialized countries, Japan and some of the EU member states wanted to ensure that any such trading was competitive and transparent so as to prevent the United States using its political leverage to gain preferential access, particularly *vis-à-vis* the likely Russian surplus; the EU was also particularly anxious that trading should not enable the United States to avoid domestic action as the main agent. It also raised concerns about domestic rules of allocation. As the deadline loomed, the text shown in Box 4.2 was hammered out. It indicates the extent of OECD agreement on principles governing emissions trading, including the establishment of an industry-level international emissions trading market, but with some key issues still unresolved.[11]

In fact, this text never even reached the full plenary discussions because Ambassador Estrada knew that more fundamental objections would be raised by the developing countries. From the ashes of the ensuing confrontation (see Chapter 3) the bare minimum of enabling language survives in the Protocol – inserted among the Articles on implementation procedures – as three bald sentences (Article 17):

> The Conference of the Parties shall define the relevant principles, modalities, rules and guidelines, in particular for verification, reporting and accountability for emissions trading.
>
> The Parties included in Annex B may participate in emissions trading for the purposes of fulfilling their commitments under Article 3.
>
> Any such trading shall be supplemental to domestic actions for the purpose of meeting quantified emission limitation and reduction commitments under that Article.

The creative ambiguity embodied in these three sentences was sufficient to allow the supporters of emissions trading to understand that it is an intrinsic and irreversible part of the Protocol, while other Parties claimed that the Protocol could not proceed until agreement on implementation details was

---

[11] This text represents the extent of agreement reached after more than a week of intensive negotiations, with differences still evident about cumulating early crediting towards compliance [2], supplementarity caps [3] and the strength of requirements on the openness and competitiveness of the system (the three variants of para 7 corresponding roughly to US, Japanese and EU positions respectively). In fact, subsequent informal discussions reached further, with a US–French compromise proposal on a para 7 referring to 'government-regulated clearing houses', but the United States then sought to link this to agreement on unresolved disputes about supplementarity.

Box 4.2: OECD proposal on emissions trading: the text that never made it

Proposed Article 6

1. For the purpose of meeting its commitments under Article 3 in a cost-effective manner, any Party included in Annex 1 ... may, in accordance with the international rules on emissions trading [to be] established in paragraphs 5 and 7 of this Article, transfer to or acquire from any Party included in Annex 1 ... any of its emissions allowed under Article 3, provided that each such Party is in compliance with its obligations ... and has in place a national mechanism for the certification and verification of emission trades.

[2. Emission levels achieved before the start of any trading system established under the Protocol can[not] be used as the basis for emissions trading.]

[3. A Party shall not transfer or acquire more than x% of its emissions allowed in any budget period.]

4. A Party whose emissions are in excess of its emissions budget in any budget period may acquire, but may not transfer, emissions allowed.

5. Each Party shall report to the Secretariat on an annual basis on all of its transfers and acquisitions under this Article, and on the functioning of its national mechanisms for the certification and verification of emission trades. The Secretariat shall publish, on an annual basis, a tabulation of such emission trades made in the previous year and of the adjustments required under Articles 3.10 and 3.11. Further guidelines for reporting may be adopted at the First Meeting of the Parties.

6. A Party may authorize intermediaries to participate, under the responsibility of that Party, in actions leading to the transfer or acquisition, under this Article, of emissions allowed. Parties shall be responsible for all transfers and acquisitions, including those resulting from transactions by intermediaries, under this Article.

7. Based on the experience in the first budget period the Parties shall review whether the emissions trading market is functioning in an open and competitive manner, and based on such a review, take appropriate action.

   For the purpose of ensuring transparent and accessible emissions trading, inter-national guidelines to guarantee competitive emissions trading shall be adopted at the First Meeting of the Parties.

   Parties shall ensure that the emissions trading system is transparent, accessible, and is functioning in a competitive manner. Trading by Parties shall take place through a clearing house system. To this end, the first Meeting of the Parties shall set up the clearing house system and the relevant principles and rules including those related to legal liability.

8. If a question of a Party's implementation of the requirements of [monitoring, reporting, etc.] is identified ... transfers and acquisitions of emissions allowed may continue to be made, provided that such emissions allowed may not be used by any Party to meet its obligations under Article 3 until any issue of compliance is resolved.

The Group of 77 and China requested – and ultimately obtained – deletion of this Article; see Chapter 3.

reached. In practice the post-Kyoto debate makes it clear that the clause has achieved the *fait accompli* that there will be some kind of emissions trading: the debate has indeed turned to the relevant 'principles, modalities, rules and guidelines', and to the meaning of making such trading 'supplemental to domestic actions'. These issues are considered in Chapter 6 of this book.

4.4.2 Joint implementation within Annex I

Article 6 of the Protocol enables emission savings or sink enhancement arising from cross-border investments between Annex I Parties to be transferred between them. This is joint implementation at the project level, in the sense that the term came to be used in the debates prior to Kyoto (Chapter 3). However, because it occurs between countries that are both subject to legally binding constraints, it does not carry many of the political and technical complexities associated with joint implementation more widely. The Article establishes that JI projects between industries within Annex I may proceed and generate 'emission reduction units'. This necessarily involves private investment, but to have legal significance under the Protocol – and hence value to the governments concerned – it must be sanctioned by the governments of the participating industries.

Agreement must be reached on the emissions saved by the investment, as compared with what would otherwise have been emitted. At this point, the emission savings agreed between the Parties become equivalent to an international emissions trade, being deducted from the allowed emissions of the host country, and added to the allowed emissions of the investing country. Because the combined emissions from the countries remain constrained, however, the accuracy of the estimated emissions savings is, from the standpoint of the environment and of the Protocol, of secondary importance; it is a matter for negotiation between the governments and industries concerned. This is discussed further in Chapter 5.

This Article on project-based joint implementation between Annex I countries was the least disputed of the transfer mechanisms, comprising mostly text already agreed before Kyoto. It makes the process strongly conditional upon fulfilment of obligations regarding the adequate and

Box 4.3: Joint implementation (Article 6)

1. [A]ny Party included in Annex I may transfer to, or acquire from, any other such Party emission reduction units resulting from projects aimed at reducing anthropogenic emissions by sources or enhancing anthropogenic removals by sinks of greenhouse gases in any sector of the economy, provided that:
   (a) Any such project has the approval of the Parties involved;
   (b) Any such project [reduces emissions or enhances removals by sinks], additional to any that would otherwise occur;
   (c) It does not acquire any emission reduction units if it is not in compliance with its obligations on [compilation of emission inventories and reporting];
   (d) The acquisition of emission reduction units shall be supplemental to domestic actions for the purposes of meeting commitments under Article 3.

2. The COP/MOP may, at its first session or as soon as practicable thereafter, further elaborate guidelines for the implementation of this Article, including for verification and reporting.

3. A Party included in Annex I may authorize legal entities to participate, under its responsibility ...

4. If a question of implementation ... is identified in accordance with the relevant provisions of Article 8 [expert review] emission reduction units [transferred by that Party] ... may not be used by [another] Party to meet its commitments ... until any issue of compliance is resolved.

credible development of national emission inventories and reporting thereof by Parties *acquiring* emission reduction units, but apparently not necessarily by those transferring them – a point of some dispute in the aftermath of Kyoto.[12]

Apart from this point, the idea of project investments leading to the generation of emission reduction units transferred between Parties with binding obligations was not felt to involve any new principles and it built readily upon ongoing AIJ activities, many of which were occurring in EIT countries. In the months after Kyoto, Japan announced a substantial programme of JI project investments in Russia; the least controversial mechanism is thus the first to take a concrete form and to start attracting real investments.

---

[12] The EU stated that it had always understood these obligations as referring to both sides in any JI transaction, but some JUSSCANNZ countries disagreed with this intepretation.

4.4.3 The clean development mechanism

In addition to these mechanisms for transfer between Annex I Parties, the Protocol establishes the 'clean development mechanism' which, in principle, enables activities similar to joint implementation to proceed with non-Annex I countries. The CDM, created at a late stage by welding two apparently opposed proposals (Chapter 3), is an amalgam of different elements of precursor ideas which were stitched together and expanded at Kyoto, into Article 12, comprising ten paragraphs that are unique in the history of international agreements.

The core elements of the CDM are summarized in Box 4.4. Its stated purpose is to help developing countries to achieve sustainable development and so contribute to the ultimate objective of the Convention, and to 'assist Annex I Parties in achieving compliance' with their specific commitments. Project activities under the CDM shall 'benefit' developing countries, and generate 'certified emission reductions' which Annex I Parties may use to 'contribute to compliance with part of their quantified commitments'. Emission reductions shall be certified on the basis of criteria (borrowing language from the Berlin AIJ decision) including voluntary participation, 'real, measurable and long-term benefits' related to mitigating climate change, and emissions additionality ('reductions that are additional to any that would occur in the absence of the certified project activity'). The CDM is not a fund, but shall 'assist in arranging funding of certified project activities as necessary', and participation may explicitly involve private and/or public entities.

Two main features distinguish the CDM from the earlier proposals on joint implementation. The first is the strength of multilateral control over the process: it shall be supervised by an executive board and 'shall be subject to the authority and guidance' of the Parties collectively, that shall at the first meeting after entry into force (COP/MOP 1) take actions to ensure 'transparency, efficiency and accountability through independent auditing and verification of project activities'.

The second distinguishing feature is that 'a share of the proceeds from certified project activities' shall be used to cover administrative expenses as well as to assist particularly vulnerable developing countries to meet the costs of adapting to climate change. This clause, which was crucial in

Box 4.4: The clean development mechanism (Article 12)

1. A clean development mechanism is hereby defined.

2. The purpose of the clean development mechanism shall be to assist Parties not included in Annex I in achieving sustainable development and in contributing to the ultimate objective of the Convention, and to assist Parties included in Annex I in achieving compliance with their ... commitments under Article 3.

3. Under the clean development mechanism:
   (a) Parties not included in Annex I will benefit from project activities resulting in certified emission reductions (CERs); and
   (b) Parties included in Annex I may use the CERs accruing from such project activities to contribute to compliance with part of their ... commitments under Article 3, as determined by the Conference of the Parties serving as the meeting of the Parties to this Protocol (COP/MOP).

4. The clean development mechanism shall be subject to the authority and guidance of the COP/MOP and be supervised by an executive board of the clean development mechanism.

5. Emission reductions resulting from each project activity shall be certified by operational entities to be designated by the COP/MOP, on the basis of:
   (a) voluntary participation approved by each Party involved;
   (b) real, measurable, and long-term benefits related to the mitigation of climate change; and
   (c) reductions in emissions that are additional to any that would occur in the absence of the certified project activity.

6. The clean development mechanism shall assist in arranging funding of certified project activities as necessary.

7. The COP/MOP shall, at its first session, elaborate modalities and procedures with the objective of ensuring transparency, efficiency and accountability through independent auditing and verification of project activities.

8. The COP/MOP shall ensure that a share of the proceeds from certified project activities is used to cover administrative expenses as well as to assist developing country Parties that are particularly vulnerable to the adverse effects of climate change to meet the costs of adaptation.

9. Participation under the CDM ... may involve private and/or public entities ... subject to whatever guidance may be provided by the executive board ...

10. CERs obtained during the period from the year 2000 up to the beginning of the first commitment period can be used to assist in achieving compliance in the first commitment period.

building sufficient G77 support for the CDM, represents an important novelty in funding sources, and starts to give concrete form to the Convention commitment in this area.[13]

---

[13] The Convention (Article 4.4) stated that the developed countries (listed in Annex II) should assist countries that are 'particularly vulnerable to the adverse effects of climate

Perhaps the most surprising feature of the CDM is the final paragraph, which states that certified emission reductions CCERs 'from the year 2000 up to the beginning of the first commitment period' can be used to 'assist in achieving compliance in the first commitment period'. In principle, therefore, CERs, envisaged in the most complex, contentious, novel and ill-defined instrument in the whole Protocol, and the one involving developing countries that had fought to make any action conditional upon developed-country leadership, may be generated sooner than anything else in the agreement – and well before the Protocol itself is likely to enter force.[14] Furthermore these credits may be *cumulated* towards meeting commitments defined over 2008–12 – though there is no reference to CERs generated *during* the commitment period, which is presumably just an omission in the drafting. The implications of early crediting turn out to be potentially quite momentous, as discussed in Chapter 6 of this book.

A subsequent dispute over the CDM emerged with the observation, after the dust from Kyoto had settled, that the CDM referred only to 'emission reductions'. The United States and several other JUSSCANNZ countries claimed that the CDM was definitely intended also to include sinks and this had been clear at the time, but Ambassador Estrada has forcefully rejected this claim, and whether to include land use and forestry in the CDM will be a topic for future negotiation.[15]

---

change in meeting costs of adaptation to those adverse effects'. However, the mandate for the Global Environmental Facility was focused upon fostering abatement of emissions, and there were long-running disputes about extending its mandate to incorporate support for adaptation that were not resolved until the Buenos Aires conference (Chapter 7). There is concern among developing countries that they may in practice have to contribute to 'proceeds from the CDM'.

[14] The EU as a whole disliked the CDM, seeing it as both a potential loophole and a competitor to the Global Environmental Facility; some of its member states particularly detested this clause on early crediting. However, the Europeans were still conferring internally when this came up for final agreement and it was declared agreed before the EU raised its flag to object. Ambassador Estrada refused to concede to the EU's belated objection – he could hardly do otherwise by then – and offered only the crumb of an obligation to 'analyse the implications' of early crediting.

[15] Ambassador Estrada expressed an unequivocal view on this topic after Kyoto. He observed that 'in contrast' to Article 6 (JI), 'Article 12 on the CDM only refers to reductions of emissions and says nothing about removals of greenhouse gases. It is only logical

Notwithstanding these continuing disputes, the CDM was an astonishing result, and the Protocol could almost certainly not have been adopted without it. It gives the Kyoto Protocol commitments the global investment scope which the United States and most other JUSSCANNZ countries had been desperately seeking, and explicitly enshrines the role of the private sector. Yet almost all the concerns that developing countries and others had raised about global joint implementation apply potentially to the CDM, so its adoption can be seen a remarkable *volte-face* on their part. This was justified presumably by the potential protection afforded by the executive board and COP/MOP, and the lure of funds to assist adaptation. Nevertheless it highlights the very real and significant problems which must be addressed in the governance of the CDM (discussed briefly in Chapter 7 of this book).

4.4.4 Conclusions on the mechanisms for international transfer

The three mechanisms in the Kyoto Protocol that enable international flexibility in theory allow the industrialized countries to meet their emission commitments through bilateral trading of assigned amounts among themselves, and through investment anywhere in the world. Chapter 5 of this book quantifies their potential impact on the practical consequences of the commitments, and finds it is huge: the meaning of the commitments is inseparable from the flexibilities surrounding them.

Such flexibility is unprecedented in an international agreement. Politically it reflects the dominance and objectives of the United States, backed strongly by several other JUSSCANNZ countries, as described in Chapter 3. Underlying this, the international flexibility specifically reflects

---

to conclude that different wording reflects different meaning ... it has been suggested there was an understanding among negotiators to make the texts of Articles 6 and 12 uniform on this point. That was never brought to my knowledge, neither during the negotiations nor after the negotiations and before formal approval by the Conference. Delegates involved were well-experienced diplomats, scientists and professional staff, and nobody should be induced to error ... the only real truth is in the political will of governments, and sequestration will be included in the CDM or not according to that will'. Raúl Estrada-Oyuela, 'First Approaches and Unanswered Questions', in *Issues and Options: The Clean Development Mechanism,* New York: UNDP, 1998. See also Chapter 7.

the general process of economic globalization culminating in the 1990s and incorporating:

- the pressures from companies whose international reach made economic mockery of attempts to confine commitments within national boundaries;
- the desire of OECD countries to give their companies full scope to seek out emission reduction possibilities internationally;
- the desire shared by East and West to use corporate investment to help clean up after communism;
- the growing acceptance of developing countries of the role of private investment and the desirability of attracting foreign companies.

Many of these ideas had been anathema a generation earlier; by the mid-1990s, when the Protocol's core ideas were born, they had become almost hegemonic in economic but not in environmental policy. The Protocol's mechanisms for international transfer had their origin at the peak of globalization, and just before the Asian financial crisis started to legitimize attendant cynicism. The Protocol is essentially an agreement to extend economic globalization to environmental policy: to establish a global emissions market to counter the global environmental consequences of global economic growth.

But the Protocol settled few of the core issues about how such mechanisms might actually work and be governed in the international context. That task is complicated by the inconsistencies between the mechanisms, which may be rationalized by understanding the process, but not in terms of outcome. The commitments are defined for 2008–12 but companies need incentives now if they are to find it profitable to take the actions that will be required to meet them. Early and cumulated crediting is allowed for the CDM, the most complex and novel of mechanisms, which will involve investment in regions where typically projects may take many years to get going; but early and cumulated crediting is not specified for JI among Annex I countries, where governments and companies are already more geared for action. Sinks are included for JI, but remain disputed in the CDM. Monitoring and certification procedures are different between the two project-level mechanisms; emissions trading, in theory the simplest mechanism but one on which a great deal hangs politically, remains wholly undefined. The unfinished business of Kyoto's mechanisms for international transfer is immense, as the subsequent Buenos Aires conference illustrated starkly (and as taken up tentatively in Part II of this book).

4.5 Additional issues relating to developing countries

4.5.1 'Advancing the commitments' of all Parties: national programmes, technology transfer and financial mechanisms

Apart from the CDM, several other parts of the Protocol cover activities that involve developing countries. Article 10, derived from the Berlin Mandate obligation to 'continue to advance the implementation' of the Convention's universal Article 4.1 commitments, can be traced almost paragraph by paragraph to that Article, slightly strengthening some paragraphs, omitting others, and reproducing others with minor wording changes.

The main outstanding dispute on this Article at the finale in Kyoto (see Chapter 3) resulted in some expansion of the role and scope of programmes by and communications from the developing countries.[16] The advance was modest, partly because of the fact that this requirement in theory has to apply to every country, from the large and populous to the tiny and poverty-stricken. It might, however, help to stimulate capacity in the developing world somewhat in the way that communications for Annex I countries began to do for them after the Convention.

---

[16] The headline language governing national programmes in the Protocol paragraph 10(b) is identical to the Convention's Article 4.1(b), but is supplemented by two sub-paragraphs – the focus of the outstanding dispute at Kyoto (Chapter 3). Bo Kjellen's compromise language deftly circumvented G77 resistance to *requirements* that such programmes address specific sectors by transforming this into statements of fact: such programmes *would* '*inter alia*, concern the energy, transport and industry sectors as well as agriculture, forestry and waste management', and stating that adaptation technologies and methods for improving spatial planning *would* 'improve adaptation to climate change'; and it expanded the scope of developing-country communications beyond the Convention's requirements on inventories and information 'that the Party considers relevant and suitable for inclusion', so that under the Protocol they 'shall seek to include … as appropriate … information on programmes which contain measures that the Party believes contribute to addressing climate change and its adverse impacts'.

In other places, the wording does lay greater emphasis upon the role of private investment and of actions by and in the developing countries themselves. Article 10(c) of the Kyoto Protocol states that Parties should

> cooperate in the promotion of effective modalities for the development, application and diffusion of ... environmentally sound technologies, know-how, practices and processes pertinent to climate change, in particular to developing countries, including the formulation of policies and programmes for the effective transfer of environmentally sound technologies that are publicly owned or in the public domain and the creation of an enabling environment for the private sector.

The opening statement that Parties should 'cooperate in' contrasts with the Convention's injunction that developed-country Parties only 'take all practicable steps'. It also expands similar language in the Convention by adding the general catch-all on 'practices and processes pertinent to climate change'. Perhaps most significantly, it formally recognizes the role of the private sector and the need for an 'enabling environment'. All this reflects significant evolution of thinking, but is practically meaningless in terms of specific commitments.

Similar remarks would apply to most of the other paragraphs in this Article, which ends by reiterating the need to 'give full consideration' to meet the specific needs and concerns of various categories of developing countries, including those most vulnerable to climate change and to the impact of response measures.

Article 11 on financing simply applies the Convention's language on the financial mechanism to the Article 10 commitments, almost verbatim.

The Protocol does not make explicit any link between the flexible mechanisms and technology transfer. JI/CDM may be an effective tool of technology transfer if investment is channelled into projects that replace old and inefficient fossil fuel technology, or create new industries in environmentally sustainable technologies, but the Kyoto Protocol does not deal with these aspects directly and treats technology transfer as a separate topic. However, it does acknowledge that technology transfer today is increasingly integrated into policy debates about investment, and that the problems of managing investment for better development are the same as for increasing technology transfer.[17]

4.5.2 Minimizing adverse impacts

A defining tension in the negotiations on climate change, ever since their inception, has been the anxiety of some developing countries (especially AOSIS) about the adverse impacts of climate change on the one hand, set against contrasting anxieties (especially in OPEC) about the adverse economic impacts of response measures. Both sets of anxieties were in fact shared more widely among the developing countries and, as noted, a major political success of OPEC was to align its fears procedurally alongside those of other 'vulnerable groups' listed in Articles 4.8 and 4.9 of the Convention. A key clause in the Protocol's Article 3 on commitments is the final paragraph, which states that:

> Each Party included in Annex I shall strive to implement the commitments ...
> in such a way as to minimize adverse social, environmental and economic
> impacts on developing country Parties, particularly those identified in Article
> 4, paragraphs 8 and 9, of the Convention ... [the Meeting/Conference of
> Parties shall] ... consider what actions are necessary to minimize the adverse
> effects of climate change and/or the impacts of response measures on Parties
> referred to in those paragraphs. Among the issues to be considered shall be the
> establishment of funding, insurance and transfer of technology.

This reinforces, and is more specific than, the similar provision in Article 2 on policies and measures. In addition, a decision appended to the Protocol at Kyoto launches a specific fast-track process 'to identify and determine actions necessary to meet the specific needs of developing countries', referring to the Convention Articles 4.8 and 4.9 on vulnerable groups. These were the key provisions which gave OPEC enough to build on to accept the Protocol.

---

[17] An extensive analysis of technology transfer and investment in the context of climate change and the Kyoto Protocol is given by Tim Forsyth, *International Investment and Climate Change*, London: RIIA, May 1999.

The paragraph thus certainly establishes the basis for subsequent negotiation on measures on 'funding, insurance and transfer of technology' relating to vulnerable groups. Arguably, it could also be used to challenge abatement measures that OPEC considers discriminate against international oil. Since political pressure within Annex I countries will favour measures which tend to protect domestic interests (such as German coal subsidies) at the expense of imported fuels, the final paragraph could open up a very large can of worms indeed.[18]

4.5.3 Amendment procedures and expansion to include new countries

The main purpose of the chairman's proposal on voluntary accession (Chapter 3) was to provide institutional procedures by which new countries wishing to take on quantified commitments could join Annex B of the Protocol. With the defeat of that proposal, the main route left for additional countries to be incorporated is by joining Annex I of the Convention and amending Annex B, which lists the quantified commitments that industrialized countries agreed to at Kyoto, to incorporate additional countries. Unfortunately there are two big obstacles to this.

First, the Protocol states that Annexes 'are an integral part' of the Protocol and can only be amended with the full amendment procedures. These, set out in Article 20, are taken directly from the Convention (see Chapter 2), and require agreement *and ratification* by three-quarters of the Parties to the Protocol. This is a high and cumbersome hurdle, which is bound at minimum to take a couple of years even if the negotiators are all agreed (amendments to Annex B also have to be agreed in writing by the Parties affected).

The other obstacle – or rather ambiguity – is that most of the Protocol's commitments apply to the Parties listed in Annex I of the Convention.

---

[18] Potentially it could set the stage for an enduring struggle between the coal industries within Annex I countries and the international oil business, with Annex I governments at the interface. In the short term this struggle will be limited by the complexity of energy trade flows (which include considerable non-OPEC oil as well as the growing international coal trade); nevertheless the potential for developing-country exporters to debate the impacts of specific Annex I energy policies is considerable and could have profound long-run implications.

Annex B is supposed just to list their commitments. *Only* Article 17 on emissions trading – at the insistence of the United States – refers to 'the Parties listed in Annex B' as a discrete group. However, whether it is possible to participate in emissions trading without joining the core structure of commitments defined in Article 3 and the rest of the Protocol, and the Convention itself, is – to say the least – unclear.

The dilemma is readily resolved for Parties willing to join the full set of Annex I commitments under the Convention, but there are obvious objections to that for most developing countries.[19] In short, in the area most crucial to the future evolution of the regime, the Protocol is a mess. The next significant legal step may have to involve a wholesale amendment to the structure and geographic scope of commitments – perhaps in the context of a second commitment period – rather than piecemeal expansion, a process which may prove significant in determining the future evolution of the regime (discussed in Chapter 8 of this book).

4.6 Compliance, future development and related issues: monitoring, reporting and review[20]

The task of ensuring that the Kyoto commitments are met and ultimately lead to strengthened action towards the objective of the Convention can be divided into three main components: the assessment of compliance with specific obligations; the enforcement of those obligations; and the review of adequacy and revision of commitments. In most respects the Protocol is remarkably stringent on the first of these, currently and typically weak on the second, and somewhat ambiguous on the third. This section summarizes the provisions on each.

---

[19] It would require countries nominally to fulfil the objective of Article 4.2 of aiming 'to return by the end of the present decade to earlier levels of emissions …', which is impossible for most developing countries, though as a non-binding aim it could arguably be overlooked. It would also require countries to take on the full panoply of Annex I reporting requirements, without the 'degree of flexibility' afforded to the transition economies.

[20] Section drafted by John Lanchbery.

4.6.1 Compliance assessment: uncertainty, reporting and review of information

The Kyoto Protocol lays great emphasis on compliance assessment. It repeatedly stresses the need for accountability and verification. In part, this emphasis reflects the fact that the Protocol contains legally binding commitments and many Parties feared that others might cheat on what may be an expensive agreement to implement. This concern especially focused on the mechanisms for international transfer, which many states either feared or did not fully understand or both.

There was also general concern, evident in the Protocol provisions relating to sinks for carbon dioxide, that many emission sources and sinks might be hard to estimate and that compliance assessment would be correspondingly hard. As noted, the qualifying emission sources were derived directly from IPCC guideline categories, and the adjustments to assigned amounts to reflect net emissions from land-use change and forestry are designed to exclude emissions or removals that could not be reliably monitored. In this and other respects, the Protocol inextricably weaves together technical and legal aspects of compliance. The interdependence of these within a treaty that depends so much on science is perhaps obvious. Nevertheless, most environmental treaties do not make such links as strongly. It may bode well for the development of the Protocol that it does so from the outset.

The Kyoto Protocol provisions for reporting and review are scattered throughout the document, making their overall effect difficult to grasp at a glance. Reporting and review are first mentioned in *Article 3* (which is ostensibly about targets) which states that forestry-related activities shall be reported in a transparent and verifiable manner, reflecting concerns about the uncertainties associated with this sector. This is linked to Article 5 (on methodologies) and stresses that any decision on including forest-related emissions and removals shall take into account 'uncertainties, transparency in reporting, verifiability, the methodological work of the IPCC, the advice provided by the SBSTA in accordance with Article 5 and decisions of the COP'.

Article 3 also summarizes how transfers under the flexibility mechanisms shall be accounted for (Chapter 3, Section 4 above); indeed, without the provisions of Article 3, the basic functioning of emissions trading would be obscure.

Article 5 provides a framework for how Parties will devise and implement methodologies. It specifies that Parties in Annex I shall institute national systems for compiling inventories of greenhouse gases no later than a year before the start of the first commitment period, i.e. by 2007. The Article clearly envisages that the Protocol methodologies will build upon the present OECD/IPCC methodologies, and states that these shall be regularly reviewed and adjusted. However, revisions (including revisions to the GWPs used for the different gases) shall be applied to commitment periods subsequent to the revision, to prevent 'shifting of the goalposts' relating to commitments already made.[21] Thus, Article 5 seeks to allow for the growth and development of methodologies over time in light of increasing knowledge, while protecting the core basis upon which the specific commitments were made.

Article 7 addresses national reporting, focusing mainly on emission inventories and other information relating to compliance with Article 3 commitments. This again reflects the Parties' focus on monitoring compliance rather than, for example, specific policies and measures.[22] Article 7 first reiterates an existing decision for Parties to submit annual inventories of greenhouse gases but adds the rider that 'the necessary supplementary information for ensuring compliance with Article 3' should also be submitted.[23] The next paragraph more or less repeats this, but with

---

[21] It was agreed prior to COP-3 that all methodological changes should be applied retrospectively to all prior years, including the baseline year. This was essential to make meaningful compliance-related data comparisons. It does, however, mean that a change of methodologies is likely to result in changes to assigned amounts and that such changes cannot therefore be allowed in commitment periods.

[22] Policies and measures are addressed in Articles 2 and 10, but unlike the Convention itself, the question of reporting them is only touched on obliquely. Of course, because reporting on policies and their implementation is covered by the Convention and COP decisions there is, strictly speaking, no need to refer to the subject again. But then, there are fairly good Convention provisions and COP decisions on inventories and yet the Protocol frequently repeats and elaborates on them.

[23] This is a wise provision because many inventories submitted so far are neither complete nor transparent, nor are the uncertainties associated with emission estimates usually made clear.

reference to the overall 'commitments under the Protocol'. These paragraphs (together with a subsequent decision at COP-4 on comparison of national inventories against independent data) thus prepare the way for a thorough system for verification of compliance.

The Article is less prescriptive about the timing and content of reports as a whole, leaving this decision to the Conference/meeting of the Parties. It also only mentions that reporting guidelines should be reviewed periodically. However, it clearly specifies that the Conference/meeting of the Parties must decide on modalities for accounting of assigned amounts prior to the first commitment period. Concerns about the effect of uncertainty in transferring parts of assigned amounts are thus, in principle, addressed.

Article 8 deals with how the information submitted under Article 7 shall be reviewed. It builds on the existing Convention system of 'in-depth review teams' coordinated by the Secretariat. The teams will continue to be selected from experts nominated by Parties and, perhaps, by intergovernmental organizations such as the IPCC or OECD. Again, specific mention is made of inventories and assigned amounts which must be reviewed annually. The provisions on how to review communications as a whole are more vague but such reviews must be 'thorough and comprehensive'. The expert teams are obliged to report back to the Conference/ meeting of the Parties, 'assessing the implementation of the commitments of the Party [concerned] and identifying any problems in, and factors influencing, the fulfilment of commitments'. The Secretariat is specifically tasked with listing the questions raised by the expert reports and submitting the lists to the Conference/meeting of the Parties for decision as required.

The Articles on the mechanisms for international transfer also contain specific reference to compliance assessment. *Article 6* on joint implementation specifically excludes Parties that are not in compliance with Articles 5 and 7 on methodologies and reporting from acquiring ERUs, though not specifically from transferring them (Section 4.4 above). The Article also suggests that the first Conference/meeting of the Parties may further elaborate guidelines for implementing JI, including verification and reporting, and introduces a 'traffic light' approach to accountability in traded

CERs.[24] Article 12 on the CDM contains a 'catch-all' paragraph, stating that the first Conference/meeting of the Parties shall 'elaborate modalities, and procedures with the objective of ensuring transparency, efficiency and accountability through independent auditing and verification of project activities'.[25] Article 17 on emissions trading contains a similar provision, that the Convention's Conference of Parties (not the COP/MOP to the Protocol) shall 'define the relevant principles, modalities, rules and guidelines, in particular for verification, reporting and accountability for emissions trading'.

Thus, despite all the provisions there remains a huge amount of work in this area. Not only is there no agreement on how to allow for the uncertainty in emission estimates, but as yet little work has been done on the issue. Similarly, the Articles on JI within Annex I and on the CDM both state that project activities shall be 'additional' to those that have occurred anyway, but the question of how to assess project benefits against what might have happened (but has not) is a thorny one that has still to be properly addressed. Finally, both the emissions trading and the developed-country JI Articles contain the provision that they shall be 'supplementary to domestic actions'; as noted, this phrase is subject to widely differing interpretations.

4.6.2 Enforcement provisions

Enforcement of commitments on a country that does not adhere to them tends to be one of the first issues raised in popular discussion about international law, and one of the last issues to be addressed in actual international treaties. This is for perfectly good reasons. The general presumption underpinning international law is that serious countries would not sign up to and ratify an agreement unless they intended to fulfil

---

[24] If a question regarding implementation (of any sort) arises about one of the Parties participating in a JI project or projects, then the ERUs arising from the projects cannot be used until the compliance-related issue is resolved. This is sometimes likened to a traffic light which switches to a warning orange when a compliance question is raised.

[25] This is because of the short time available and the involvement of developing countries. The latter necessarily raises additional reporting and review issues, over and above those in the other mechanisms, and was deferred for further attention.

their commitments: far better not to sign than to undermine the whole basis of international law by flouting an agreement. Discussion of specific enforcement mechanisms also necessarily involves procedures that could impinge upon national sovereignty and are hence very sensitive.

Specific legal processes concerning compliance are dealt with in Articles 16, 18 and 19. Article 16 is simply a provision to allow a 'multilateral consultative process' being negotiated under Article 13 of the Convention to be adopted, possibly in modified form, into the Protocol. At the time of the Kyoto COP the content of this long-running effort had still to be agreed and so Article 16 is necessarily short and a little vague. Indeed, in June 1998 in Bonn, the Parties failed to conclude a two-year process devoted to elaborating the Convention's Article 13 when the United States objected to the composition of the committee that would oversee its operation. Apart from this question, the fleshed-out content of the Article had been agreed. Prior to the conclusion of the Convention it was envisaged as a non-compliance provision, most probably along the lines of the Montreal Protocol's non-compliance procedure, but (given in part the lack of quantified commitments) it evolved as a mechanism to help developing countries with their reporting commitments. At first this might seem rather trivial, but at the heart of any compliance process is a good reporting process and many countries have considerable problems, both technical and financial, in compiling reports that will be adequate for compliance assessment.[26]

Unlike Article 16, Article 18 is directed at overall compliance with specific commitments (and so the two Articles could be seen as complementary). Like so many of the Articles relating to verification and compliance, it basically provides some framework for more negotiations. It states that at its first meeting the Conference/meeting of the Parties shall approve effective procedures and mechanisms to determine and address cases of

---

[26] If the composition of the committee can be resolved, then Article 13 will certainly be useful to the Convention. Whether it will be worth adopting it into the Protocol is another matter. Developing countries could receive assistance in reporting via the Convention so there is no need, at present, for them do so via the Protocol. On the other hand, if developing countries are to become increasingly active participants in Protocol commitments regarding targets, a specific assistance mechanism for them in the Protocol might be beneficial.

non-compliance with the Protocol. It then goes on to say that it should develop an 'indicative list of consequences' for non-compliance – possibly, penalties. This makes it a potentially stringent mechanism, far more so than in most other international agreements. It is clear that the Parties intend, if necessary, to try to enforce compliance by imposing a range of penalties. Although some environmental agreements have penalized their Parties this has usually been done in an ad hoc manner rather than in the routine and systematic way implied by Article 18.

Penalties for countries are normally a feature of economic agreements. They are not automatic even in most arms control agreements, which are generally regarded as more politically important than environmental treaties. Article 18 could thus represent a major step towards increased environmental protection at the international level. However, a potential drawback is its final sentence which stipulates that any mechanisms or processes that entail binding 'consequences' must be adopted by an amendment to the Protocol; in other words, a new, formal agreement would have to be signed and ratified.[27]

Article 19 of the Protocol simply allows the settlement of disputes procedure in the Convention to be applied to the Protocol. These provisions are similar to those in many other environmental agreements and are wisely regarded as almost useless – a view supported by the fact that, as far as the author is aware, such provisions have never been used in environmental agreements. This is basically because they ultimately result in a sort of trial in which one side must lose. States are generally reluctant

---

[27] The need for a formal amendment has already caused problems in elaborating the Protocol. In the United States, for example, the constitutional separation of powers means that when the administration (led by the President) and the legislature (the Congress comprising the Senate and House of Representatives) are of opposing political persuasions, as they are now, the Senate is reluctant to ratify any international agreement signed by the President. The United States, represented by the administration, thus shies away from linkages between Article 18 and other compliance-related processes, especially those that will be needed for the transfer mechanisms. This is because the United States wants the transfer mechanisms to become operational as soon as possible, and fears that linking them to a separate ratification process via Article 18 will result in long delays. The EU, on the other hand, wishes to negotiate the compliance mechanisms holistically. It is hard to refute the logic of this argument and privately most US negotiators agree with it in theory, but their current constitutional gridlock makes it difficult for them to agree in practice.

to involve themselves in international processes over which they have no control and which they could lose.

4.6.3 Review and strengthening of commitments

Article 9 on review procedures for the Protocol is quite brief and essentially says that in line with new information and reviews of implementation, the Conference/meeting of the Parties shall review the Protocol and decide on 'appropriate action'. A first review of the Protocol shall take place at the second Conference/meeting of the Parties and subsequent reviews shall be regular and timely. This is weaker than many Parties in Kyoto wished, because the Conference/meeting of the Parties can fulfil these review functions anyway. There is no direct reference to the overall objective (of ultimately stabilizing atmospheric concentrations), but reviews will be 'coordinated with pertinent reviews under the Convention'.

When linked with Article 3, paragraph 9, which states that the consideration of commitments for subsequent periods should begin no later than 2005, the process is stronger than it otherwise appears.

In addition, Article 13 sets out the functions of the Conference/meeting of the Parties, which shall

a) Assess ... the implementation of this Protocol by the Parties, the overall effects of the measures taken pursuant to this Protocol, in particular environmental, economic and social effects as well as their cumulative impacts and the extent to which progress towards the objective of the Convention is being achieved;

b) Periodically examine the obligations of the Parties under this Protocol, giving due consideration to any reviews required ... in the light of the objective of the Convention, the experience gained in its implementation and the evolution of scientific and technological knowledge ...;

c) Exercise such other functions as may be required for the implementation of this Protocol, and consider any assignment resulting from a decision by the Conference of the Parties.

Thus, although the provisions on reviewing the adequacy of commitments are not individually very forceful, taken together they may prove fairly stringent – if, that is, the governments of the world do fulfil existing commitments as a step towards ultimately strengthening and expanding the Kyoto Protocol.

4.7 Conclusions

The Kyoto Protocol is a complex and path-breaking agreement in many ways. Its central achievement is the definition of legally binding quantified constraints on greenhouse gas emissions from each industrialized country – an achievement that some experts had declared politically impossible. The quantified commitments for the major OECD countries, furthermore, appear considerably stronger than many analysts expected. Central to the agreement was the inclusion of various mechanisms for international transfer that in principle should enable the effort implicit in the headline commitments to be dispersed much more widely and efficiently.

These mechanisms for international transfer are the most novel and complex aspects of the Protocol and many issues remain to be resolved concerning their operation, implementation and implications (issues that are considered in Part II of this book). The requirement to strive to minimize the adverse impacts both of climate change and of the response measures adopted under the Protocol could also open difficult new ground in international relations. The Protocol is suffused with requirements for ensuring that compliance with the quantitative commitments can be verified, but the clear intent to develop stronger mechanisms for enforcement is not yet matched by means.

The Protocol is weaker in areas not directly linked to the specific quantified commitments, but does provide ground to build on. Indicative policies and measures are listed and agreement must be pursued to cover international bunker fuels that are omitted from the Protocol's quantified commitments. The Protocol strengthens – but not greatly – the Convention's 'soft law' commitments involving developing countries, and associated issues of finance and technology transfer. Provisions addressing the adequacy and strengthening of commitments are limited, but sufficient to move forward if and when countries are willing to do so.

The most obvious omissions concern procedures for the accession of additional countries into the specific quantified commitments, as well as any detail at all on emissions trading. In addition, the Protocol offers no potential for specific treatment of the very long-lived industrial trace gases, and no direct reference to collaboration on new technologies, which had also been suggested (though this omission is partly on account of efforts already being pursued through the International Energy Agency's Implementing Agreements, which can encompass many considerations other than just climate change).

Despite such limitations, set in the context of the usual glacial pace and substantive constraints of global negotiations, the Protocol is a remarkable achievement. Furthermore, the concerns of each of the major groups presented at the beginning of Chapter 2 are also apparent. The EU obtained the binding regime with quantified reductions below 1990 levels for which it (and AOSIS) had striven since the inception of global climate change negotiations in that year, though in a very different form than it had sought. In return the United States and other OECD countries achieved every single dimension of flexibility that they had sought (with the exception of borrowing). Central and east European countries maintained their status as part of the industrialized world's institutional structures, while gaining additional flexibility and commitments that will probably yield net economic benefits for most of them.

The bulk of developing countries in the G77 achieved their twin core aims of strengthening Annex I commitments while avoiding new commitments themselves, though they had to yield to the globalization of industrialized-country efforts through investments under the CDM. Within the G77, AOSIS, African and other groups anxious for help in adapting to the impacts of climate change made some advances, while OPEC and others worried about losing revenues established a general principle of minimizing adverse impacts and gained footholds for pursuing their specific concerns.

While almost every sentence in the Protocol has a rational explanation and history that can be traced, there are some striking inconsistencies in the package overall. Except in terms of process, there is no logical reason to have emission credits from the CDM starting in the year 2000 against industrialized-country commitments defined over 2008–12. Indeed the core commitments will clearly require early action in the industrialized countries (which the Convention already requires) if they are to be met, but the clause on 'demonstrable progress by 2005' is the closest the Protocol gets to formalizing this. Emissions trading and JI within Annex I shall be 'supplemental' to domestic actions, but credits obtained under the CDM may 'contribute' to 'part of' Annex I commitments. The relationship between the CDM and the established financial mechanism is unclear.

Overall, however, the Protocol succeeds in building specific commitments and a clear structure upon the general foundations laid by the Climate Change Convention. Part II of this book seeks to unravel what it all may actually mean.

Part II

Analysis of commitments, mechanisms and prospects

Chapter 5

Environmental and economic implications of the Kyoto commitments

As the dust settled on Kyoto, reaction to the commitments adopted spanned the full panoply. Some environmentalists declared that the commitments were so inadequate as to be a sham, rendering the whole Protocol irrelevant to the real task of safeguarding the planet. At the opposite extreme, some industrialists and energy economists compared the headline emission reduction commitments against emission projections and declared that the commitments simply could not be met, or at least that the attempt to meet them would impose immense and unacceptable economic burdens. This chapter sets out the environmental and economic implications of the Kyoto commitments.

5.1 Environmental consequences of the Kyoto commitments

In terms of overall emissions, the impact of the Kyoto Protocol goal to reduce Annex I emissions to at least 5% below 1990 levels appears rather modest. As illustrated in Chapter 3, by the mid-1990s emissions in the EU were just below 1990 levels, those in the rest of the OECD were generally 6–10% up, and emissions in the EITs had collapsed by 15–50%. In *aggregate*, these changes mean that in the mid-1990s, greenhouse gas emissions from Annex I countries were 5–6% below 1990 levels. The collective commitment of the Kyoto Protocol, therefore, is essentially to stabilize over the subsequent 15 years the aggregate emissions from the industrialized world – or rather, to allow some growth, allowing for the effect of sinks and credits from the CDM.[1]

---

[1] The inclusion of LUCF emissions for Australia (and any others with positive LUCF emissions in 1990), set against declining trends, adds marginally to the increase available for other gases.

Non-Annex I countries accounted for about 31% of global emissions in 1990, but their emissions have grown very rapidly during the decade; taking IEA projections for developing countries and assuming they are unaltered by the Asian financial crisis or the Kyoto Protocol, *global* emissions could still rise to 31% above 1990 levels[2] during the commitment period. This emphasizes that even if all commitments are met, the Kyoto Protocol is but the first substantive step in the much wider and longer-term challenge of controlling climate change.

The main environmental results of all this are summarized in Figure 5.1, in terms of projected emissions, concentrations, temperature and sea-level changes.[3] Since the environmental consequences are spread over a long time, assumptions need to be made about subsequent commitments. The analysis summarized in Figure 5.1 assumes that no other countries make subsequent commitments, exploring only variants in which Annex I countries take on different degrees of additional action. Assumed global emission paths are shown in Figure 5.1(a).

This analysis shows that the Protocol commitments themselves could reduce CO_2 concentrations in the middle of the twenty-first century by about 10–25 ppm (parts per million), and by 20–80 ppm by the end of the twenty-first century, depending on the degree to which subsequent Annex I commitments are strengthened. The corresponding reduction in global average temperature increase, trailing the concentration trend, is around 0.05–0.1°C by mid-century, and 0.08–0.3°C by the end of the century. These represent reductions in the rate of temperature increase by 4–14% across the different scenarios (a more robust result than the absolute temperature change, which depends on the basic climate sensitivity assumed).

The response of sea level is even more delayed. Sea-level rise is reduced by only about 1cm in the middle of the twenty-first century on these scenarios, and by a few centimetres at the end of the century, span-

---

[2] Data from Table 5.3. Non-Annex I emissions will grow by 114%. Implementation of the Kyoto Protocol by the Annex I countries, however, means a reduction of 10% from the reference scenario.

[3] T. Wigley, 'The Kyoto Protocol: CO_2, CH_4 and Climate Implications', *Geophysical Research Letters*, 1998.

Figure 5.1: Potential impact of the Kyoto Protocol on future emissions, concentrations, temperature and sea level

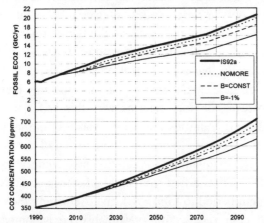

(a) Global emissions and concentrations, 1990–2100, under three post-Kyoto scenarios

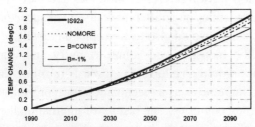

(b) Global mean temperature changes, 1990–2100, under three post-Kyoto scenarios

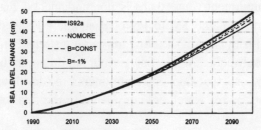

(c) Global mean sea-level rise, 1990–2100, under three post-Kyoto scenarios

Notes: The three post-Kyoto scenarios are:
(I) No further emission reductions (emission growth resumes in proportion to reference case);
(II) Constant Annex I emissions after 2010;
(III) Annex I emissions decline at 1% per year compound from 2010 levels.
Non-Annex I emissions grow without restraint in all these scenarios.
Source: T. Wigley, 'The Kyoto Protocol: CO_2, CH_4 and Climate Implications', *Geophysical Research Letters*, 1998.

ning a 2–13% reduction in the rate of sea-level rise. These studies confirm that the Kyoto commitments, in and of themselves, will have only a very modest impact on the rate and pace of climate change. Even if countries achieve the commitments of the Kyoto Protocol, it will be but a small first step towards stabilizing the atmosphere, and it is certain that a great deal of attention needs to be paid to adaptation to the climate change that appears increasingly inevitable.

There are, however, obvious reasons why this is far from the whole story. The developing countries are taking various actions that have the effect of limiting emissions (including those fostered through the GEF) even in the absence of quantified commitments. The Kyoto Protocol as agreed is only intended as a first step, and includes specific reference to a second commitment period to follow the first. Achieving the commitments in industrialized countries would presumably be accompanied by the development of new infrastructure, technologies and industries that would set their energy economies on a different course over subsequent years and decades. These changes would also tend to spread globally.

Moreover, if the industrialized countries ratify and implement their commitments, they would go a long way towards fulfilling their moral and legal leadership requirements as a precondition for more substantive actions in and by developing countries. The history of international environmental agreements has almost entirely been one of increasing the scope and strength of commitments, and there is no fundamental reason why the future development of the Kyoto Protocol should be any different. The different scenarios in Figure 5.1 allow for varying strengthening of Annex I commitments subsequent to the first period, but do not include any additional action by, or expansion of commitments to, developing countries (the specific options for which are considered in the final chapter of this book).

This offers a very different perspective on the Kyoto commitments. Figure 5.2 shows an estimate of 'optimal global CO_2 emission trajectories' required to move towards atmospheric CO_2 stabilization at 450 ppm and at 550 ppm (which would equate to total atmospheric changes in the range 500–650 ppm-equivalent, taking account of other greenhouse gases). The main and distinguishing feature of this analysis is that it focuses upon the

Figure 5.2: A view of optimal global CO_2 trajectories for stabilization at 450 ppm CO_2 and 550 ppm CO_2

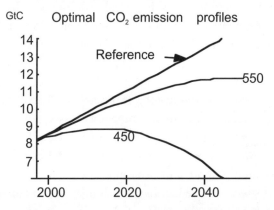

GtC   Optimal CO_2 emission profiles

Source: M. HaDuong, M. Grubb and J. C. Hourcade, 'Influence of Socioeconomic Inertia and Uncertainty on Optimal CO_2 Abatement', *Nature*, No. 390, 20 November 1997.
Notes: Optimal global CO_2 emissions from energy + deforestation, assuming a 'characteristic time' of 50 years for transitions in the global energy system, for atmospheric stabilization at 450 and 550 ppm CO_2. The reference case is a 2% per year increase in the absence of a climate problem, assumed to be economically optimal (i.e. including 'no regret' reductions); marginal abatement costs increase quadratically for abatement below this.

dynamics of change, assuming extensive inertia in the global system, and seeks to optimize the *pressure* for change over time, recognizing that technologies and systems could be expected to adapt to changing circumstances.

On this basis, with emissions allowed to start diverging from the 'reference' case in 1997, the optimal emissions trajectory would require *global* CO_2 emissions in 2010 to be about 15% below the reference case to stabilize ultimately at 450 ppm, while for 550 ppm optimal global emissions in 2010 would be around 5% below the reference.[4] With action concentrated in the industrialized world, its required reductions would obviously be correspondingly greater, and 'no regret' reductions could be expected to add perhaps an additional 5–10% reductions by 2010 in each

---

[4] The reference case is assumed to be a 2% per year linear increase, roughly tracking the IPCC's IS92a scenario used as reference in note 3. The reference is assumed to be economically least-cost in the absence of climate change.

case as compared to business-as-usual trends (see below).

This suggests that the Kyoto 'assigned amounts' for OECD countries equate to a strength of action that, if implemented domestically, might set them on a course commensurate with stabilization somewhere in the range 450–550 ppm CO_2. Of course, this is not the situation: the rest of the world will not be taking commensurate action, and OECD countries will not be confining their action to within their borders. Collectively, the action required of *Annex I* countries under the Protocol to achieve the commitments is probably much closer to the trajectory associated with global stabilization at 550 ppm. The future is uncertain, and the Kyoto commitments themselves will entail very modest environmental consequences but they can be considered a good start. The more detailed implications, and the prospects for sustaining reductions, of course depend in part upon assessment of trends and the economics of abatement – topics to which we now turn.

5.2 Economic consequences of the Kyoto commitments

5.2.1 Emission trends and cutbacks required

A first indication of the task facing industrialized countries in implementing the Kyoto commitments may be observed simply by considering the reductions required from the levels already attained during the 1990s. Emissions data to 1995–6 were given in Chapter 3, and, as noted, the *aggregate* Kyoto commitment is to hold Annex I emissions roughly at the levels of the mid-1990s fifteen years hence.

In terms of *distribution* the story is of course radically different. In the EU, emissions had declined – albeit not primarily due to climate change policies – and the most recent projections suggest that member states will – just – achieve the Convention's aim of holding emissions of CO_2 and other greenhouse gases below 1990 levels in the year 2000. In fact, it is likely that the modest abatement policies adopted in some European countries during the 1990s have tipped the balance in this respect. The rest of the OECD is commited to bigger reductions from the higher levels they had reached in the mid-1990s, while the EITs have scope for a big increase.

Figure 5.3 illustrates the specific changes from 1995 levels for different

Box 5.1: Countries and groups used for regional analysis of Annex I commitments

Group	Comprising	Share in Annex I (%)		Notes
		1990 emissions	Assigned amounts	
USA	USA	33.7	33.1	
EU	15 member states	23.8	23.1	Germany and UK account for about 50% of the EU's 1990 emissions
Japan	Japan	6.5	6.5	
Canada	Canada	3.3	3.2	
Australia	Australia	2.9	3.3	
Other OECD	Iceland, Norway, New Zealand, Switzerland	1.1	1.1	
Russia	Russia	16.2–17.5*	17.1–18.4*	
New EU	Poland, Hungary, Czech Republic, Slovenia, Estonia	5.2	5.1	The EITs on fast track for EU membership (excluding Cyprus)
Other CEEC	Other central and east European countries (other Baltics, Slovakia, Romania, Bulgaria, former Yugoslav republics, Ukraine)	7.3	7.4	Ukraine accounts for about half this group's 1990 emissions

Source: Data compiled by authors from national communications and other sources as available.
* The higher values use data from Russian Second National Communication made available shortly before going to press.

groups within Annex I (the regional grouping is shown in Box 5.1, and maintained in analysing emissions trading later in the chapter). This confirms that the Kyoto commitments represent a big reduction from 1995 levels for all the OECD countries except Australia, and a big increase for all the central and east European countries, especially the ones less advanced in their economic transition that are not currently on track for EU membership.

In absolute terms, the cutback implied for the United States from 1995

Figure 5.3: Gap between Kyoto commitments and 1995 emission levels in principal Annex I groups

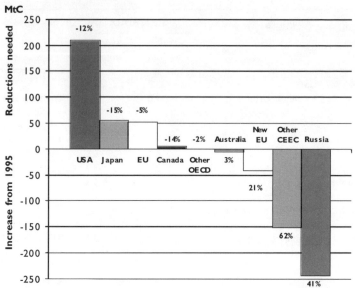

Source: Authors

Note: The bars show the difference in absolute terms, together with the percentage difference, between reported emissions in 1995 and the Kyoto commitments. Negative data indicate scope for increase from 1995 levels. For details of the groupings see Box 5.1. LUCF emissions are not included: they would increase the allowed scope for emissions growth in Australia especially.

levels accounts for more than half of that required for the OECD overall and almost exactly equals the increase allowed to Russia. The increase allowed to Ukraine is greater than the cutback required in the EU or Japan individually. In terms of percentage changes, Japan and Canada face the stiffest cutbacks, followed closely by the United States. Apart from Iceland, Australia is the only OECD country with a growth allowance from 1995 levels, that would be further increased by the (last-minute) addition of LUCF accounting in the base year (though estimates of

---

[5] Indeed these emissions appear to have changed from being a huge source to being a sink and back to being a source of CO_2 emissions: Compare FCCC/SBI/1997/INF.4 (22 October 1997) with FCCC/CP/1998/11/Add.2 (5 October 1998) and FCCC/CP/1998/INF.9 (31 October 1998).

Australian LUCF emissions in 1990 have proved highly variable).[5] The EU accession countries have scope for growth of 18% (about 1% per year compounded over the 15-year period), while Russia can increase by more than 30% and the other EITs (in aggregate) by even more.

5.2.2 Economic costs of the Kyoto commitments

Given these divergent trends, how much might the Kyoto commitments cost? This is an intrinsically difficult question to answer. As set out in Appendix 2, fundamental economic debates continue concerning ways of defining and estimating abatement costs, and accounting for factors such as apparent 'no regret' options that appear already cost-effective (but have not been taken up), technological progress and secondary benefits of emissions control. Many engineers, for example, maintain that the scope for cost-effective improvements in energy efficiency especially mean that the Kyoto targets could be achieved with net economic benefits.

Table 5.1 shows results from a wide range of somewhat more conventional economic modelling studies (though two of the models are more engineering based) presented to an OECD workshop. These show wide divergences in costs of achieving the Kyoto commitments domestically in the different regions. The marginal cost of hitting the target tends in most of the models to be highest in Japan and Europe and somewhat lower in the United States; across the full range of models estimates of marginal costs vary by a factor of ten *within* each region. Neglecting the two outlier models, marginal costs for implementing the commitments domestically are mostly grouped in the range $80–$200/tC. These reflect modelling estimates of meeting the commitments without any international flexibility, using CO_2 taxes alone, without allowance for possible 'no regret' reductions from other approaches. In contrast the biggest national economic study of the cost of the Kyoto commitments, assuming full flexibility, estimated that the actual marginal costs to the United States would lie in the range of $14/tC to $23/tC.[6] The study estimated that this was about a

---

[6] J. Yellen et al., 'The Kyoto Protocol and the President's Policies to Address Climate Change', Council of Economic Advisors, Washington, 1998.

Analysis of commitments, mechanisms and prospects

Table 5.1: Estimated costs of achieving Kyoto commitments from various economic models

Model	Marginal cost of target achieved domestically, $/tC			% GDP loss domestic implementation[a]	% GDP loss with full trading
	USA	Europe	Japan		
SGM	163			0.4	
MERGE	274			1.0	0.25
G-Cubed	63	167	252	0.3–1.4[b]	'Decline significantly'
POLES	82	130–140	240	0.2–0.3	
GTEM	375	773	751	0.7–2.0	0.3
Worldscan	38	78	87	–	
GREEN	149	196	77	0.4–0.9	0.1–0.5
AIM	166	214	253	Less than 0.5	

Source: Adapted from Tables 1 and 2 in OECD, *Economic Modeling of Climate Change: OECD Workshop Report*, Paris: OECD, October 1998.
Notes:
[a] Note that the exact welfare measure varies between the models and these estimates may not be directly comparable.
[b] GNP: in G-Cubed the US GDP increases slightly due to competitiveness gains: GDP stated as a gain of 0.1% for the United States, 1.6% loss in Japan and 1.5% loss in other OECD.

quarter of the costs without international flexibility.

The total costs, expressed in terms of estimated GDP losses by the year 2010, also span a large range across the OECD studies. If no allowance is made for the international flexibility in the Protocol – that is, if the commitments are implemented entirely domestically – the estimated GDP losses range from 0.2 to 2% of GDP, or 0.3 to 1.4% neglecting the outlying models. International flexibility (modelled as trading within Annex I) greatly reduces these costs in all the modelling studies, to under 0.5% of GDP across Annex I. Furthermore, and quite apart from technical debates about modelling approaches, these figures probably substantially overestimate the actual costs of implementing the Kyoto Protocol because most of them do not include the non-CO_2 gases, sinks or the CDM, but assume rather that the commitments apply to CO_2 alone.

There are of course various ways of presenting such numbers. Something under a 0.5% loss in Annex I GDP by 2010 is equivalent to forgoing a few months of GDP growth and is hardly discernible compared to the projected overall growth and uncertainties therein. In absolute terms of course it is more impressive: projected Annex I GDP by 2010 is around $17,000 billion, 0.5% of which is $85 billion. To put this in perspective, however, even this maximum estimate of the cost of implementing the Kyoto Protocol is less than the value wiped off shares on the London stock exchange on the day of the Brazilian currency crash in January 1999. As indicated by the other models and other considerations noted, the actual cost is likely to be much lower.

The aggregate economic consequences of the Kyoto Protocol, like the environmental benefits, can thus be reasonably described as rather modest. But this does not mean that the commitments will be easy to achieve. To see this, one need look no further than the record of climate change policy in the 1990s and current emission projections.

5.2.3 Emission projections and the Kyoto commitments

During the negotiation of the Convention itself, it was recognized that emissions in most OECD countries were on a rising trajectory; hence the reference to 'returning emissions to 1990 levels'. The fact that emissions had risen by the mid-1990s is thus not in itself inconsistent or surprising. However, countries had committed themselves to taking steps to reverse the rising trend, and more troubling is the very limited extent of policies adopted. Policies for carbon/energy taxation in the EU and the United States have mostly failed, energy efficiency programmes have not been greatly extended, and coal production has remained supported in several countries. The rapid conversion to natural gas and progress in renewable energy programmes in some countries do not seem sufficient to reverse the rising trends.

Table 5.2 shows official projections of emissions up to 2010, based on the most recent data officially submitted by Parties in their National Communications. These present a sobering picture. Japan, the United States, and Australia project emissions as growing by more than 20%, with Canada projecting only a little lower. EU projections vary widely, but the aggregate official EU projection – hinged upon ongoing reductions

Table 5.2: Emission projections up to 2010 submitted by Parties (% from base year)

Country	GHG	CO_2	LUCF	CH_4	N_2O
EU					
Austria	n.a.	−6	n.a.	−2[a]	n.a.
Belgium	+12[b]	+15[b]	0	−22[b]	+16[b]
Denmark	−25	−26	−123	−15	−18
Finland	12–29	4–32	10–39	−22	33–39
France	−6	8	−84	−22	−41
Germany	n.a.	−16	n.a.	−51	−31
Greece	n.a.	27	n.a.	n.a.	n.a.
Ireland	10	33	−88	+3	−11
Italy	−7	−4	2	−32	−10
Luxembourg	−39	−43	0	−7	+7
Netherlands	5	9	−13	−43	+9
Portugal	n.a.	68	n.a.	−12	+3
Spain	n.a.	25	n.a.	+10	0
Sweden	55	10	35	−13	+37
UK	−7	3	−58	−35	−55
United States	26	23	13	−11	−6
Japan	24	20	33	−6	+24
Australia	28	40	−36	+10	+24
Canada	n.a.	19	n.a.	+18	−6
Other OECD					
Iceland	n.a.	35	n.a.	−1	+25
New Zealand	12	43	−3	−8	−4
Norway	8	33	−57	−23	+10
Switzerland	−6	−3	−17	−21	−2
Russia	−11	−3	−40	−9	−20
EU accession					
Czech Republic	−1	−1	−150	+7	+4
Estonia	n.a.	−51	−2	n.a.	n.a.
Hungary	n.a.	−23[a]	n.a.	n.a.	n.a.
Poland	n.a.	4	n.a.	n.a.	n.a.
Slovenia	n.a.	n.a.	n.a.	n.a.	n.a.
Other CEEC					
Bulgaria	11	7	−35	+12	+62
Croatia	n.a.	n.a.	n.a.	n.a.	n.a.
Latvia	−75	−50	−27	−39	−26
Lithuania	−39[a]	−15 to +27	13	−12	−58 to −55
Romania	n.a.	n.a.	n.a.	n.a.	n.a.
Slovakia	−14	−11 to −5	−87	−44 to −8	−32 to +10
Ukraine	−21	−16	−36	−26	+7

Source: FCCC/CP/1998/11/Add.2

Notes: n.a.=not available.   [a]Data for 2000.   [b]Data for 2005.

particularly in Germany – is roughly for stabilization (which contrasts sharply with all the international economic modelling studies, which project renewed growth).

Recent developments further aggravate the situation. International oil prices, having remained soft during the 1990s, finally collapsed during 1998 to around $10 per barrel, the lowest for 25 years. In Germany, the Green Party in the newly elected 'red–green' coalition government started to promote plans for a rapid phase-out of the country's nuclear power production. Although the Green Party obviously intends that the effect on emissions should be offset by its other proposed policies, including removing subsidies on coal, the strength of support for the coal industry in the 'red' part of the coalition, and the constraints on expenditure, call other parts of the package into doubt. A rapid nuclear phase-out would almost certainly be accompanied by considerable emission increases, possibly adding up to 10% to German emissions and vastly complicating the difficulty of meeting the EU target overall. However, by the time of going to press it already appeared unlikely that these plans would be implemented, and even the Green Party was split, partly because of the implications for climate change.

The difficulties appear even greater when contrasted with the energy projections by the International Energy Agency, whose aggregate projections are shown in Table 5.3. The most striking discrepancy concerns Europe, where national reports to the UN FCCC project a figure close to emissions stabilization in aggregate but the IEA projects an increase of almost 30% in CO_2 emissions from 1990 levels. The IEA projections in fact represent a reversal of long historic trends towards decarbonization, partly because of projected reductions in nuclear power generation especially in Europe, even *without* an accelerated nuclear phase-out in Germany.

Considerable caution needs to be exercised in using such projections. There is the semantic but important question of what is meant by 'business-as-usual' projections, since one of the few things known for certain about the future is that it will not be 'as usual'. Furthermore, projections are

---

[7] T. Baumgartner and A. Midttun, *The Politics of Energy Forecasting*, Oxford: Clarendon Press, 1987.

Table 5.3: CO_2 emissions increases and required cutbacks by major Annex I groups: IEA projections (MtC)

	1900	BAU projection 2010	Projected emissions increase 1900–2010	Reductions below BAU projection in 2010 required to meet Kyoto commitments	
OECD					
• Europe	998	1,258	260	340	27%
• Pacific	370	484	114	126	26%
• North America	1,456	1,920	464	566	29%
EITs	1,207	1,051	-156		
Annex I total	**4,031**	**4,712**	**681**	**883**	**19%**
• China	658	1,451	793		
• Rest of world	1,045	2,191	1,146		
World total	**5,836**	**8,506**	**2,670**		

Source: *World Energy Outlook*, Paris: IEA, 1998, Table 4.1, p. 53.

inevitably influenced by perceptions, hopes and political objectives.[7] The low EU governmental projections undoubtedly reflect both hope and intent to constrain greenhouse gas emissions in their 'reference' scenario. IEA projections have historically tended to be very high. But the projections are also driven by more fundamental factors, including continued population growth in the New World countries, the end of nuclear construction in most of the OECD, and the sheer inertia in energy infrastructure, institutions and policies.

None of this is to deny the extensive technical opportunities available for limiting emissions. Indeed there is a vast technical literature on how greenhouse gas emissions can be reduced with net economic benefit. To European and Japanese visitors, the opportunities seem particularly large in North America, where the continued lack of some basic housekeeping measures of energy efficiency help (but only in part) to explain CO_2 emissions almost twice as high per person, and significantly higher also per unit of GDP, than in Europe or Japan. American negotiators in turn tend to be more conscious of the waste in east European societies and in developing countries.

The fundamental point is that technical opportunities do not readily translate into rapidly achievable emission reductions, for a whole host of reasons beyond the scope of this book. Even if the economic costs are low, it is implausible that any US administration could implement the scale of changes required for the country to achieve its assigned amount purely in terms of domestic greenhouse gas emissions, even if it wanted to. The same is true in practice for Japan, Canada and several other OECD countries.

The situation in central and eastern Europe – and particularly Russia and Ukraine – is quite different (and highly uncertain). According to official projections, the desperately sought economic growth will inevitably result in emissions rising back towards their levels before the transition from communism, albeit to widely varying degrees. In reality this seems debatable. Most of these countries are still much less efficient in their use of energy than the EU or Japan, and Russia at least remains more energy intensive than the United States.[8] Since 1992, despite economic growth in several EIT countries including the Czech Republic, Hungary, Poland and Slovakia, emissions continued to decrease before roughly stabilizing in the mid-1990s. It is arguable that CO_2 emissions need not grow, and could even decline, on account of the efficiency improvements associated with renewed economic growth. The uncertainties are particularly large – and important – in Russia and Ukraine, given their size and chaotic economic situation (see Box 5.2).

The inclusion of non-CO_2 greenhouse gases does not fundamentally change the *domestic* picture for the big emitters. As noted in Chapter 4, these 'other greenhouse gases' are estimated to have comprised about 20% of total 1990 greenhouse gas emissions in the EU and Canada, about 15% in the United States, and a little over 5% in Japan. These figures would be

---

[8] Economic structures dependent upon energy-intensive industries, often with old technologies, have resulted in high energy consumption per unit of GDP and high carbon intensity. This in turn has led to high per capita and absolute emissions of CO_2, generally between those of the EU and the United States. National carbon intensities (CO_2 emissions per unit of GDP) are declining, but most remain well above the levels in the EU. More relevant indicators based on purchasing power parity estimates suggest that by 1993 carbon intensities in Hungary, for example, were comparable to those in the UK and Germany, while Polish carbon intensity was below that of the United States. By 1993, most central European countries were less carbon intensive than the United States, while Russia (and probably other FSU) remained above US levels. See K. Simeonova and F. Missfeldt, *Emission Trends in Economies in Transition,* EEP Climate Change Briefing, No. 8, November 1997.

Box 5.2: Which way for Russian and Ukrainian net greenhouse gas emissions?

Russia and Ukraine between them account for over 20% of the total 'assigned amount' of emissions under the Kyoto Protocol.* Their likely emissions are thus particularly important within the context of EIT uncertainties. Ukrainian emissions are estimated to have declined by more than 50% since 1990 and as yet they show no sign of upturn. In Russia, emissions since 1993 appear to have been fairly static at about 30% below 1990 levels. However, Russian industrial production has declined by almost 60%.

One reason why Russian emissions have declined less than those elsewhere in the former USSR appears to be that it has not been politically possible to reform the energy sector. Vast amounts of energy are still delivered to customers who do not pay: in October 1998, following the financial crisis, the giant gas supplier Gazprom stated that only 6% of customers had paid bills in September, and surveys suggest that less than half of industrial customers *could* pay; the rest would simply go out of business if payment were enforced. The possibility needs to be recognized that Russian emissions will go down, not up, if and as economic reforms succeed. The economic projections used in most global models simply ignore these factors.

In addition, Russia and Ukraine are countries with huge land areas, including some potentially very productive soils; Ukraine used to be the 'breadbasket of Europe'. The potential for sinks must be enormous, though no data were found for Ukraine. The Russian 2nd National Communication estimates total net sinks from forests in 1994 at about a third of its CO_2 emissions.

In short, there is a distinct possibility that Russia and Ukraine will have far more surplus assigned amounts under the Protocol than any quantified analysis (including ours presented in this chapter) has so far considered possible. If this does turn out to be the case, the implications for the Kyoto regime could indeed prove dramatic.

Note: * The Russian assigned amount accounts for about 18% of the Annex I total (see Box 5.1). Estimates for Ukraine, including non-CO_2 gases, are uncertain in the absence of a national communication; the data gathered by the authors amount to 4.3%.

a little bigger as a percentage of the assigned amount, because for the trace industrial gases 1995 emissions can be used, but the difficulty of limiting some of these emissions means that they probably ease the task of achieving emission goals domestically by the equivalent of 'only' a few percentage points in Japan, the EU and the United States. For certain other countries – notably Australia, New Zealand, the Scandinavian countries and most of the former Soviet countries – the other greenhouse gases are much more important.

Emissions trading and joint implementation in effect give the big OECD emitters access to these emission reduction opportunities in other countries, including the EITs. In principle, the CDM may also allow them to offset their commitments against emission savings in the developing world, where some of the 'other gases' are also more important. The analysis in the rest of this chapter attempts to look more closely at the implications of all these factors.

5.3   The potential impact of flexibility within Annex I

The Kyoto Protocol is a complex agreement. It defines specific commitments for each industrialized country, but there are so many dimensions of flexibility that its real quantitative implications are unclear. This section analyses numerically how the specific emission commitments could combine with the flexibility mechanisms within Annex I, focusing upon the possible implications for plausible emissions of CO_2 and other greenhouse gases during the commitment period that would be consistent with the Protocol. It also deals with some economic aspects.

5.3.1 Analytic framework

The analysis uses a simple computer-based model of emissions from each industrialized country during the commitment period. Given data on likely emissions in the absence of abatement and on the costs of control, the model presents behaviour that minimizes total abatement costs during the commitment period among the Annex I countries. In principle, the result could be obtained by using any mix of the Annex I mechanisms for international adjustments and flexibility (bubbling, emissions trading and JI), in combination with optimal choices by each country about the mix of different gases and sources to be limited. Thus, for example, it means that Japan can contribute to meeting its formal commitment by trading 'assigned amounts' or by investing in Russian methane reductions, if that is cheaper than limiting its domestic CO_2 emissions. In terms of a purely static economic analysis, bubbling and the two Annex I transfer mechanisms are almost identical, and this chapter seeks simply to explore the possible overall implications of such mechanisms for the level and distribution of emissions and costs.

Box 5.3: Abatement cost modelling in ITEA

The basis of ITEA is a straightforward marginal abatement cost function for each of the involved Parties. This cost curve is defined in reductions relative to the BAU emissions in 2010 and broken down into two separate components. The first component comprises reductions which incur negligible economic costs ('no regret' reductions), but which are not achieved in the BAU scenario for political or other reasons. The second part of the curve represents linearly rising marginal costs after the 'no regret' reductions are exhausted. The marginal abatement cost curve is given in Figure 5.4(a). A linearly rising marginal cost curve results in quadratically increasing total costs for reductions; this is displayed in Figure 5.4(b).

ITEA includes the non-CO_2 greenhouse gases, which have very different abatement costs and 'no regret' options. Therefore we have created a second marginal cost curve for each country representing the other GHGs, grouped together according to their global warming potential (GWP). In a scenario with multiple gases, as is the case with the Kyoto Protocol, we have summed the two marginal cost curves of each country.

The specific assumptions for the analysis are given in Table 5.4, and are explained in Appendix 3.

Figure 5.4: Standard marginal (a) and total abatement (b) cost curves for the ITEA model

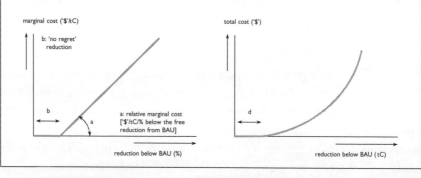

The model assumes competitive, least-cost behaviour during the commitment period; Russia, for example, 'sells' its methane reductions to the highest bidder (which also defines the average price of emission quotas). Given the commitments, each Party trades on any difference in abatement costs, to mutual benefit. In effect the model assumes a competitive market for emission quotas, completely transparent for participants and without costs for the trading itself. Obviously this is a crude representation of reality for such a nascent system, but the results are instructive for understanding the potential implications of what has been created by the Kyoto Protocol.

Table 5.4: Base assumptions in the ITEA analysis

Region	Relative marginal cost	'No regret' (% from BAU)		BAU CO$_2$ growth (% from base year)	
	('$'/tC/%)	Base	High	Base	High
European Union	1.20	6	3	2.3	2.3
United States	1.00	10	6	23.2	23.2
Japan	1.40	5	3	14.9	20.0
Australia	1.00	12	8	42.1	42.1
Canada	1.00	10	8	18.7	18.7
Other OECD	1.50	4	0	26.5	26.5
Russia	0.75	14	10	−17.8	−4.7
New EU	0.75	10	8	−18.4	9.0
Other CEEC (excl. Ukraine)	0.75	12	8	−23.6	−12.6
Ukraine	0.75	14	10	−17.7	−9.2
Other GHGs (all regions)	0.50 (ref.) 0.75 (high)	20	10		

Source: See Appendix 3.

The International Trading of Emissions Allowances (ITEA) model is designed to be simple, transparent and readily adaptable to explore the implications of different assumptions about emission trends and abatement costs. It aims to be comprehensive in its coverage of gases and regions, and to relate to official projections and real-world phenomena, giving some scope for 'no regret' potentials, rather than pursuing economic sophistication. However, the model omits sinks and the CDM, because of the huge uncertainties involved; these are considered in Section 5.4. The basis of ITEA is a simple function defining the cost of abatement relative to 'reference' projections, further explained in Box 5.3 and Appendix 3. Given its commitments, each Party trades on any differential in abatement costs. Although the model separates fossil fuel CO$_2$ emissions from other gases, all trade takes place in the full basket of greenhouse gases.[9]

---

[9] Since it is commitments that would be exchanged, not physical emissions, it would be meaningless to talk of trading CO$_2$ when the commitments themselves are defined in terms of a CO$_2$-equivalent basket of gases. See Chapter 6.

5.3.2 Composition and assumptions

The Annex I Parties are grouped for a clearer presentation of the results, as defined in Box 5.1 above, based on scale and similarity between the countries in economic, regional and political terms:

- the countries of the European Union, a signatory in its own right, are grouped, although the European countries can be viewed independently in order to test the influence of the 'bubble';
- the United States and Japan are considered separately; together with the EU these represent over two-thirds of current Annex I GHG emissions;
- Australia is treated separately because of its unique situation and treatment under the Protocol.

This leaves two separate groupings, Canada and the other OECD countries, in the West. The latter, though regionally diverse, all share rapidly growing economies and a particularly high share of non-fossil energy supply.

The EITs, the former communist economies, are divided in three groups:

- the Russian Federation, with a sixth of total Annex I GHG emissions in the base year;
- the countries that are on a 'fast track' to join the EU early in the twenty-first century (New EU), which are advanced in economic transition and whose economies and (mostly) emissions are growing again;
- other central and eastern European countries (Other CEEC), including the former Soviet republic of Ukraine. This third group of countries had the biggest decrease in their economies and emissions, and they are still in the early stage of economic transition. The Ukrainian base year emissions are around half of the total for this group, but the emissions paths of all the countries are similar.

In the model *assigned amounts* for all Parties are calculated exactly in accordance with the provisions of the Kyoto Protocol, namely base year emissions exactly as defined in the Protocol adjusted by the prescribed percentage commitment.[10] The base year emissions data for all Parties are

---

[10] Base year emissions are thus defined as reported 1990 emissions of the basket of the six greenhouse gases, except for:

acquired from the National Communications (the Second National Communications, where available), except for Ukraine (no Communication: 1990 base year assumed with data taken from IEA).

Reference emission projections for the year 2010, which are used in ITEA to represent the average annual reference emissions during the commitment period of 2008–12, are mostly taken from the National Communications where available.[11] However, because of incomplete data and intrinsic uncertainties, we analyse two variants of reference emission projections for the EITs,[12] and also for Japan whose emission prospects appear the most uncertain of all OECD countries because of widely divergent views about its nuclear programme:

- for the EITs that have presented a credible range of scenarios for 2010 (mostly the New EU) we take as reference their 'low' projections as our 'base' case, and their 'high' projections for our 'high' case;
- for other EITs we assume annual growth from the most recent year for which data are available (including projections up to 2000) at rates of 1% (base case) or 2% (high case);

- the Economies in Transition that have chosen to use a different historical base year as agreed at COP-2 (see Table 4.1);
- the three industrial trace gases for which 'any Party included in Annex I may use 1995 as its base year'; in ITEA the choice of each country is automatically the year with the highest emissions, and therefore easier compliance;
- the 'Australia clause' that includes 1990 land-use emissions where these are positive (Australia and marginally for the UK and Estonia).

[11] Among the OECD, results are particularly sensitive to assumptions about US emissions because they comprise such a large share. The US National Communication projects CO_2 growth to 23% above 1990 levels by 2010 and we use this. Claims that US CO_2 emissions could grow by up to 30% by 2010 were circulated in autumn 1997, but we doubt their analytic strength as they appear in part to be over-reactions to the jump in US emissions in 1996 – which was subsequently itself revised downwards and followed by much weaker growth in 1997. Excessive projections, and long-term extrapolation of recent trends, have historically been common errors in forecasts of energy demand. CO_2 projections are lowest in the EU due to existing reductions, low population growth, and continuing decarbonization even assuming BAU; on the other hand, the EU includes the least developed and fastest-growing regions in the OECD and faces early nuclear phase-out in Sweden, so its projections may be optimistic and costs more evenly distributed than presented here.

[12] The projections used in the model for the EITs are given and discussed in Appendix 3.

- for Japan we assume reference CO_2 emissions to be 14.9% above 1990 levels in the base case, and 20% above 1990 levels in the 'high' case.[13]

Where not available in the National Communications emission projections for 2010, the non-CO_2 GHGs are assumed to be equal to the most recent data available or the latest year in the projections. The specific data are given in Appendix 3.

Economic assumptions are made on the basis of a comparison of important energy indicators between the regions, estimated as summarized in Appendix 3. These are inevitably based partly upon judgment. Our analysis is almost unique among economic models in incorporating a modest quantity of 'no regret' or cost-free reductions relative to the reference case.[14] The scope of this and the relative marginal costs for the different countries in Annex I define the abatement cost curve (see Box 5.3). Table 5.4 gives the assumptions for ITEA, for the base and 'high' cases. The 'high' assumptions test the sensitivity of results to more pessimistic assumptions that involve higher reference emissions growth in key countries, and much less scope for 'no regret' reductions, reflecting a more conservative view of the ability to limit emissions at low cost.

The marginal cost slopes have been defined relative to costs in the United States. In all analyses we present the marginal cost slope for the United States as 1. All costs are given in '$'. We use '$' in quotes deliberately, because our focus is upon relative rather than absolute costs. When we treat 1'$' as 1 US$ (or 1 euro or ¥100) it would yield costs towards the low end of most economic studies; treating it as US$10 represents input costs exceeding the high end of the spectrum in the peer-reviewed literature.

---

[13] Projections submitted by Japan to the IEA in 1995 projected near stabilization of CO_2 emissions due to very rapid nuclear construction, but it is already apparent that this will not occur. Other projections assuming minimal new nuclear development project rapid CO_2 emissions growth. We have taken a mean of these for our base case projection.

[14] Most economic models are optimizing models in which such possibilities are excluded by construction, or, more formally, are assumed either to be incorporated automatically in the reference case (if genuinely no cost) or not really 'no regret' due to various hidden costs; see Appendix 3 for economic discussions. In our view this is inadequate and some climate change policies would bring additional real benefits that would not be realized in the absence of pressures to limit greenhouse gas emissions.

In general there are no restrictions on trading in the analysis. We do, however, also analyse one important variant, concerning the issue of 'hot air' for EITs. To the extent that these countries' 'business-as-usual' emissions are still below their assigned amounts, this gap represents a surplus that could be sold to another Party (which could then increase its emissions accordingly). Various proposals have been put forward for eliminating such 'hot air' trading, as summarized in Chapter 6. In our numerical analysis here, rather than place any caps upon trading in general, we explore the implications of allowing countries only to trade emission reductions below their projected reference emissions levels. The countries are thus not constrained; nor are they allowed, however, to make windfall profits in the event of their allocations turning out to be genuinely surplus to BAU requirements.

5.3.3 Aggregate results

The aggregated results are displayed in Table 5.5. With unrestricted trading, total Annex I emissions in the commitment period equal the sum of the quantified commitments at 94.7% of 1990 levels. This is almost exactly the level attained by the mid-1990s, due to the collapse of the EIT emissions. The influence of 'hot air' trading is visible in the fact that emissions with unrestricted trading are higher than without any trading. In the base case, the surplus allowances amount to several percentage points of the total allowed Annex I emissions, and eliminating this surplus brings aggregate emissions down to 10% below 1990 levels.

The most startling result is that with the base case assumptions, unrestricted trading enables the commitments of the Kyoto Protocol to be met without aggregate resource cost. This does not imply that nothing is done. Countries are exploiting most of the 'no regret' potential, which, for some, would itself imply radical changes in energy policy. In addition, there is a large volume of East–West emissions trade. Since the model assumes emissions trading to be competitive and no country has to take 'positive cost' measures, the model assumes that such trading is free. In practice, of course, the sellers would extract some price, reflecting for example the value of enabling some importing governments to avoid difficult political

Table 5.5: Aggregate implications of Kyoto emission commitments

	Base case			High case		
	No trade	No restrictions	No hot air	No trade	No restrictions	No hot air
Average emissions from base year (%)	−10.00	−5.23	−10.00	−7.08	− 5.23	−7.08
Average costs per capita ('$'/cap)	1.37	−	0.05	3.12	0.49	0.90

Source: Authors, ITEA.

steps required to realize 'no regret' measures, so there would probably be some transfer payments.

Nevertheless, given the observations earlier in this chapter that the head-line commitments for the major OECD countries would mean extremely onerous reductions if the targets had to be implemented domestically, this is an extraordinary conclusion, explored further below.

Under the base case assumptions, if hot air is excluded then countries have to move just beyond 'no regret' measures. The costs are still small, with large savings arising from emissions trading. Under the 'high' assumptions, costs are greater, but are again greatly reduced by trading, and the vast majority of these savings are still realized even when 'hot air' is excluded. Imperfect trading, and the difficulty of distinguishing imple-mentation difficulties from actual economic costs in the real world, mean that the results concerning costs should be considered only as a crude guide.

5.3.4 Implications for trading and emissions in different regions

The various Annex I Parties will be affected by the Kyoto Protocol in very different ways. As noted before, the abatement costs, commitments and expected emissions growth vary widely across Annex I: the extraordinary divergence in emission commitments relative to 1995 levels was illustrated in Section 5.2 above. This regional variation explains why international flexibility has such a large impact. Large volumes are traded; and this has

Table 5.6: Traded volumes under the high case (no constraints: moderate trading)

	QELRC	Volume bought/sold		Resulting domestic emissions	
	MtC/yr	MtC/yr	% QELRC	% QELRC	% 1990 emissions
EU	1035	−34	−3	97	89
US	1485	241	16	116	108
Japan	291	46	16	116	109
Australia	148	−14	−10	90	98
Canada	146	12	8	108	102
Other OECD	49	3	7	107	105
Russia	768	−166	−22	78	78
New EU	231	−12	−5	95	89
Other CEEC	333	−76	−23	77	75

Source: Authors, ITEA.

striking implications for emissions in different regions, as summarized in Tables 5.6 and 5.7.

Table 5.6 shows the absolute volumes of trade and the implications for emissions under the 'high' case, with no restrictions; this probably corresponds most closely to official projections, including those for eastern Europe. In this case, aggregate trading between the regions amounts to about 300 million tonnes per year of carbon-equivalent, the great majority of which is acquired by the United States. This country adds about 16% to its QELRC by imports, and its emissions end up at 8% above 1990 levels. Japan's position is similar relative to its commitment and 1990 levels. In this scenario the EU and Australia join the EITs in exporting assigned amounts.

Table 5.7 shows domestic emissions relative to the base year under the different scenarios. Trading is greatest in the base case with no restrictions, with over 500 million tonnes traded, more than 10% of the total 4,500 MtC Annex I assigned amount. For reasons already discussed, we consider this scenario more plausible, and quite possibly still too high in terms of EIT trends, suggesting that trading could be even greater.

The most striking and consistent result to emerge from these analyses is

Table 5.7: Emission distribution in Annex I after trade (% relative to base year)

Region	Commitment	Base case		High case		CO_2 gap
		No restrictions	No hot air	No restrictions	No hot air	% (%)
Annex I	**–5.23**	**–5.23**	**–10.00**	**–5.23**	**–7.08**	**+4–5**
EU	–8.0	–8.6	–12.7	–11.0	–12.5	+7–8
US	–7.0	+11.4	+6.0	+8.1	+6.1	+3–5
Japan	–6.0	+7.5	+4.1	+8.7	+7.3	+4–5
Australia	+8.0	–0.1	–6.8	–2.4	–4.5	+30–31
Canada	–6.0	+6.6	+1.2	+1.6	–0.4	+4–6
Other OECD	–1.8	+6.4	+0.8	+4.9	+3.1	+20–25
Russia	0.0	–26.5	–31.3	–21.6	–23.6	–2–1
New EU	–6.5	–26.5	–30.9	–11.4	–13.6	+1–3
Other CEEC	–3.5	–29.1	–33.3	–25.5	–27.4	+0–1

Source: Authors, ITEA.

Note: The CO_2 gap is given in percentage points relative to overall GHG emission levels; that is, the GHG emissions in Annex I without restrictions will be about 5% below base year, CO_2 emissions could end up 4–5% higher, at 0–1% below base year.

the large gap between domestic emissions (particularly CO_2) and the overall 'Kyoto commitment' in some OECD countries. Most of the political debate about emission targets concentrates on domestic CO_2 emissions, and assumes that these will be more or less the same as the Kyoto commitment. Our results show wide differences. Emissions in the EU do end up close to the Kyoto commitment in aggregate (or a little lower if 'hot air' is excluded). This reflects the low BAU projection. Aggregate CO_2 emissions would be higher, quite close to 1990 levels if 'hot air' trading is included. Within the EU, emission levels could be widely distributed, with emissions in Germany, the UK and others in the range of 10–20% below 1990 levels in order to accommodate growth in some of the less developed member states. Countries faced by high costs to limit domestic CO_2 emissions will take the cheaper options of action on other gases and emissions trading, leaving domestic CO_2 emissions at higher levels.

We find that by 2010, under either set of assumptions, comprehensive gas coverage and emissions trading leaves domestic CO_2 emissions in

Figure 5.5: US assigned amount and possible emissions under the Kyoto Protocol

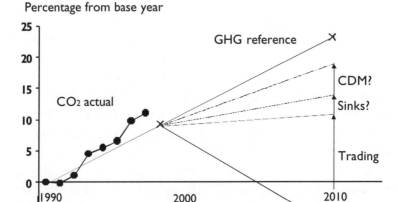

Source: Authors, ITEA.

Note: 'Least cost' trading in the commitment period, for the base case used in RIIA's emissions trading model, would leave US greenhouse gas emissions about 11% above 1990 levels. CO_2 emissions would be considerably higher assuming greater abatement in other greenhouse gases and use of sinks and the CDM. Excluding 'hot air' would reduce US emissions by about 4%.

Japan and the United States at 15 to 20 percentage points higher than their Kyoto commitment. For some of the smaller OECD countries like Norway and New Zealand, domestic CO_2 emissions may exceed the Kyoto commitment by up to 30%. This is a huge gap, and it illustrates most starkly both the dangers and the opportunities offered by emissions trading. Taken at face value, the Kyoto agreement could still leave CO_2 emissions in the United States and Japan around 10% above 1990 levels (and even much more for some other countries) without breaching the letter of the formal agreement – and that is before considering additional flexibility from sinks and CDM investments outside Annex I. Figure 5.5 displays these results for the United States, with illustrative additions for sinks and the CDM (considered further in Section 5.4).

This is not the intended outcome of the Kyoto agreement. Particularly if some of this international exchange represents 'hot air' trading, such an outcome would risk bringing mechanisms for 'flexibility and efficiency' into serious disrepute, and delay still further the time when developing countries accept that they too need to adopt quantified commitments.

These results – in contrast to the absolute costs associated with any given commitment (see Section 5.3.3) – are not very sensitive to differing assumptions within plausible ranges, given the official BAU projections as a basis. This can be seen by comparing the two cases. This is because certain fundamental characteristics are almost universally accepted: across Annex I it is cheaper to achieve a given reduction relative to 1990 levels in EITs than in the major OECD countries, and it is easier to do so on multiple gases than on CO_2 emissions alone. Inevitably, therefore, flexibility means that OECD CO_2 emissions will be above the level of the Kyoto commitment. More pessimistic assumptions about the costs of limiting CO_2 emissions only widen the gap between domestic CO_2 and the Kyoto commitment further.

5.3.5 Cost and related implications

Table 5.8 illustrates the approximate cost implications of the commitments and the impact of emissions trading on different regions. In this analysis, the United States incurs higher costs than Japan or the EU, which is contrary to the results of the economic models discussed earlier (Section 5.2.2, Table 5.1). This highlights the fundamentally different basis of the analysis, in working from official reference emission projections submitted to the Climate Change Secretariat. Most global economic models project emissions growth in the EU, sometimes to a degree comparable to that in the United States, but the official EU projection is for approximate emissions stabilization, contrasting sharply with the projected US growth of 26% (23% for CO_2) above 1990 levels. This swamps the higher 'no regret' potential and slightly lower marginal cost slope assumed for the United States.

Under the reference assumptions, commitments in the EU and Australia (as well as most of the EITs) can be met through 'no regret' measures only, given the large scope for such reductions, particularly for non-CO_2 gases, combined with the modest baseline projections for the EU and the

Table 5.8: Impact of trading on the cost distribution for the Kyoto targets ('$'/capita)

Region	Base case			High case		
	No trade	No restrictions	No hot air	No trade	No restrictions	No hot air
Annex I	+1.37	–	+0.05	+3.12	+0.49	+0.90
EU	–	–	–0.07	+0.02	+0.02	+0.02
USA	+5.00	–	+0.62	+10.80	+3.14	+4.21
Japan	+1.78	–	+0.18	+4.96	+1.11	+1.49
Australia	–	–	–0.66	–	–0.86	–1.16
Canada	+1.64	–	+0.33	+3.85	+1.70	+2.28
Other OECD	+0.24	–	+0.10	+2.10	+0.93	+1.25
Russia	–	–	–0.43	–	–2.01	–1.57
New EU	–	–	–0.23	+0.02	+0.06	+0.22
Other CEEC	–	–	–0.25	–	–1.40	–0.89

Source: Authors, ITEA, with EU in a bubble, Kyoto commitments, all GHGs.

very large contribution of non-fossil fuel sources in Australia. Most other OECD countries incur costs, which emissions trading (excluding 'hot air') reduces by a factor of up to 10, because of the access it gives to the large pools of low-cost reductions in EITs and the smaller OECD countries; Australia joins the EITs in benefiting significantly from exporting assigned amounts ('quotas' in the following economic discussion).

For our 'high' assumptions (which include less scope for cheap reductions in all greenhouse gases), abatement costs for North America and Japan are again substantially reduced by trading; their average annual abatement costs probably lie in the region of $1 to $10 per capita.[15] The EU and Australia are again marginal exporters of quotas. The model assumptions

---

[15] The numbers in the table are relative: multiplication by a factor in the range 1 to 10 would yield estimates in US$ corresponding to relative optimism or pessimism in the underlying aggregate assumptions: see Section 5.3.2. US$10 per capita corresponds to about US$10 billion across Annex I, which is somewhat under 0.1% of projected annual GDP in the industrialized countries for the commitment period. These costs are substantially lower than most of the results reported in Section 5.2 principally because the ITEA model includes the full range of gases and commitments exactly as signed in the Protocol, works from officially submitted reference projections, and includes some allowance for 'no regret' possibilities. See Appendix 3 and the *caveats* in Section 5.3.6.

are set so that revenues equal additional costs (i.e. the model assumes that political and practical constraints act to prevent OECD countries from excess profits from quota sales), hence the EU does not benefit significantly and Australia benefits only from modest sales of 'no regret' quotas. The major benefits accrue to the EITs.

Excluding 'hot air' (mainly from Russia and Ukraine) raises the price of quotas: the costs to Japan and North America increase by around 30% (this is still far lower than the costs without trading), revenues to the non-FSU EITs increase, and the benefits accrue much more evenly across the transition countries overall.

According to our model, emissions trading could reduce costs by 70% compared to domestic action in a restricted scenario (high growth, no 'hot air' trading) or by much more still with no restrictions. It is important to note that big cost reductions from trading are not contingent upon 'hot air' trading. On top of this, the CDM and sinks, offer scope for additional cost reductions.

During the negotiations in and before Kyoto much effort was put into differentiating the commitments to spread costs more evenly among Annex I Parties. The differentiation formulas proposed by various Parties did not survive the negotiations, and our results suggest that the commitments are largely independent of abatement costs or effort needed, but reflect political will. The ITEA model indicates that the mechanisms for international transfer, such as the emissions trading modelled here, will not only reduce but also spread the costs of the Kyoto Protocol more evenly between the Annex I countries.

In this context it is important to recall that the Kyoto commitments were adopted in tandem with the mechanisms for international transfer, which were absolutely central to agreement particularly by the United States and other JUSSCANNZ countries. The authors recall a finding they made before the negotiations in Kyoto: 'For a given cost to the major OECD countries, the two forms of flexibility [multiple gases and emissions trading] could enable an increase of up to 5 percentage points each in the strength of the target, and the effects are more than additive; thus, full flexibility should enable a commitment about 10 percentage points greater than could be expected from a flat-rate CO_2-only commitment without trading.'[16] The evidence is clear that the mechanisms for international transfer did lead to much stronger commitments than would have been agreed without them. It is less clear that negotiators realized the extraordinary scale of flexibility that is potentially available in the agreement as signed.

5.3.6 Caveats

It must be emphasized that, like all such analyses, the results depend on model and input assumptions. The underlying assumption of perfect, competitive trading with no transaction costs is obviously an idealization, albeit one shared by all economic studies currently available. Nevertheless, although the 'absolute' costs resulting from the model may vary widely for different assumptions (which is not a problem for ITEA since the costs are defined in a relative manner), the general patterns of results concerning the impact of flexibilites appear reasonably robust.

Obviously the pattern and results could change in some respects. There is indeed a notable contrast between some of the official emission projections presented to the Climate Change Convention, and those presented in some other economic studies (including some by other government departments). This is a major part of the reason why aggregate ITEA results are very different from those of the economic models cited in Section 5.2, and it points to an important issue. Advocates of the general economic models tend to suggest that the official projections reflect politicization of forecasts and wishful thinking. The government departments concerned tend to defend their projections as more realistic because they can reflect a variety of 'no-regret' policies and structural changes (often including some liberalization of the electricity sector) that are beyond the scope of the more generalized economic models. Obviously there is no test for the future and the evidence from the past is mixed, but at least concerning the EU in aggregate, most economic models have for the past two decades projected emission increases that have failed to materialize in reality.

---

[16] Michael Grubb and Christiaan Vrolijk, *Defining and Trading Emission Commitments: the Implications of Flexibility*, EEP Climate Change Briefing, No. 10, December 1997.

The difficulties that inadequate source data pose should also not be underestimated. Not only have some countries (notably Ukraine) not yet submitted national reports, but data in some others are incomplete or inconsistent, internally or as compared to other sources. Most significant, estimates for Russian greenhouse gas emissions in 1990 vary by around 10% between different sources. Estimates of LUCF vary widely for almost all countries at present. The Secretariat of the UN FCCC has a major task in ensuring quality and consistency in data.

Concerning the ITEA results presented here, the most obvious possibilities for change could involve:

- altered assumptions regarding the highly uncertain situation in the EITs – though this might well revise their projections downwards;
- higher 'reference' emissions in the EU;
- nuclear power in Japan or Europe, with higher costs if their official BAU emission projections are over-optimistic (too low) for this or other reasons;
- incorrect assumptions on non-CO_2 gases, on which emissions and economic data are much more sparse.

In addition, the United States is so big that significantly altered assumptions concerning its emission and abatement prospects could have a large bearing on aggregate results. Probably the most important factor differentiating the actual emission implications of the Kyoto commitments from the analysis here, however – apart from the outstanding questions concerning possible constraints on trading – is the inclusion of sinks and the CDM.

5.4 Potential impact of sinks and the CDM

The inclusion of sinks (through land-use change and forestry) and the CDM in the Kyoto Protocol adds further dimensions of flexibility to those analysed above. These are impossible to quantify, given not just data uncertainties, but the extent to which their scope will depend on definitions and procedures yet to be negotiated. However, general qualitative features are clear. Both, by definition, add further dimensions of flexibility and both indeed are intended to offer lower-cost options for meeting commitments than exist within the scope of Annex I emissions abatement itself.[17]

As noted in Chapter 3, the inclusion of sinks in the Kyoto Protocol was an area of both political controversy and great technical uncertainty. Currently it remains impossible to estimate in detail the quantitative impact of allowing countries to offset their emissions against 'direct human-induced land-use change and forestry activities, limited to afforestation, reforestation and deforestation since 1990', plus potential 'additional human-induced activities ... in agricultural soils and land-use and forestry categories' that may subsequently be negotiated.

A few crude estimates have been advanced, however. The net emissions from land-use change and forestry in 1990, as summarized in Chapter 3, were around 10% or more of gross emissions in many Annex I Parties, and much more in some. Only a small fraction of total LUCF changes may actually *qualify* as valid, additional and verifiable anthropogenic sinks under the terms of the Protocol – the detailed interpretation of which, as noted, remains to be negotiated. On the other hand, of course, LUCF emissions in 1990 are prior to activities designed specifically to enhance sinks. Estimates of potential absorption from forest-related activities during the commitment period are reported for a few countries in Table 5.9, together with results of one wider analysis of the potential specifically for reforestation.

These suggest that for New Zealand and the Scandinavian countries, on some interpretations sinks could account for virtually all of the 'reductions' required (compared to projected trends), and could constitute a substantial contribution in Canada. In the larger countries the contribution may be proportionately smaller, but would still equate to a few percentage points of the Kyoto commitments, compared with the project requirements to reduce emissions by 10–30% compared to reference projections. It should also be noted that Russia and Ukraine both have large areas of potentially very productive but poorly utilized land.

---

[17] The only exception would be for Annex I countries in which the qualifying LUCF categories resulted in net positive emissions during the commitment period that exceeded the country's total 1990 land-use change emissions, a situation which seems highly unlikely.

Table 5.9: Potential sinks from anthropogenic LUCF activities

	Assigned amount	Possible net sink from forestry during 2008–12		Notes
Government estimates	MtC/yr	MtC/yr	% Assigned amount	
Finland[a]	18	3–6	17–33	Reported as 12–23TgCO$_2$/yr
New Zealand[a]	22	> 7	> 30	Reported as > 35 MtC over period
Sweden[a]	19	0–2	0–10	'Anything from 0 to 200,000 ha depending on definition of reforestation'; converted at 10t/ha
Estimates from Nilsson and Schopfhauser[b]				
Canada	145	10.4	8	
Europe		15	1–2	
USA	1483	45	3	
Former USSR	c. 1000	30	3	
South America	–	140		
Asia	–	180		
Africa	–	14		

Notes:
[a] Reports to UNFCCC Workshop, September 1998.
[b] Derived from S. Nilsson and W. Schopfhauser, 'The Carbon-Sequestration Potential of a Global Afforestation Program', C*limatic Change*, 30, 1995, pp. 267–93. Estimated by author on basis of estimated annual average uptake into biomass over 1995–2050. The original authors estimate that soil uptake would add on average 25%, though other literature demonstrates soil uptake to be extremely complex and variable according to conditions.

Total potential LUCF changes in the developing countries are much larger. Overall the IPCC estimated that global reforestation projects might absorb 12–15% of projected emissions over the period 1990-2050. Thus potential absorption in the developing world – if sinks are to qualify through the CDM – would be larger still. Overall, the potential impact of land-use change and forestry, including sinks, under the terms of the Protocol remains impossible to project both because of intrinsic data uncertainties and because many of the defining issues remain subject to future negotiations. Clearly, however, sinks will be very important for a few countries' ability to meet their commitments, while in aggregate they may significantly ease the pressure to curtail emissions.

The impact of the CDM overall, even if it remains restricted to emission reduction projects, is even harder to assess. It is bound up with very complex debates about the criteria for projects and crediting under the CDM (discussed in Chapter 7) and the terms of transfer including charges for adaptation and administration (Chapter 6). To some extent, indeed, Western governments may view the CDM as a 'safety valve', offering emission credits as a way of offsetting difficulties in meeting domestic commitments. Needless to say, developing countries are unlikely to take a similar view.

In the aftermath of Kyoto, debate started about the likely scale of resources available to developing countries through the CDM. Various economic modelling studies have been released that ignore the non-CO_2 greenhouse gases, the potential for 'no regret' reductions in all gases, and sinks within Annex B; several have also taken somewhat higher reference projections for emissions, particularly in the United States and EITs. Such studies, summarized in Table 5.10, have suggested that the CDM could attract resource flows potentially up to $20 billion or more, with a traded volume of credits up to over 700 MtC, or more than 50% of all reductions needed from BAU emissions.

Our analysis suggests this to be highly implausible. Because of all the uncertainties about the cost and scope of reductions available from sinks and the CDM, and corresponding uncertainty about the impact on the price of traded credits, consistent analysis is very difficult. But even with our 'high' scenario assumptions, without any restrictions on Annex I trading we find a volume of less than 150 MtC, which would represent a value of the CDM credits of below $5 billion annual-equivalent during the commitment period.[18] Excluding the 'hot air' from trading within Annex I could double CDM value and raise the volume at maximum to just over 200 MtC. It is also hard to see how many more projects could be developed given the complexity of the CDM (Chapter 7).

One other important observation here relates to the collective aim stated in the very first paragraph on commitments: Annex I countries have adopted these commitments 'with a view to reducing their overall emissions of

---

[18] The cost figure is derived from the cost savings achieved in the ITEA model and the estimated absolute costs of $120 billion in the MIT study for domestic action and CO_2 only.

Table 5.10: Estimates of the size of the CDM

Study	Cost ($billion)	Emission credits (MtC)	Implied Annex I emissions (% change from 1990)
Haites[a,e]	1–21	27–572	–4.7 to +6.9
MIT[b,e]	2.5–26	273–723	+0.5 to +10.0
Austin[c]	5.2–13	397–503	+3.2 to +5.4
US administration[d]	4.2–7.9	100–188	–3.1 to –1.3
ITEA	<5[f]	67–141	–3.8 to –2.3

Source: Christiaan Vrolijk, 'The Potential Size of the CDM', in *UNCTAD Newsletter*, update of *Short Paper Series*, No. 9 of the Ad Hoc International Working Group on the Clean Development Mechanism, organized by UNCTAD, UNEP, UNIDO, UNDP.
Notes:
[a] Erik Haites, *Estimate of the Potential Market for Cooperative Mechanisms in 2010*, Toronto, 11 September 1998.
[b] A.D. Ellerman, H.D. Jacoby and A. Decaux, 'The Effects on Developing Countries of the Kyoto Protocol and CO_2 Emissions Trading', *MIT Joint Program on the Science and Policy of Global Change Report No. 41*, November 1998, Cambridge, MA.
[c] Duncan Austin et al., *Opportunities for Financing Sustainable Development via the CDM: a Discussion Draft*, 7 November 1998.
[d] US Administration, *The Kyoto Protocol and the President's Policies to Address Climate Change*, Washington, July 1998.
[e] The low values in these estimate ranges are for scenarios with supplementarity restrictions.
[f] Cost estimates in ITEA are relative only; using the MIT estimate for domestic action only, ITEA's results are below $5bn.

such gases by at least 5 per cent below 1990 levels in the commitment period'. This clearly does not include absorption by sinks or offsets under the CDM. Consequently extensive use of sinks or the CDM would violate this aim. The economic studies that involve large-scale use of the CDM result in total Annex I emissions being well above 1990 levels, more than a 10% growth from the levels of the mid-1990s (Table 5.10). Such an outcome would hardly seem in keeping with either the letter (Article 3.1) or the spirit of the Kyoto Protocol.

One interesting observation, however, is that excluding the trade of genuinely surplus entitlements ('hot air') could be one way of 'making room' for the use of sinks and the CDM without violating the collective Annex I commitment – while also shoring up the value of emission credits and flows through the CDM.

5.5 Conclusions: inflation, efficiency and competition between the mechanisms

The Kyoto Protocol commitments, from one perspective, are both environmentally and economically rather modest. Without further extension they do little to constrain the growth in concentrations, temperature or sea level. Economic analyses vary widely, but even the more pessimistic economic studies suggest that in aggregate the costs of meeting the commitments will be hard to discern against the scale and uncertainties of economic growth.

From other perspectives, however, the commitments are anything but modest. The headline commitments by themselves, without any international flexibility, would require radical change in OECD energy consumption – so radical that the commitments would never have been adopted but for these mechanisms, and some countries would certainly not ratify the Protocol without them.

This chapter underlines just how crucial are the Protocol's mechanisms for international transfer in determining its real implications. They reduce the costs of compliance hugely, in all cases, even with restrictions. But under some circumstances we find that the scope for flexibility may be so great as to reduce the economic costs to near zero, particularly if the EIT commitments do turn out to represent a substantial surplus ('hot air') that is traded, or if definitions of sinks and the CDM offer extensive additional possibilities. Under these circumstances, the action required within the OECD countries may be very limited.

This reflects the multiple flexibility in the Kyoto Protocol. These flexibilities were driven by national fears about the cost implications of strenuous commitments by the JUSSCANNZ countries – which do indeed shoulder most of the effort under the Protocol in our economic analysis – and their efforts to secure anything that might lower these costs. But probably only the United States really understood (and it may not have done so fully) just how much flexibility might be built in, and countries sought to protect themselves further with additional derogations (the different EIT base years, different base years for the three industrial trace gases, the gross–net approach on sinks, the 'Australia clause', exclusion of bunker fuels, and continuing attempts by Iceland to gain derogations for large projects).

Such pressures will continue to exist in the further negotiations over rules on sinks and the CDM and over possible constraints on trading. One very real danger facing the Kyoto regime is thus that of inflation – the accretion of so many dimensions of flexibility that the root 'currency' of emission reductions becomes almost worthless.

A classical economic perspective would suggest that, providing the reductions are real and not fraudulent, this is not a problem: it simply reflects that the flexibilities can uncover a huge potential for low-cost reductions, and this is good, not bad. Unfortunately things are not so simple. In the first place, trading of 'hot air' does not represent emission reductions and has nothing to do with least-cost abatement. Also, the question of 'additionality' under the CDM – and possibly sinks – is so complex that it cannot be assumed that all emission reductions under these mechanisms will be real and additional. Jacoby et al. commented that 'Kyoto is likely to yield far less than the targeted emission reduction. That failure will most likely be papered over with creative accounting, shifting definitions of carbon sinks, and so on'.[19]

But also important, the classical economic perspective – like the quantified analysis in this chapter – is static. It takes no account of induced innovation or a host of other issues associated with developing long-term efficient trajectories towards atmospheric stabilization. As evidenced by the gap between modest economic costs and the tremendous difficulty experienced in implementing emission constraints, to an important degree the real obstacles lie in the inertia of habits, institutions, infrastructure and industries. If the OECD countries really did achieve most of the commitments set out in the Kyoto Protocol domestically, even allowing for some use of sinks, this would represent a radical redirection of their energy systems. It would be accompanied by the development of technologies and industries for abatement that could spread globally. If, however, their efforts are so wide and thin that the commitments can be met largely by actions that involve minimum effort – and perhaps with credit for things that would have happened anyway – then real progress towards tackling climate change will be far less.

---

[19] Henry D. Jacoby, Ronald G. Prinn and Richard Schmalense, 'Kyoto's Unfinished Business', *Foreign Affairs*, July/August 1998.

There is, in short, a potential tension between cost minimization in the first commitment period, and the generation of sufficient – and efficient – pressures required to change course towards long-term stabilization. These complex dynamic questions thus combine with political debates about the appropriate degree of flexibility in implementing the Kyoto Protocol, in which some parties fear that *too much* flexibility would enable the richest countries to avoid making the changes and innovations that are central to establishing leadership and long-term resolution of the climate problem.

In general, the dominant political pressures in Kyoto were to expand flexibilities without limit. In implementing Kyoto, those pressures will remain, for example in debates over the rules governing sinks, trading and the CDM. But a countervailing factor may emerge, namely a growing awareness of the dangers of inflation, and consequent competition between the mechanisms. Our analysis shows clearly that even in purely economic terms, all do not benefit from maximizing flexibility. Trading of 'hot air' reduces the value of transferring legitimate emission reductions, for example through joint implementation or sector-level trades. An expansive CDM would reduce the value of all emissions trading and JI within Annex I. Global inclusion of sinks would reduce the resources flowing through the mechanisms to the energy sector in both EITs and developing countries, and so forth. Just as countries had to learn that printing money does not solve fiscal problems, they will learn that generating additional flexibilities does not generate more revenues – or solve the climate problem.

Thus the Kyoto Protocol has constructed the basis for what may become an extraordinary multi-dimensional struggle over the scope and design of what are in essence interrelated and artificial markets. It is a struggle that has no parallel in the history of international politics.

Chapter 6

Implementing international transfers under the Kyoto mechanisms

Chapter 5 provided a numerical analysis of the possible implications of the Kyoto Protocol's mechanisms for international transfer. Clearly they are hugely important in determining the implications of the Protocol's commitments. It has also been noted that these mechanisms are unprecedented in international legal agreements. How can they be implemented, and what are the key issues to be addressed?

Unlike some other studies, this book does not simply consider each mechanism in turn, because there are important common themes and interrelationships. This chapter discusses some of the policy options and outstanding questions specifically related to the *acquisition* and *transfer* of emission credits and allowances: credits from JI and the CDM, and the trading of assigned amounts. The analysis starts by considering common themes underpinning the mechanisms, and concludes by examining the relationships between them from the standpoint of Annex I emission commitments, including cross-cutting issues such as supplementarity. Chapter 7 then discusses the outstanding issues governing the operation of the CDM itself as a discrete and unique mechanism encompassing the concerns of the developing world.

6.1 General principles governing the creation and transfer of emission credits and allowances

6.1.1 Roots

The three mechanisms for international transfer share a common root in the definition of commitments, Article 3, paras 10–12 (reproduced in Chapter 4, section 4). This establishes that the three mechanisms shall involve nominally different 'currencies', but shall all be added together to define a government's assigned amount in meeting its commitments:

- CDM projects under Article 12 shall generate 'certified emission reductions' (CERs);
- joint implementation projects under Article 6 shall generate 'emission reduction units' (ERUs);
- emissions trading under Article 17 shall involve the direct transfer of 'parts of assigned amounts'.

Thus each of the three mechanisms is institutionally separate, involving different systems of governance, monitoring, validation, etc.; but ultimately each generates an asset that comes to the same thing, namely an addition to (or subtraction from) the overall emissions of greenhouse gases allowed from an Annex I Party. To alter its allowed emissions, a country could choose any mix of the three, unless the Parties decide on specific constraints in pursuit of the clauses on 'supplementarity' (discussed in Section 6.5).

6.1.2 Corporate involvement

The two project-level mechanisms (JI and CDM) have clear references to corporate involvement: 'a Party included in Annex I may authorize legal entities to participate, under its responsibility' in JI (Article 6), and 'participation under the CDM ... may involve private and/or public entities' (Article 12). In other words, these project mechanisms are intended to involve the private sector, using international corporate investment as an engine for the generation and transfer of 'emission credits' that Annex I Parties may then acquire towards fulfilling their commitments.

The article on emissions trading is too brief to make specific reference to the role of the private sector, but it is clearly the intent of most OECD countries at least – as reflected in the draft text they almost agreed at Kyoto (Chapter 4, Box 4.2) – that there should be substantial corporate involvement.

Under the Protocol, companies cannot be held directly accountable to the meeting of Parties, at least for exchanges operating under Articles 6 or 17: their accountability can only be expressed through that of the countries in which they are incorporated. Any corporate transfer of credits or emission permits internationally would have to interface with the governmental commitments of assigned amounts to have any value to the governments under the Kyoto Protocol.

There are different possible approaches to this. The common theme is that for companies to have a real incentive to engage internationally, they have to face a domestic incentive to control emissions that can be offset by international investment (or purchase of emission permits) under the agreed terms and procedures.

The form of domestic legislation in Annex I countries could be quite varied. Governments could create domestic emissions trading systems upon whatever basis they wished concerning participation, allocation, liability, etc.; and credits or permits acquired internationally would be considered as exchangeable with domestic emission permits. Other governments might develop domestic carbon taxes, but allow credits or permits acquired internationally to be offset against the tax base. In each case, the companies in the national system would be held accountable under national law, which would also then stipulate the qualifying procedures for obtaining credits or permits from abroad that would be recognized under the domestic legislation. Other approaches may also be possible, for example some of the quantified 'negotiated agreements' prevalent in continental Europe could be extended to state that credits or permits acquired internationally would be considered to count towards the agreed emissions target.

In each case, the domestic procedures for validating credits or permits acquired internationally would of course have to include coherence with the specific procedures for certification under the Protocol, without which the transfers would be worthless to the government in meeting its obligations. Some more specific aspects for each of the mechanisms are consider later.

6.1.3 Gas and sectoral coverage

One reason why governments may wish to keep the institutions and units of corporate exchange separate from the governmental commitments derives from concerns about monitoring and verification. As noted in Chapter 4, the Protocol is laced with references to the need for activities related to the quantified commitments to be monitored and verified adequately. Concern about verification is expressed particularly with regard to the mechanisms for international transfer.

Emissions trading itself, in terms solely of the exchange of assigned amounts between governments, probably does not raise any major issues of monitoring and verification. Despite the name, it is commitments, not emissions, that would be traded. The problem of monitoring compliance with commitments is very real, but it is a problem of monitoring compliance with assigned amounts, and that issue is unaffected by intergovernmental trading provided that the two governments concerned agree on the amount being traded and both register the trade with the institutions of the Protocol.

For the same reason, it is meaningless to talk about confining intergovernmental emissions trading to particular sectors or sources. The unit being traded under the Protocol would be a governmental commitment denominated in terms of the full basket of gases listed in Annex A, plus LUCF activities qualifying under Article 3, and it would not belong to any particular source or sector.

The situation is very different for the project-level mechanisms, *and* for corporate emissions trading between national systems. It might, for example, be far easier to estimate total national methane leakage from energy production and distribution than to monitor methane emissions separately for all the different sections and companies in the energy system. Thus to facilitate monitoring and verifiability in the face of uncertainty, one or other government might confine a domestic quantitative regulation (e.g. a domestic trading system) to particular sectors or gases – for example CO_2 from power stations in the first instance, as a monitorable source with large corporate entities. There is no reason why such companies should not then engage in international trade, with trading of CO_2 at the corporate level being directly translated into an equivalent exchange of GHG-equivalent assigned amounts between their governments under the Kyoto Protocol.

Similar issues would apply to projects under JI and the CDM, with the added complexity that in these mechanisms other criteria must be met, including that of ensuring that emission reductions are 'additional to any that would otherwise occur'. For this reason it is almost certain that qualifying activities would in some way be constrained to particular sectors, sources or even project types that met established criteria including those of monitoring and verification at the project level. Nevertheless, the incentive driving such activities would be the acquisition of emission credits that could ultimately be translated into allowed greenhouse-gas-equivalent emissions from some Annex I Party.

6.2 Acquisition and trade of emission credits from JI and the CDM

Under the Kyoto Protocol, emission credits may be generated from specific projects that result in 'additional' emission reductions and that meet certain other criteria. Chapter 7 notes different models by which CERs might be generated and made available from the CDM and shows that it is bound to be a complex business because of the difficulty of establishing whether emission savings are 'additional' and fulfil the range of other criteria. The generation of ERUs from JI will be much simpler for various reasons (see Box 6.1). Both mechanisms, however, face some common issues concerning the acquisition and transfer of credits.

6.2.1 Tradability of credits

There is some debate about whether the credits should be tradable, or should be available to the investor only as a fixed asset that is owned by or can be passed on to only the host or investor government.

For the CDM this question appears to be easily resolved. At least if the CDM allows companies to acquire credits directly from bilateral activities, then CERs will have to be tradable for one simple reason. Since CERs only have value to companies or countries subject to emissions control, a ban on CER trading would immediately put developing-country industries at a disadvantage: investors from Annex I countries would get something that was not of value to local developing-country investors, who would thereby be disadvantaged. Even if developing-country representatives in the climate negotiations agreed to such a system (perhaps seeking a way

Box 6.1: Credit generation from joint implementation

Especially since Kyoto, joint implementation between Annex I Parties has attracted far less attention and debate than the CDM. Article 6, governing this, is the most developed of the Kyoto mechanisms and was the least controversial – notwithstanding the *contretemps* over whether the Article was intended to prevent Parties not in compliance from transferring emission reduction units (Chapter 4, Section 4.4.2). It is also the main one for which there is already significant investment activity: just a few months after Kyoto, the Japanese government announced plans for up to 20 'JI' projects with Russia.*

This is because the concept of joint implementation is far easier, politically and technically, among industrialized countries that are bound by an emissions target. It is politically easier because it only involves countries that are perceived as part of the industrialized world, so there is no root G77 concern about the legitimacy of involvement or denigration of the Convention's principle of industrialized-country leadership, and the exchanges all occur within the Annex I collective commitment. Technically it is easier for several reasons.

Assessing additionality of emission savings from JI projects is easier than for the CDM for two reasons. In most cases, JI projects can reasonably be expected to lead to an actual reduction of emissions by the replacement of old plant with something more modern and efficient. This is intrinsically easier to assess – and contain – than an investment that is claimed to displace some other projected new construction, which may be more common under the CDM. More important, since the transfer of an ERU is subtracted form the host-country assigned amount, that country has a direct interest in ensuring that transfers are plausible estimates, and not excessive. However, there are exceptions to this, notably countries such as Russia and Ukraine that potentially have considerable surplus assigned amounts; in these cases there may be little or no cost to them in being too generous in estimating the 'additional' emission savings.

Eligibility criteria are also far simpler. Because JI is concerned primarily with project-level exchange of credits relating directly to shared commitments in the first period, Article 6 does not specify project criteria relating to 'real, measurable and long-term benefits', and it does not state an explicit governing purpose of sustainable development. Thus, overall, the issues relating to project eligibility and validation – including additionality – would appear to be a subset of the far more extensive criteria required to govern projects qualifying under the CDM.

Institutionally, the procedures are also simpler. A JI investment by a company in another Annex I country would result in it receiving a 'credit'; the size of the credit would have to be agreed by the two governments, and would be transferred between them as the trade of emission reduction units under Article 6 (JI). This aspect is very similar to the exchange of emission permits between companies under a 'case-by-case' approach to emissions trading (Figure 6.2, p. 208).

* *Joint Implementation Quarterly*, Paterswolde, The Netherlands, June 1998 and subsequent issues.

of attracting foreign investment while maintaining control) their counterparts in the World Trade Organization would promptly throw it out.

CER trading can only be avoided if the CDM entirely rejects the possibility of investors gaining CERs directly from investment activities. This 'portfolio fund' model, outlined in Chapter 7, would only allow CERs to be obtained from a multilateral fund that in turn invested in projects in developing countries. But it is still hard to see how, or why, secondary markets in CERs could be prevented, especially since Article 12 explicitly states that the activities 'may involve private and/or public entities'.

Similar logic ultimately applies to ERUs generated from joint implementation under Article 6. In theory it would be possible to state that investing companies, having acquired ERUs, could only pass them to their governments and could not trade them on to other entities. But since an ERU might be worth far less to one Annex I Party than another, this would again introduce objections about competitiveness, and it is hard to see any rationale for such a restriction – or indeed how it could be enforced.

Consequently, it is hard to avoid the conclusion that a general market would emerge in emission credits.

6.2.2 Who should acquire credits?

Another topic of debate concerns who should receive credits from a project in the first place. This has particularly been a topic of debate in the CDM, with different proposals emerging about how to share credits between investor and host.

The debate seems spurious. If the CDM really were to act purely as a portfolio fund the question is irrelevant, since the fund itself would generate all the credits for selling on to any purchaser. For bilateral projects also it seems spurious for the international community to set rules on credit sharing. For any substantial project, the distribution of CERs among the entities involved (including the host government and companies) is likely to be an intrinsic part of the project negotiations and it is hard to see why multilateral rules should constrain that.[1]

---

[1] See Chapter 7 for further discussion of CDM projects, credits and governance.

To put it simply, JI and the CDM are project mechanisms and it is the project itself that should receive any credits. The way such credits are divided between those involved could be resolved by the participants, together with the host-country government that sets the contractual framework for joint ventures. There are likely to be valuable lessons here from more familiar types of commodity-generating projects, including the often tortuous negotiations over production-sharing legislation for oil and gas joint ventures; but it is hard to see why the international community should get involved.

The question of who gets credits does, however, point to an issue that has received little attention in joint implementation. JI is generally assumed to apply primarily to East–West investment, but of course this is not specified: it could equally be applied to investment between any Annex I countries, for example between two members of the EU or any other OECD countries. Foreign investment flows between OECD countries are huge and include a great deal of cross-ownership; OECD countries have developed highly transparent economic borders and companies may invest quite freely, including in projects that clearly lead to reduced emissions. Under what circumstances should they receive ERUs? Should Denmark receive ERUs when Danish wind energy companies invest in the United States – or the UK? Should US gas and engineering companies receive ERUs as part of their contribution to European gas power or co-generation?

The determining factor here – assuming that additionality can be adequately demonstrated – is whether the host government wants to offer ERUs. It is of course under no obligation to do so, but this raises other important issues. If a foreign investor can clearly offer emission reductions more efficiently than any domestic entity, or is willing to offer an incremental cost/investment greater than the value of the ERU to the host government, it would be hard to justify not offering ERUs; this may be particularly relevant in many central and east European countries. But since an ERU would have a tangible economic value, it would not be credible to offer it to foreign investors if it was not available domestically: like trying to prevent the tradability of CERs, it would contravene WTO rules but in reverse of the normal way of favouring domestic industries.

The logic is inescapable that if governments did want to sell ERUs to attract foreign investment, then they would also have to offer ERUs – or the exact equivalent – to their own industries as well, with the same criteria applied to domestic and foreign companies.

6.2.3 Early crediting in the CDM and in joint implementation

This raises another issue in the credit mechanisms, namely early crediting. The provision for early crediting in the CDM – that CERs obtained 'during the period from the year 2000 up to the beginning of the first commitment period' can be used towards compliance – is unlike anything else in the Protocol. It will obviously amplify the CDM, for both good and bad, including the CDM's impact in spreading Annex I effort over time as well as space.[2] This is arguably offset by the varied economic and political benefits of engaging industry and developing-country investments early on – at least if the criteria are rigorous.

But early crediting has another important consequence, in that it creates a curious and specific imbalance between the developing countries and industrialized regions that may have far-reaching consequences for the Kyoto regime. It could lead to preferential location of industries in developing countries, not just to avoid domestic emission legislation but also to generate credits that would not be available for exactly the same action within Annex I. This could raise important trade issues. In particular, if CDM projects result in tradable products, then producers of the same products within Annex I would have a valid complaint under the fundamental WTO rules of non-discrimination, since the international regime would be subsidizing their direct competitors. That situation is unlikely to be tenable.

---

[2] Any CDM credits generated between 2000 and 2007 translate directly to a weakened abatement commitment required for compliance during 2008–12. Since achieving compliance would inevitably require action before 2008, the availability of credits for doing so lessens the action required during the commitment period (unless it is so effective that there is considerable overcompliance and consequent banking). The same does not apply to early crediting from JI. Of course, both may be useful ways of establishing incentives for early action – and for enabling compliance overall.

Figure 6.1: Early crediting of projects

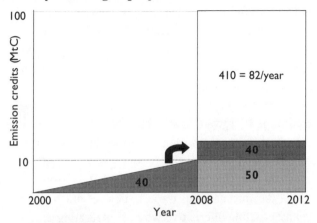

Note: The figure shows internal transfers with early crediting. The hypothetical country has an assigned amount of 500 MtC (100 MtC/yr) and offers credits for early action to its industry. Credited emission savings and advanced permit allocations build up to 10 MtC/yr by the beginning of the commitment period, with 40 MtC cumulated over the eight years 2000–07 that the industries can use during the commitment period. These amounts are taken out of the government budget for the parts of the economy not included in the credit/trading system. Similar accounting could apply to international transfers of ERUs with early crediting, with the cumulated credits in this case being subtracted from one government's assigned amount and transferred to another's.

One solution would be to abandon early crediting, but this is unlikely to be either politically tenable or desirable. Certainly there are valid questions about whether it is possible to establish credible crediting systems as early as 2000, especially since the rules are now only scheduled to be agreed at COP-6 around the end of 2000. Such is the desire for early crediting, however, that governments may well aim to agree rules that apply from shortly afterwards.

It is fortunate, therefore, that there is another obvious solution at hand – one which is already under consideration for other reasons. Article 6 on joint implementation does not establish early crediting – but does not preclude it either. There is absolutely nothing to prevent a country that wishes to offer ERUs for investment under Article 6 from doing so before the commitment period. These credits, obviously, would have to be bankable, counting towards emissions in the commitment period, and would have to be deducted from the Party's assigned amount so as to preserve its total allowed emissions during the commitment period (Figure 6.1); it would thus not carry the risk of inflation inherent in CDM early crediting.

This links directly with an emerging debate about how to stimulate early action towards meeting the Protocol's commitments, which has led to proposals that industries be given credits for early action. It suggests that one of the keys to the Kyoto Protocol will be early crediting – early domestic crediting, to reward companies that take early action, by, in effect, giving them a legal entitlement to certain emissions during the commitment period (part of the country's assigned amount); and internationally, by carrying out transfers of ERUs from the year 2001 onwards. Given the international trade issues set out here, it is not clear that there is any credible alternative way of integrating and implementing the Kyoto Protocol's provisions on the project mechanisms.

6.2.4 Timing and liability in credit generation: ex-post or ex-ante?

Early crediting in JI may also be desirable because of the demands of the certification process. The clean development mechanism is absolutely clear that its units (certified emission reductions) shall be generated *from* the results of specific project activities, i.e. *after* the activity occurs (*ex-post*).[3] Project plans, decisions or commitments are not the defining point: projects must happen and the results be measured, certified, etc. before any CERs are generated under the terms of the Protocol. This is also implied for joint implementation, which refers to 'emission reduction units resulting from' projects.

This is far from universally accepted. It is widely argued that companies will need advance (*ex-ante*) crediting to give them confidence to proceed with investments. Indeed, while developing projects, investors must judge the availability of credits, and it is not reasonable that they should be expected to second-guess the certification process itself. Consequently, there is likely to be a need for a pre-crediting approval process so that project developers can proceed with some degree of assurance that their

---

[3] 'Parties ... may use the CERs accruing from such project activities ... emission reductions resulting from each project activity shall be certified ...'

project will be deemed eligible for crediting when and if it is in operation. This could go further, with an official estimate of the credits likely to be available if the project performs as projected.

However, investors themselves – and not the international institutional machinery – must bear the risk of project non-performance, and of price risk concerning the value of credits. There is a general principle at stake here, relating to questions of liability. The transfer of ERUs or CERs will be an international transaction, outside the scope of specific national jurisdictions. If things go wrong, there is simply not the same kind of machinery available to enforce corrective or punitive action as there is within a national jurisdiction. Furthermore, investors in the project mechanisms, primarily those from OECD countries – and the governments behind them – will generally be the most powerful parties in a project. Consequently it is important that *they* bear the risk of inadequate performance: otherwise they will have no incentive to avoid bad projects or to ensure integrity in project assessment and management regarding emissions performance.

Therefore, while the international machinery might guarantee whether a project *as submitted* should qualify for crediting, and might even make an estimate of the volume of credits to be generated, it must not actually issue advance credits. Of course, private markets might emerge in 'future credits', in which pre-crediting certification might be very important, but this is a different matter.

This does not entirely resolve the issue of liability. However, it is reasonable to argue that while pre-crediting certification cannot be taken as a legal obligation, once credits have actually been issued on the basis of project performance, companies cannot be held liable if a subsequent assessment should raise questions about the validity of credits for the project – providing of course that the information they supplied was accurate. The complication arises if a *country* fails to comply, either with reporting obligations or its specific commitments.[4] If the *project* fulfils all criteria,

---

[4] 'Will the EPA (or the EU for that matter) hold harmless the actions of a private sector firm that helps an emitter in a third-world country reduce emissions even if that country fails to meet its target?' (P.K. Verlege, 'Entitlements Redux', *Petroleum Economics Monthly*, XIV, No. 10, October 1997). While the example relating to a developing country is inappropriate in the context of the Protocol as it stands, the issues raised are clear.

but the host *country* does not comply with some other aspect of the Protocol, again it does not seem reasonable to hold the investor liable, at least if there was no advance warning (cf. the provisions of Article 6.4 outlined in Chapter 4). The situation is less clear regarding the use of ERUs by the acquiring government, and this relates to questions concerning accountability in emissions trading that are discussed in the next section.

6.3 Governing international emissions trading

Emissions trading, as described in Chapters 3 and 4, was, for the United States and some other Parties, the crux on which Kyoto stands, and on which it very nearly fell. All that could be agreed in the final hours was that the Conference of Parties would have to define the relevant principles, rules, modalities and guidelines governing emissions trading, that Parties could engage in trading, and that it should be supplemental to domestic action. This section considers a few key issues; then the final part of this chapter considers the overall question of supplementarity and related issues.[5]

6.3.1 Principles governing emissions trading

As noted, the allocation of assigned amounts involves important issues of principle. These are rather different from questions about how the mechanisms for trading agreed assigned amounts might work, and are touched on in Chapter 8. Another important principle is that the structure of emissions trading should be compatible with the project mechanisms, and should allow for corporate involvement in some form, the mechanics of which are outlined below. Other possible principles could be advanced, for

---

[5] An extensive literature has developed on emissions trading, particularly relating to domestic experience and its possible applicability to greenhouse gas trading. A principal review and reference is that by T. Tietenberg, M. Grubb, A. Michaelowa, B. Swift and Z. Zhang, 'International Rules for Greenhouse Gas Emissions Trading', UN Conference on Trade and Development, Geneva, 1999. This includes extensive references to the literature on emissions trading that are not repeated here. See also 'Further reading and sources of information'. The literature is growing rapidly and this chapter makes no claim to be comprehensive in its coverage.

example concerning transparency and competitiveness in emissions trading; these are discussed briefly below.

One principle common to both the project mechanisms is that of emissions additionality. A corresponding principle for emissions trading could be that the system should be implemented in ways that ensure that no trade of assigned amounts can lead to collective emissions being higher than in the absence of that trade. This is essentially a principle of emissions conservation: trading should not under any circumstances be a vehicle for weakening the overall degree of limitation as compared with the situation in the absence of those trades. At first sight this is almost tautological: it appears self-evident, if trading works as the textbooks say it should. Unfortunately, as shown in this section, there are ways in which this principle could be violated if appropriate rules and guidelines are not established; that is why it is another useful principle to keep in sight.

6.3.2 Corporate involvement

As noted in the introduction to this chapter, the driving force behind corporate involvement in emissions trading must be domestic legislation. Though this could take many forms, the simplest to consider is when governments establish a domestic emissions permit system. The question is then how this might interface with international exchanges of assigned amounts under Article 17.

At one extreme, governments might simply transfer title deeds for part of their assigned amount to corporations and authorize them to trade these internationally subject to government tracking of trades. For various reasons (including those of monitoring noted in the introduction) this is probably not attractive. Another approach would be to create discrete domestic systems for trading of governmentally defined emission permits, but to enable companies to 'cash in' these permits with governments against assigned amounts specifically designated for international exchanges.

A third option would be to authorize companies to trade between national systems subject to such trade being reported and approved by the governments concerned; the governments would simultaneously register a directly equivalent trade of assigned amount between the countries

Figure 6.2: Accountability in international emissions trading

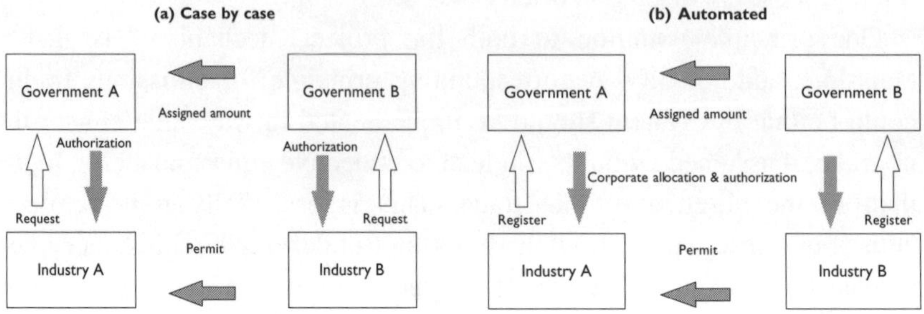

Source: Authors.

concerned, which then assume responsibility for the trade under the Protocol. There would appear to be two variants of this (Figure 6.2). The simpler but more bureaucratic approach (a) is to require direct case-by-case authorization of any such trades. To avoid this however, two governments could in principle agree to mutual recognition of domestic systems and sign an agreement governing automatic recognition of trades between the two systems (b). Parallels with financial and other markets are obvious.

On either variant, each company would be accountable to its government for registering the trade; and each government would in turn be accountable to the institutions of the Protocol for enforcing this and for registering it as an equivalent exchange of assigned amounts. With modern technology this is probably not as complicated as it sounds, once appropriate procedures are in place. What is important is that *equivalence* can be established between the corporate trade and the equivalent intergovernmental exchange of assigned amounts.

6.3.3 Transparency, competitiveness and trade implications

Once granted to industry, greenhouse gas permits would represent a financial asset. If given out by the government to industry, they would affect competitiveness between firms. One concern about emissions trading that emerged in the late stages of debate in Kyoto and after, expressed particularly by France, was that some countries might use allocation of emission permits to confer advantage on domestic companies. This might, for example, conflict with the French government's desire to tax emissions or to auction permits.

Brack et al. argue that emission permits would probably be considered as a service under the terms of the General Agreement on Trade in Services, and that domestic allocation of such permits would confer a corresponding financial benefit falling under the provisions of the Uruguay Round agreement on Subsidies and Countervailing Measures (SCM).[6] Allocation would thus become subject to GATT-type principles of non-discrimin-ation. This would imply that any enterprise operating within a government's jurisdiction should have the same rights of access to permits as any other, regardless of the nationality of its ownership. Brack et al. note that this need not create any particular problems, but it *would* rule out any favourable treatment of domestic-owned industry in permit allocation, and it could rule out any arbitrary treatment of different sectors within industry. In other words, although governments would be quite free to 'grandfather' emission permits to industry if they wished, they would have to respect SCM principles and would have to demonstrate non-discrimination – implying also a considerable degree of transparency in the allocation process.

The demand for transparency and competitiveness also applies to the trading process itself, and this might drive a GHG trading system further than the bilateral exchange model indicated in the previous section. As noted, Japan and some other JUSSCANNZ countries were concerned about the possibility of preferential access by some countries; that is, they were worried that the United States might use its political and market power to obtain assigned amounts from Russia on preferential terms.

The simplest way of ensuring competitiveness in emissions trading would be to require that assigned amounts, or equivalent carbon permits, must only be traded through registered and regulating trading houses. Governments, or industries, could at any time place permits for emitting greenhouse gases on their local Board of Trade, national Stock Exchange or Bourse. Trading prices would be advertised, and permits would be

---

[6] D. Brack with C. Windram and M. Grubb, *International Trade and Climate Change Policies*, London: RIIA/Earthscan, forthcoming 1999.

bought and sold competitively and transparently – and perhaps specu-
latively, according to beliefs about the real need for and costs of emission
permits during the commitment period. Once the trading mechanisms
were in place, there would be nothing to prevent ERUs and CERs being
traded in exactly the same way. The Kyoto Protocol would have created
what might well emerge as one of the world's largest traded assets – an
entitlement to emit greenhouse gases – as an entirely artificial construct:
a market to control the biggest by-product of the world's market
economies.

6.3.4 Accountability and compliance

Where exactly would accountability lie in such a market? Since emissions
trading means trading of an *allowance* to emit during the commitment
period, it will almost certainly be subject to futures (*ex-ante*) trading by
Parties or other entities that want to adjust in advance the amount they *will*
be allowed to emit during the commitment period. This immediately raises
important issues about the status of traded assigned amounts in the event
of non-compliance by a selling party.

The fact that the system is an artificial construct means that observance
of the rules and agreed constraints is critical to the Kyoto system.
Emissions trading injects a new dimension into the issue of compliance
that is without precedence in previous international agreements. If one
country were simply to sell excessive quantities of assigned amounts, so
that it could not comply at the end of the commitment period, the
ramifications would go far beyond that country's default and could
degrade the whole system.

Obviously, if any Annex I Party fails to meet its commitment – its
emissions exceed its assigned amount remaining after transferring assigned
amounts and ERUs – it will be subject to whatever non-compliance
mechanisms are negotiated under the Protocol. As noted in Chapter 4, the
specific compliance provisions are not strong: all that it was possible to
agree is that the Conference/meeting of Parties shall develop 'procedures
and mechanisms ... including the development of an indicative list of
consequences, taking into account the cause, type, degree and frequency

of non-compliance ... Any procedures and mechanisms entailing binding consequences shall be adopted by means of an Amendment to this Protocol.'

This highlights the fact that compliance is a complex and graduated issue, to be treated sensitively according to causes and consequences. In this there is little or no analogy with domestic trading systems, which are under the authority of national law and enforcement agencies.[7]

The specific issue this raises for emissions trading is the question of accountability for traded units: whether units already transferred from a Party that ultimately fails to comply should retain their full amount. Contrary to some claims, the Protocol does not say anything about this explicitly.

There are two basic options. With *seller-only* accountability ('seller beware'), units already traded would retain their full value. This would have the following consequences:

- in the event of non-compliance by any transferring Party, total Annex I emissions would be increased as compared with the overall Kyoto Protocol commitments, violating the principle of emissions conservation under trading;
- the value of a trade would be unaffected by the prospect of non-compliance by the selling Party; thus an acquiring (buying) entity would face no disincentive to trading with countries that take a lax attitude to compliance (or internal enforcement).

As a result, potential buyers would have no incentive to promote strong compliance procedures; on the contrary, all Parties (except the 'honest sellers', who could find themselves undercut) could gain economically from weak procedures and the participation of weak or irresponsible states. For this reason, seller-only accountability could encourage a regime

---

[7] Under the US SO_2 system, for example, the trading entities (companies) are accountable under US law: if a company does not comply it is fined; if it does not pay the fine the company directors may be jailed. The security of the system resides in the US government, and not surprisingly there have been no cases of non-compliance in practice. The parallels with international law are so remote that only those completely immersed in the US system seem able to draw them without further consideration.

of weak compliance and default, to the detriment of the atmosphere and of the integrity of the Kyoto regime itself.

In consequence, there is increasing acceptance that seller-only accountability would not be an acceptable basis for the international trading regime. The alternative is a system with some element of 'buyer beware' accountability (sometimes called 'shared' accountability, since the seller is still just as liable to non-compliance procedures as before). This could be expressed as a rule that any transferred assigned amount should only contribute towards compliance by the acquiring Party to the extent that the transferring Party also complies. One variant would be to discount all units transferred from such a Party in proportion to the degree of its shortfall. Another variant would be *sequential* discounting, which would render the most recently traded units invalid in the event of non-compliance.

These options would ensure that collective emissions do not exceed the commitments at Kyoto, and would create an incentive for all acquiring entities to assess the reliability of transferring Parties. Thus, units from unreliable Parties would carry a risk and so would trade at a discount, which is a highly desirable feature. Fundamentally, the consequences of default would remain within the sphere of human institutions, rather than being disowned to the atmosphere.

This complicates the system administratively, since the units are no longer necessarily automatically interchangeable. In practice, however, verification would anyway require that units transferred under emissions trading be tagged according to the source and date of transfer, so it is not such a big obstacle. As with other traded services, market services would emerge to evaluate the associated risks.

There will clearly be political resistance to a rule of buyer liability, particularly from companies which would prefer that a unit acquired carries no risk of default. A particularly complex relationship could emerge between project-level activities and state compliance. As noted previously, it is highly debatable whether investors in JI projects should be held liable if the host country does not comply with its quantitative commitment. But if the resulting ERU is used by the acquiring government towards its obligations, this carries some of the same problems as with seller-only liability in intergovernmental trading. Essentially the same problem arises for corporate trade of emission permits – and shared accountability at the government level might be undermined if the entities could avoid that discipline by funnelling assigned amounts through corporate allocations in domestic emission permit systems.

It might, however, be possible to make governmental shared account-ability compatible with the corporate desire for certainty regarding the status of ERUs or permits. One possibility might be for trading to occur through government-regulated clearing houses that themselves would absorb the risk: buying on a 'buyer beware' basis, aggregating a portfolio, and selling on a different basis to domestic clients. Another possibility might be for insurance companies to play a similar role, acquiring a portfolio of emission units, perhaps primarily from governments considered certain to comply, with which to insure companies against default by the country of origin. A third option would simply be for governments to absorb the risk, and make sure that they carry sufficient surplus assigned amounts from reliable sources. What is increasingly plain is that shared accountability in trading basic assigned amounts, and possibly the transfer of ERUs, under the Kyoto Protocol is both possible and desirable.

6.3.5 Additionality and 'hot air'

The principle of emissions conservation can also be violated if any Parties are allocated assigned amounts that exceed what their emissions would be even in the absence of any limitation, and transfer this surplus to other Parties that use it to lessen their degree of abatement. This is the 'hot air' problem. With the Kyoto negotiations adopting pledge-based rather than rule-based allocations, it is understandable that some countries in the midst of radical economic transition and with highly uncertain emission projections demanded a wide safety margin in their allocations. The hot air issue concerns the implications of their transferring to other Parties what may turn out to be a substantial surplus.

The total amount of hot air in the Kyoto allocations is by its nature uncertain. However, as discussed in Chapter 5, and depending particularly upon trends in Russia and Ukraine, it could amount to several hundred million tonnes of carbon-equivalent – perhaps as much as 10% of total Annex I allowed emissions.

The transfer of this hot air would not represent any real emission reductions by the exporting country, and, by transferring the surplus to a country that may use it, it makes emissions higher than they would be in the absence of trading. It thus violates the principle of emissions conservation.

Is hot air trading really a problem? Most JUSSCANNZ countries regard it simply as a feature of the deal struck at Kyoto. However, as noted, the political basis of the 'deal' on emissions trading is ambiguous and delicate, and furthermore the most relevant EITs claimed that they needed their full assigned amounts for further growth: they argued that there would be no hot air to trade, and that they only expected to have assigned amounts to sell as a result of policy-driven changes that lead to emission reductions compared with business-as-usual trends. The whole question of what to do if there is hot air is thus both open and difficult.

Hot air trading undermines the economic rationale for emissions trading as an instrument of efficiency, and also brings more direct and serious problems. Politically, assigning to some countries large surpluses that are then transferred to the wealthiest countries, enabling them to avoid substantive action, will be seen by developing countries as violating the spirit of the Kyoto agreement and the principle of the Convention itself by violating the aim of developed-country leadership; it would also depress the value of any transfers under the CDM. Publicly, it would be seen quite simply as using emissions trading to cheat, and it would bring the system rapidly into disrepute.

Potentially the most fatal aspect of hot air trading is its implications for the expansion of the regime. If countries presently within Annex I are allowed both to obtain and then to trade assigned amounts that turn out to be far in excess of their needs, there would appear to be no politically plausible way of preventing other countries from joining on a similar basis in the future (at least for as long as allocations remain based on ad-hoc national bargaining). Failure to tackle 'hot air' trading could thus lead to the whole Kyoto system ballooning; it would be hard to prevent new countries joining with inflated or too cautious targets that would inject an expanding volume of hot air, and result in the collapse of any incentive to meaningful action.

The really difficult question, however, is whether anything can be done about the hot air issue. In outline, it is easy to suggest a *guideline* that follows from the above discussion: any Party with a surplus assigned amount should transfer only that part which represents actual policy-driven emission reductions as compared to business-as-usual emission trends. This does *not* imply that the Kyoto assigned amounts necessarily need to be renegotiated; the problem is one of transferring amounts that turn out to be genuine surplus. Nor does it imply opposing 'growth targets' for countries that face genuine pressures for emissions growth. Nor does it imply that the countries concerned would be unable to benefit from trading. Notably, most transition economies and developing countries have relatively inefficient energy systems, and they could expect to benefit substantially from measures to remove these inefficiencies and to sell resulting tonnes irrespective of the guideline on only trading 'real limitations'. Implementing such a guideline is, however, technically and politically complex and problematic.

Approaches to dealing with 'hot air' The feasibility of excluding 'hot air' trading will depend upon the method proposed for doing so. The most blunt approach would be to try to persuade Russia and Ukraine to accept, at some future date, revised and much lower quantified commitments and get others to agree to this as an Amendment to the Convention's Annex. This is unlikely to be politically feasible, and indeed has various drawbacks and dangers. Constraining trading itself ('ceilings') has been advanced as one possible way of excluding hot air; but this in reality is a separate issue, discussed below under 'supplementarity' (Section 6.4). A 'supplementarity' ceiling upon countries acquiring tonnes, for example, would not necesarily solve the problem of hot air; it could simply mean that countries with real surplus assigned amounts would spread their hot air more widely among acquiring countries.

A direct ceiling on sales from transferring countries – for example, allowing countries to sell only 10% of their allowed emissions – could be somewhat more effective. However, it would simply mean that countries

Box 6.2: A 'review and assessment' approach to tackling 'hot air' trading

A politically credible approach could combine a statement of principle with an ongoing review and compliance assessment mandated to identify and discount any hot air trading. This would include:

1. a guideline that Parties should transfer only that part of their assigned amount that represents policy-driven emission reductions as compared with business-as-usual emission trends;

2. a procedure by which ongoing review and compliance assessment could include a quantitative analysis of a country's likely emissions, net of any policy-driven reductions, in a business-as-usual scenario; this could be according to agreed and representative procedures conducted under one of the official Subsidiary Bodies to the Convention;

3. a rule that any tonnes assessed at the end of the compliance period to have represented hot air trading should be discounted in the compliance assessment of all Parties.

The assessment process (2) would be an ongoing procedure, and ultimately retrospective at the end of the compliance period; as with other aspects of liability assessment, therefore, there would be progressive indications of hot air problems, and such tonnes would be increasingly heavily discounted in trading.

Such an approach would not involve renegotiation of assigned amounts based on long-term (and disputed) business-as-usual emission projections, nor would it involve singling out particular countries. However, application of the final and critical step does hinge upon establishing a principle of shared liability with 'buyer beware', so that tonnes traded can subsequently be discounted if they are found and agreed to represent hot air.

Obviously, such assessment would be a complex procedure. Given that most climate change policies are in fact adopted for multiple reasons – and to ensure that those countries potentially with hot air would still be net beneficiaries – a fairly inclusive definition would probably be appropriate. But changes such as UK electricity privatization, German reunification, the primary economic transformation in eastern Europe and the Asian financial crisis clearly would not qualify as 'climate policy-related reductions' in this sense. The primary aim would be not precision, but the provision of a mechanism for minimizing windfall profits and inflation of the trading system arising from allocations that turn out to be excessive for reasons that are clearly wholly unrelated to climate change policies.

with a big surplus would sell their hot air but would then not face any incentive to real abatement. There may be legitimate grounds for considering caps on trading, but such ceilings would not solve the hot air problem and we do not consider them further here.

Rather, the guideline suggests that hot air needs to be tackled at source, recognizing it first and foremost as a problem of the *combination* of allocation and trading, and of uncertainty regarding major economic trends. Consequently a different and more sophisticated approach is needed. Box 6.2 summarizes a retrospective approach that would seek to discourage and ultimately discount any hot air trading once the situation has become clearer.

If the hot air problem is to be tackled seriously, both in relation to existing allocations and potential future entrants, the inherent uncertainties in emission projections probably leave little choice: an ongoing and ultimately retrospective assessment may be the only viable approach. It could perhaps only be invoked after a formal charge of hot air trading; this would have the side benefit of giving countries an incentive to avoid seeking and transferring obviously excessive assigned amounts. Even if such a system were not formalized, discussion of it would help to focus attention upon a fundamentally important question: in a wholly artificial market constructed through intergovernmental negotiation, what should and what should not constitute legitimate allocations and trades?

6.4 Supplementarity, balancing and charges under the Kyoto mechanisms

The requirement that JI and emissions trading should be 'supplemental to domestic actions for the purposes of meeting commitments' reflects the widespread concern that international flexibility may be used to avoid adequate domestic action. The CDM statement that Annex I Parties could use CERs to 'contribute to compliance with part of their commitments reflects the same concern.

'Supplementarity' was one of those creative ambiguities which allowed countries to agree while fundamentally disagreeing. Some developing countries indeed wished 'supplemental' to mean 'supplemental to the domestic actions required in industrialized countries to achieve their commitments domestically', so that all the international action would indeed be additional to domestic action. This, however, is clearly inconsistent with the basic mechanics of exchange defined in Article 3. At the opposite extreme, the United States maintains that any action abroad will be 'supplemental' to whatever domestic action it adopts, so that the phrase is in effect redundant.

Indeed, most of the US literature on the Kyoto mechanisms appears almost to ignore the fact that the supplementarity clause was one of only three sentences that survived as an essential part of the Kyoto compromise on emissions trading. It was the compromise that spanned the gulf between those on the one hand who instinctively believed that the commitments should be fixed as an index of how the industrialized countries individually should change their emissions, and, on the other, those (especially economists) who could see no reason at all for forcing a country to reduce emissions domestically that could be reduced more cheaply abroad.

6.4.1 Rationales for supplementarity constraints

The arguments advanced for supplementarity have been mainly political in nature. It is argued that using international transfers as an alternative to domestic action would undermine the Convention's principle of leadership by the industrialized countries, and there is a fear that the transfer mechanisms would somehow enable the rich world to use its economic weight to transfer the burden of action onto other countries: it would simply buy its way out of doing anything difficult, an option not available to the poor. The 'supplementarity' requirement is intended somehow to constrain this.

The EU supported supplementarity, crudely, as a way of ensuring that the United States would be forced to take some action at home; it was also a (somewhat ungainly) response to the problem of hot air.

In the authors' view, these arguments have little validity for transfers between OECD countries, and far better approaches towards hot air are available (as discussed above). There is no credible reason why the United States should be forced to do more at home if emission reductions are available more cheaply in the EU, for example. But the issues are genuinely more complex when transfers occur between countries at radically different stages of economic and institutional development. In these circumstances, cost is not a good indicator of welfare and negotiations on the terms of transfer occur between very unequal parties: it is no accident that the rich and powerful generally have the greatest enthusiasm for wholly unconstrained markets. Ambassador Estrada echoed common fears about the supposed economic efficiency of untrammelled flexibility when he remarked after Kyoto that 'of course everything is cheaper in developing countries – including life'.

A different economic justification for considering constraints on the use of international flexibility stems from the observations at the end of Chapter 5 about the dynamics of change. Stabilizing the atmosphere will require far-reaching changes in technology and patterns of economic behaviour. Particularly if the Kyoto commitments with full flexibility do turn out to be as lax as the studies in Chapter 5 suggest, there is a risk that the OECD countries will not need to take much action at home; they might shoulder the burden of abatement globally, but only by spreading established techiques and never having to learn about other ways of doing things. If the richest nations never have to change course seriously, the prospects for long-term solutions are very remote. By contrast, if they are forced to implement some difficult measures at home, they might generate solutions that could then spread globally.

6.4.2 *'Concrete ceilings': proposals and problems*

However, the most difficult issue surrounding supplementarity – as with 'hot air' – is how to define and implement any such requirement. Some EU countries have proposed a 'concrete ceiling' – of 50% – on how much of a country's commitment can be met from abroad. In fact, although this sounds specific it is quite ambiguous: what, for example, would constitute 50% of New Zealand's commitment, which is an initial quantified commitment (QELRC) equal to its 1990 levels?

A more tangible formulation suggested that a country could import no more than a given percentage (e.g. 10%) of its initial QELRC. This also poses several big difficulties. One is technical. Companies involved in international transactions would not necessarily know in advance whether a specific trade might surpass the supplementarity ceiling. This difficulty would be compounded by the existence of the different mechanisms.

Generally, discussions of whether 'concrete ceilings' should apply to the mechanisms individually or collectively have led to proposing both: individual limits to preserve equity between the mechanisms, and a collective limit to prevent collective flexibility being too great.

A political concern is that 'concrete ceilings' could affect countries very differently. They might load most of the cost onto a few countries, and break the international market up into discrete pockets each bound by constraints at different costs: a splintered market subject to tremendous pressures to bend or break the rules by those most constrained. In effect, a tough flat-rate 'concrete ceiling' would face exactly the same pressures as those that broke the EU's attempts to maintain flat-rate commitments at Kyoto, but this time, the United States and Russia too might be fundamentally hostile rather than the EU's only tentative allies. Ceilings might of course be differentiated, but this would probably result in a situation in which countries simply insisted on ceilings they felt confident they would never breach.

Another approach is to place ceilings on the *supply* of emission transfers, for example by not allowing any Annex I country to sell more than 10% of its assigned amount. This has been considered as a way of limiting hot air, but has drawbacks noted in that discussion, and is a rather indirect approach that, moreover, would do nothing to limit the extent to which countries might avoid domestic action through the CDM.

In March 1999 the EU Environment Council debated its position on supplementarity and agreed that it was an important principle but could not reach agreement on the best way of defining it. The discussion has broadened to consider other options as well.

6.4.3 Policies and measures criteria

An alternative approach to emerge has been the suggestion that 'supplementarity' should be assessed with direct reference to domestic policies and measures. One variant would be to accept that international trading is 'supplemental to domestic actions' if a country is assessed as having undertaken specified domestic policies and measures. This of course links to the long-running debate about policies and measures in the Protocol itself, summarized in Chapter 3, and the avenues that the Protocol itself opens to 'consider ways and means to elaborate the coordination of such policies and measures' as listed in Article 2.

The main problem is that demonstrated by the outcome of the Protocol debate on these topics: the refusal of most countries to allow international interference in their specific domestic policies on grounds of national sovereignty and political feasibility. In particular, it is hard to see the US Congress ever accepting a treaty that made access to international trading contingent upon the United States binding itself to adopt specific policies and measures that it had to negotiate internationally.

A final option might be a hybrid approach, in which the amount that countries could import would be tied to the aggregate effects of their domestic policies and measures. This could take quantified form as a 'matching' principle in which the amount imported should be no greater than the emission savings shown to arise from domestic policies and measures. Alternatively, other indicators of the adequacy of domestic efforts could be considered, and the extent to which such constraints are 'hard' or 'soft' would also be debated.

There are, in other words, many options that remain to be considered in detail and the issue of 'supplementarity' and how to implement it will remain on the post-Kyoto negotiating agenda. The agenda may be further broadened by another issue that emerges from a different direction, but which also bears upon the question of supplementarity – namely that of equal treatment between the mechanisms, particularly in relation to the CDM's surcharge for administration and adaptation.

6.4.4 International transfer charges and price bounds

The CDM's statement that 'a share of proceeds from project activities' shall be used to cover administration charges and to help vulnerable developing countries to adapt to the adverse effects of climate change in effect establishes a charge on using the CDM. Developing countries have been quick to argue that this will discriminate against the CDM in favour of JI and emissions trading, and some have suggested that an equivalent charge should also be levied on these mechanisms.

Such a charge could also have the consequence of reducing the volume of international emission transfers. Establishing a charge on *all* international emission transfers under the Kyoto Protocol could thus advance the objective of ensuring adequate domestic action. It would also be far more equitable than a 'concrete ceiling'. It would not leave just a few countries severely affected, facing very different marginal costs; the international market would still be homogeneous, but all countries would face an equal charge for each unit that they acquired from abroad.

Is such a charge feasible? One apparently related idea – the proposal by the Danish economics professor, Tobin, to dampen financial speculation by taxing international financial transactions – never made headway for two main reasons. First, the volume of international financial transactions is so huge that a tax raising politically credible amounts of money would be too tiny to affect behaviour significantly, and would probably just add an administrative burden of little benefit. Second, many international financial transactions are by private businesses and may often include internal transfers within multinational companies simply to balance risks against currency fluctuations. The Tobin tax was also seen as intergovernmental interference with the normal business of private commerce, and never really made much progress.

None of these objections would apply to a charge levied on *intergovernmental* transfers of AAs, ERUs or CERs under the Kyoto Protocol. As noted, these transfers would ultimately have to be registered to have legal significance under the Protocol, and this would be the obvious point at which a fee could be levied. Since the transfers would only have legal significance insofar as they represented exchanges of governmental commitments, the process need not interfere with day-to-day industrial transactions. Governments anyway have to design the accounting interface between corporate exchanges of intergovernmental commitments, and this would be the point at which to design the details of how (and whether) to pass on the charge to aggregate corporate transfers.[8]

---

[8] One option under the project mechanisms might be to apply a charge at the point of issuing the credit, without any charge on subsequent transfers. For emissions trading, not all corporate transfers would have to be subject to a levy, though some kind of accounting of aggregate international transfers of emission permits – perhaps annually – could be used as a basis for passing charges on to reflect governmental payments.

One obvious use of the proceeds from such a charge would be to fund the institutional requirements of the climate change process itself, including ramification and effective participation of developing-country delegates. A second obvious use would be to help particularly vulnerable developing countries adapt to the effects of climate change, as with the CDM.

But other uses are also arguably relevant, particularly in the context of Annex I transfers. Chapter 4 noted that one issue never explicitly discussed during the Kyoto Protocol negotiations was technology research and development. While industries play a dominant role in the commercialization of technologies, government R&D is an essential part of more basic technological innovation, and international collaboration obviously makes sense. Yet after peaking at almost $2 billion in the early 1980s, overall energy R&D expenditure by OECD governments declined steadily (though the proportion spent on renewable energy, for example, has grown, reaching $887m in 1995). State-based funding also has the drawback that governments may be tempted to use their money to promote domestic R&D ventures instead of best international practice, and restrict international access to the results (though IEA implementing agreements have sought to extend technology collaboration beyond the OECD). A substantial international funding source could make a great difference to the long-term prospects for the development of new technologies for addressing climate change.

It is instructive to contrast these data with the possible revenues from a charge on international emission transfers under the Kyoto Protocol. Total transfers under the 'high' scenario amount to about 300 MtC per year during the commitment period between the major Parties. Obviously, the rate of the charge would have to be negotiated, but it is reasonable to note that a figure of $1/tC would be too low to affect behaviour significantly, while one in excess of $10/tC might be rejected as too heavy. A charge of $5/tC levied on transfers of 300 MtC per year would raise about $1500 million annually during the commitment period or, for example, about $750 million annual funding over the period 2003–12. Properly designed, it could in effect set a minimum price on international transfers and encourage companies and governments to think twice about engaging in large-scale allowance trading in preference to setting domestic houses in order first. It is, at least, an option worthy of further exploration in the ongoing negotiations.

Conversely, some proposals for domestic legislation have suggested setting an upper bound to the price. One independent proposal in the United States is to introduce domestic emissions trading, with the cost limited by a guarantee that the government would issue more permits if the price rises above a certain level – proposed as $25/tC initially in 2002, rising over time.[9] This would make the system more palatable domestically, and might interact with the international system in interesting ways. If it were regarded as an example of an 'adequate' domestic action, and the price did rise to the ceiling set, the government could reasonably obtain assigned amount from abroad to cover the additional permits it issues domestically, to remain in compliance with the Kyoto Protocol. This is one example of how an increase in the flexibility allowed to a country could be arranged if domestic measures prove to be costly. The wide flexibility ultimately available through the Protocol's mechanisms – including through sinks and the CDM – can be used to help limit the maximum cost of the Kyoto agreement, and to allow the effective strength of the commitments to be adjusted in the light of experience and the limitations of what is deemed economically reasonable and politically feasible.

6.5 Conclusions

The international transfer mechanisms of the Kyoto Protocol are a leap into *terra incognita*. Some parallels can be drawn from domestic experience with emissions crediting and trading systems, though parallels are limited and mostly more relevant to the generation of emission credits. There is no doubt that the mechanisms can be made to work, but many issues have yet to be resolved.

Emission credits from joint implementation and from the CDM should be generated *ex-post* as part of the stream of assets from a project, though various aspects will need pre-certification assurances. The resulting credits should be freely tradable among companies. The imbalance of

---

[9] R. Kopp, R. Morgenstern, W. Pizer and M. Toman, 'A Proposal for Credible Early Action in US Climate Policy', http://www/weathervane/rff.org/features/feature060, September 1998.

early crediting in the CDM implies that JI too should be structured to offer early crediting, including to domestic industry, taken out of a country's assigned amount during the commitment period. Emissions trading will operate primarily as a futures market.

Market mechanisms need a strong compliance regime to be effective. Because international enforcement mechanisms are weak, accountability for traded units will need to be shared in the event of non-compliance. In fact, the transfer mechanisms of the Kyoto Protocol may well offer ways of strengthening compliance overall, including through the use of private law governing firms operating under the mechanisms.[10]

'Hot air' in emissions trading is likely to be a serious problem; tackling it is not easy. This chapter has suggested an approach based upon progressive assessment and discounting, following the principle that traded amounts should represent policy-related emission savings. The requirement that international transfers must be 'supplemental' to domestic action is similarly not easy to implement quantitatively in credible ways. Trying to impose specific ceilings raises obvious difficulties: again, establishing some linkage to the assessed adequacy of domestic policies and measures may offer the most promising approaches. More serious attention could also be given to the possibility of a charge on intergovernmental transfers under all three mechanisms of the Kyoto Protocol, with the proceeds directed to administration, adaptation and technology development.

Critics of international flexibility have generally sought to constrain the mechanisms as far as possible. Supporters tend to argue that there should not be any constraints and that little management is required. This chapter has argued that the distinctions must be more subtle, and the mechanisms must be developed but with care and attention to the detailed implications, keeping in clear sight the objective and principles of the Convention and the climate change regime, rather than economic or environmental ideology. The Kyoto Protocol opens fertile new ground, but what springs forth will need to be managed carefully if it is to be ultimately successful.

---

[10] J. Corfee-Morlot, *Ensuring Compliance with a Global Climate Agreement*, Paris: OECD, 1998; J.Werksman, *Responding to Non-compliance under the Climate Change Regime*, OECD Information Paper, Paris: OECD, 1998.

Chapter 7

The clean development mechanism

7.1 Introduction

The clean development mechanism, the 'Kyoto surprise', is the most enigmatic and complex of all Kyoto's innovations. As noted, from one perspective it appears as a form of global joint implementation, and to this extent it shares the possibilities and problems noted in Chapter 3. However, so much is left open that it could take many forms.

The CDM has already spawned a rapidly growing literature and a frenzy of specialist meetings around the world. Quite simply, industrialized countries scent a way of alleviating their commitments, and developing countries scent money. This section does not attempt to survey all the literature, but it does attempt to provide a guide to the key issues and to offer some core observations and conclusions about the CDM.

The conventional view of the CDM in most OECD countries is that it should be primarily a way of minimizing the costs to them of the Kyoto commitments, by allowing investment in emissions-avoiding activities to generate credits wherever it is cheapest to do so. One major conclusion that emerges from the discussion in this chapter is that this is unlikely to be how the CDM can – or even should – work in practice. Rather, it is concluded that the *primary* role of the CDM should guide foreign corporate investment in developing countries towards goals of sustainable development. This will involve social, political and procedural judgments that do not simply regard the CDM as a way of generating emission credits from 'additional' projects primarily to minimize the costs of Annex I commitments.

The reasons for these conclusions are linked not just with the contrasting political priorities endemic in the North–South divide, nor with the possibility of inflation of the Kyoto system leading to insufficient action and innovation in the OECD countries. It is also argued that a CDM that is based primarily upon visions of least-cost emissions crediting is a chimera that is both technically and politically impossible. Why this is so will become clear after some rather complex issues have been examined.

There are so many elements in the CDM that they can conveniently be grouped alphabetically from A to G. Finally the elements are brought together and conclusions drawn in the context of global investment flows and the history of experience with technology transfer and crediting systems.

7.2 A: additionality, and adaptation and administration charges

7.2.1 Additionality

The problem of defining and measuring the 'additional' emission savings arising from a CDM project, as compared with what would have happened otherwise, is one of two Achilles' heels in the conventional OECD view of the CDM. The Protocol explicitly states that emission savings must be 'additional'. It does not explicitly state that financing must be 'additional' to what would have been spent otherwise, though most economists would consider such 'financial additionality' as synonymous: additional financing would be required to achieve additional emission reductions. In fact the link is not so simple, and financial additionality is considered separately below.

The fundamental difficulty, that of having to estimate emission savings relative to a 'baseline' estimate of 'what would have happened without the project' that is by definition unobservable, was noted in Chapter 3 as one of the core objections to joint implementation. Work over the last few years, including in the pilot phase of JI, has produced guidelines for assessing additionality that are probably a reasonable guide for some kinds of projects in some circumstances. It is, for example, not too hard to estimate the additional emission savings made by a wind energy project in a power system where coal is always the marginal fuel and another coal station was the obvious investment alternative. Certain kinds of reforestation projects, for example supporting forestry on lands that would otherwise have remained deserted grass or scrubland, can also be plausibly assessed at least for above-ground carbon accumulation, if that is allowed in the CDM. But these are exceptions, rather than the norm. For most projects, including many of those most actively promoted, quantified emissions additionality will remain elusive.

The reasons for this are basically simple: the future is uncertain and decision-makers are human. Indeed, businesses make their money by exercising managerial judgment in the face of uncertainty. Frequently, different companies make wildly divergent assessments of the same project – particularly of competing project proposals in which they have a stake. Is the bureaucracy of the CDM really supposed to assess competing claims about long-term project viability that are contingent upon perhaps confidential projections of costs and performance, and inherently unknown prices?

Even when there is a consensus, it is frequently wrong. In the UK in the mid-1980s, almost all electricity experts would have agreed that coal-powered stations were cheaper than gas-powered stations and would be the default investment. They were all proved wrong as the UK system was privatized and the 'dash for gas' took place. Examples in the energy world are so common that students are taught to be wary of all projections – especially when there is a consensus on them.

For the CDM there may be also the vicious paradox that the more cost-effective the project, the more uncertain the additionality. In purely economic terms, any project that would only require a small incremental benefit (such as a CER at low cost) to make it proceed would also only require a small shift in market conditions to make it viable without crediting. This problem is less stark if there is consensus to credit projects which are 'cost-effective' but which clearly would not otherwise be implemented due to other barriers; but it is enough to throw a big question mark over the supposition that the CDM should be explicitly oriented towards 'least cost'. It also suggests that the system should *explicitly* address uncertainty, for example by weighting CERs according to the degree of confidence about the additionality of emission savings.

Another important reality is that most projects and proposals are the result of a *consortium* of actors and interests that want the project anyway, opposed by others that do not want it. Cost-effectiveness is only one criterion. Any large project has to go through a process of internal negotiation and external planning procedures. In reality there may be no way of knowing whether a specific project would have gained approval without a CER. The idea that some bureaucracy in the CDM could assess this, and quantify additionality with precision, amounts to asking it to play the role of an all-knowing and all-seeing God, an entity that even most theologists reject as inconsistent with human free will.

The idea that companies and governments will start doing radically new things given the incentive of credits is also erroneous. The reality is quite different. Every government and every company that is actively considering the CDM is also actively considering which of their current projects, or desired proposed projects, might be able to gain crediting under it. This is in danger of becoming one of those truths that is unmentionable: everyone is doing it, but is not really supposed to. Yet from the authors' own experience it applies to US (and almost certainly Japanese and other) reviews of their existing foreign aid programmes; to some developing-country assessments of their programmes on renewable energy and in some cases natural gas; to projects under the existing AIJ pilot phase; and to already committed corporate programmes on cleaner energy investments.

There is not necessarily anything wrong with much of this – it is not bad if those who have already started taking climate-friendly action are rewarded for it with credits – but it does show the impossibility of measuring or even defining savings that are additional to those that would have occurred in the absence of emission credits. It also shows the danger that the CDM could become a mechanism for weakening action in the industrialized world *without* any offsetting *additional* activity in the developing countries.

Another danger of the conventional approach to 'additionality' is that it gives perverse policy incentives. Environmentally sustainable projects are least likely where the policy environment is least encouraging to them. Therefore, additionality is greatest, and most easy to prove, in the worst policy environments; this is hardly something one would wish the CDM to encourage. Ambassador Estrada focused upon this aspect in the context of whether forestry should be included in the CDM.[1]

---

[1] '[Another aspect] relates to forest management to reduce emissions which, without management, would be generated by deforestation. A couple of developing-country governments are offering "carbon certificates" for a price in exchange for sustainable management of

A final observation is that market conditions evolve. Particularly since the world is supposed to be moving towards sustainable development, one would expect lower-emitting projects to become intrinsically more attractive over time. This indeed is the case in many respects: the cost of natural gas and of renewable energy sources, for example, is generally declining relative to coal-based power, and 'clean coal' technologies are also improving. Similarly, it is obvious that most cities will have to develop and improve public transport infrastructure; this will have the side-effect of reducing greenhouse gases. The most that can be said in many cases is that crediting under the CDM might enable certain kinds of projects to proceed earlier than would otherwise have been the case.

Several important implications follow. First, if real quantified additionality in the form of assessing 'what would have happened without the crediting' is so uncertain, then real policy should not be based primarily on the chimera of accurate quantification. The negotiators will rather have to ask: how can the concept and criteria of emission reductions crediting be used as a lever for achieving desirable public policy goals? In fact the CDM text states that emission reductions should be certified 'on the basis of ... reductions in emissions that are additional to any that would occur in the absence of the certified project activity'. Certified emission reductions are relative to the absence of the *project,* not of the *crediting,* and are to be *on the basis of* estimated savings, not *an estimate of* emission savings. So there is a lot of freedom for manouevre, though obviously crediting should not be totally divorced from the goal of ensuring that the CDM results in overall additional emission reductions.

Such freedom could be used to quantify 'additionality' relative to something other than projected local conditions. For example, defining additionality relative to global or regional norms of technology and investment would remove the potential incentive for perverse public policy in recipient countries and the scope for inflating baselines. Another approach is to assess additionality relative to *observable* characteristics of

---

areas at risk of deforestation ... it is difficult to understand how the baseline on such projects can be defined ... is it not part of the normal responsibility of governments to protect their own natural resources?' R. Estrada-Oyuela, in J. Goldemberg (ed.), *Issues and Options: the Clean Development Mechanism,* New York: UNDP, 1998.

the system at the time of the investment, in ways that unambiguously drive investment towards lower emissions on average, with less attention to the 'real' additionality relative to projections.[2]

Finally, the observation that the CDM may in practice best be viewed dynamically, as a way of accelerating clean investments that might well have occurred eventually anyway, points to one way of at least containing the problem of spurious crediting. It implies that emission reductions should only be considered 'additional' for a limited duration – not beyond the end of the first commitment period at most. This also removes the spectre of credits simply expanding *ad infinitum* during the twenty-first century as projects cumulate and investors retire on the ongoing proceeds. Doubtless, ceasing crediting for current projects in or after the first commitment period would be decried for limiting incentives, but in the real world very few investment decisions would be swung by the availability of credits of uncertain value from 2013 onwards. The question can be revisited for projects initiated after negotiations start in 2005 on a second commitment period.

7.2.2 Adaptation and administration charges

The provision that 'a share of proceeds from certified project activities' should be used to cover administrative expenses and to assist vulnerable developing countries to adapt to the adverse effects of climate change was crucial to building sufficient consensus behind the CDM, but all details remain to be addressed in subsequent negotiations. Suggestions that fees should be extracted as part of investment appear both undesirable and unworkable. They would deter investment by adding to up-front costs while the degree of crediting might be uncertain. In addition, developing countries are concerned that extracting proceeds at this point – for projects that could involve considerable domestic investment – could mean that developing countries themselves pay much of the charge.

What distinguishes a CDM project from any other investment is the gaining of a CER, and since in most projects consortia of investors make

---

[2] R. Hamwey and F. Szekely, 'Practical Approaches in the Energy Sector', in J. Goldemberg (ed.), *Issues and Options*.

contributions on terms and under contractual arrangements that may well be confidential, the fees must necessarily be extracted as part of the overall crediting process. Since it is the industrialized-country investors (and governments) that would most often be seeking the CER, they would then also tend in practice to bear most of the charge.

How such fees should be defined and disbursed raises another thorny set of complications. Funding to support adaptation in developing countries, possibly also with an element for disaster relief, is likely to become more important as adverse impacts accumulate, but this takes us beyond the scope of this book.

7.3 B: bilateral or portfolio structure?

Should the CDM operate as a *bilateral* or a *portfolio* (sometimes called 'multilateral') mechanism? These options are illustrated in Figure 7.1.

Under the bilateral approach, companies or governments from Annex I Parties would invest directly in projects in developing countries and receive credits for so doing, very much as originally conceived for joint implementation but under the rules agreed and administered by the executive board. In the portfolio approach, the CDM would be a discrete entity between investor and host, aggregating projects submitted by developing countries and selling the CERs on to Annex I purchasers. In its pure form, it would indeed be an international fund financed by the selling of CERs. Yamin describes the differences and associated interests succinctly:[3]

> The bilateral approach stresses the similarities of the CDM with JI under Article 6 and appears to be favored by many Annex I Parties. The alternative, the portfolio approach, focuses on its multilateral character and possible linkages with emissions trading, and is drawing greater attention from developing countries. These approaches are not rigidly defined, or even articulated openly, but form what might be called two kinds of 'mind set' around which opinions about the future evolution of the CDM appear to be emerging. Parties' views are also influenced by experience of the AIJ pilot phase, including Costa Rica's decision to issue 'Certifiable Transferable

---

[3] F. Yamin, 'Operational and Institutional Challenges', in Goldemberg, *Issues and Options*.

Figure 7.1: Portfolio/multilateral and bilateral structures for the CDM

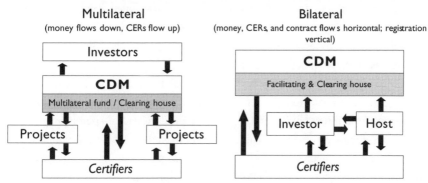

Source: M.A. Aslam, 'International Trading Aspects of the CDM', paper for the UNCTAD International Working Group on the CDM, ENVORK, Islamabad, Pakistan, 1998.

Offsets',[4] and by developments outside the Convention process such as the initiative by the World Bank to establish a prototype Carbon Fund.[5] The *bilateral approach* to the CDM emphasizes the needs of investor Parties and the interests of their private sectors. The CDM would define certain rules and exercise scrutiny over certification but the development and implementation of CDM projects, and distribution issues concerning benefits and risks, would be dealt with in a contractual manner on a project by project basis by the Parties and entities involved in the project. Under this approach, the CDM would be designed to ensure that investor and host countries (and their respective private sectors) are given the maximum amount of choice to determine the nature of CDM projects, their financial contributions and the resulting sharing of CERs with minimal interference from a centralized, international bureaucracy. This approach therefore favors the very minimal use of CDM institutional machinery which, in essence, need only consist of a clearing-house mechanism putting investors in touch with interested hosts to reduce transaction costs, and an independent certification process to generate environmental integrity and business confidence in the system.

---

[4] A. Michaelowa and A. Dutschke, 'Joint Implementation as Development Policy: The Case of Costa Rica', paper presented at OECD Forum on Climate Change, Paris, 1998.

[5] This would amount to a mutual fund for offsets that allows investors to pool their money to support a portfolio of projects developed and financed by the World Bank with a 'green' element supported by contributions to the fund.

[In this form] the CDM is seen as one of the 'flexible mechanisms' contained in the Kyoto Protocol whose purpose is to reduce Annex I Parties' compliance costs by generating cheap emission reduction 'offsets' overseas. Loading additional responsibilities on to it, such as assistance in funding, may reduce its attractiveness to Annex I Parties.

The *portfolio approach*, by contrast, stresses the multilateral character of the CDM. The basic idea behind the portfolio approach is to 'shield' host countries from direct buying and selling of CERs. Instead of approaching host countries directly, investors would buy CERs from the CDM itself, which will channel monies received to host countries that have submitted 'bundles' or portfolios of projects to the CDM for certification. Financial contributions to the CDM are not mandatory but simply the receipts from CERs sold to Annex I Parties. These receipts are channeled back to the countries providing the CERs.

The two approaches (summarized in Box 7.1) are very different, and Yamin goes on to discuss the pros and cons of each more fully, as do other authors in the same volume and increasingly elsewhere.[6] Aslam[7] also notes a 'unilateral' possibility, in which developing countries themselves invest in projects that generate credits which they can then sell internationally; this could be compatible with either bilateral or portfolio structures.

The portfolio approach clearly allows for greater multilateral control and attention to the collective concerns of developing countries, a number of which are explained below. The biggest perceived risk is that separating foreign investor from host – and including concerns about equity, sustainable development criteria, etc. – could make a hugely bureaucratic and inefficient structure which would hardly operate as an international market mechanism at all. It would also then appear to be a direct competitor to the GEF, which it would rapidly displace by virtue of generating credits.

---

[6] E.g. E. Fages and A. Michaelowa, 'Designing the Clean Development Mechanism', submitted to *Energy Policy*, forthcoming 1999. These authors argue that there are two distinct variants of the CDM's role under a bilateral approach: a straightforward information exchange, providing a database of projects proposed and possible investors; and a more active clearing house which would accept and evaluate projects, and invite tenders for funding. There are also of course many variants of the portfolio approach.

[7] M. A. Aslam, 'International Trading Aspects of the CDM', paper for the UNCTAD International Working Group on the CDM, ENVORK, Islamabad, Pakistan, 1998.

Box 7.1: Bilateral versus portfolio approach to the CDM

Bilateral approach	Portfolio approach
Project by project	'Bundling' of projects in portfolios
Investor-led	Host country-led
Private-sector emphasis	National sovereignty emphasis
Contribution to emission reductions emphasis	Contribution to sustainable development emphasis
Proceeds for adaptation unnecessary, seen as additional costs to achieve Article 3 compliance	Proceeds for adaptation seen as necessary to benefit all developing countries to increase global participation in Protocol
May concentrate on countries already benefiting from foreign direct investment	Could allow equity considerations to tailor portfolios to benefit all developing country mitigation efforts
Primary purpose of CDM is clearing-house function	Primary purpose of CDM is to obtain best price for CERs, shield hosts from undue pressure; clearing house function is a necessary feature

Like the debate on additionality, the debate on these different forms of the CDM would be clearer if actual specific projects and other experience were considered. As noted already, actual projects do not arise by simply matching a host opportunity to an investor's money; they may arise as a result of a long process of proposal, evaluation and negotiation repeated many times as the project proposal is refined with potential investors.

At one extreme, the proposed Northeast Asian natural gas pipeline (linking the Yakutsk and Irkutsk natural gas fields in Siberia to Chinese markets), might be regarded as potentially the largest CDM project on earth. It could displace massive amounts of inefficient coal-based power production in China. It would cost at least $10–20 billion, involve consortia encompassing many of the world's major international energy and engineering companies – as well huge domestic interests such as the China National Petroleum Company – and take at least ten years to come to fruition; the early stages have already been under negotiation for many years. The idea that China would place this project on a CDM bulletin board so that potential investors could just offer money and then they could all go ahead is fantasy. So is the idea that the CDM would interface between 'investors' and 'host', with China going ahead on the basis of the CDM approving and helping (how much?) to fund the project, later to sell the CERs from the project on the open market? Rather, the CDM and potential credits would have to be just another element in the overall consortia negotiations, influencing (probably marginally) project design, feasibility and organization – essentially a trilateral model not quite like any of those so far proposed for the CDM.

At the opposite extreme, a local utility in a developing country might well be willing to offer a wind energy proposal to its government's CDM office to add to the international CDM portfolio, and wait to see if CDM finance is forthcoming, perhaps along with a foreign project sponsor. That could well be more effective, as well as more equitable, than hundreds of local utilities subject to hundreds of foreign corporations, each trying to sell its preferred wind turbine (or other technology), and each making competing claims about crediting prospects as well as project performance. A portfolio fund could also help to aggregate and offset risk and give access to innumerable sources of Western finance that would never get involved in direct project finance.

In short, it is likely that the CDM will need to offer both possibilities, and indeed may act as a hybrid in some circumstances. Here there are surely also lessons from other commodity markets, for CDM credits will be a commodity of some form. Oil investment and trading, for example, used to be based almost entirely upon long-term bilateral contracts. Eventually this proved impossible to maintain and a spot market emerged, aggregating available oil in a portfolio from which purchasers could buy. Most companies now engage in a mix of fixed contracts, spot purchases and various futures options. The gas industry used to be based entirely upon long-term contracts (often with 'take-or-pay' clauses giving absolute security to the investors) to finance the basic infrastructure investment; spot markets have emerged as the industry matures and competition is introduced. The analogy is far from exact, but lessons are relevant.

Thus the CDM is likely to need a mix of bilateral and portfolio structures, a conclusion also reached by Siniscalco et al.,[8] who conclude that 'a CDM structure which allows for both multilateral and bilateral options seems more likely to foster investments, promote innovation, and

---

[8] D. Siniscalco, A. Goria and J. Janssen, 'Outstanding Issues', in Goldemberg, *Issues and Options*.

at the same time guarantee that structures are competitive'. A portfolio structure could emerge from private-sector initiatives if and as a market develops, but this of course would not necessarily meet other specified goals of the CDM, so there is a strong case for a multilaterally managed portfolio from the outset. But the defining point of multilateral control will probably not in fact be this: it will lie in the crediting process.

7.4 C: crediting, certification and the control of corruption

C is for the *crediting* process and the various issues surrounding it, including *certification* and the *control* of *corruption*. There are many dimensions to crediting, including debates about whether CERs should be openly tradable; 'credit sharing' between host and investor; and the whole range of issues surrounding monitoring, auditing and verification – as well as specific issues concerning eligibility criteria, discussed under E below.

As noted in Chapter 6, it is likely that credits themselves could only be generated as a result of assessed project performance, though uncertified futures/options based on project submissions and evaluation might well also become a discrete asset. The issues concerning tradability and credit sharing were also discussed in Chapter 6, and, as argued there, they really appear quite spurious. If there is no bilateral element to the CDM, and all credits are issued by a portfolio fund, then the issue is irrelevant. If there is any bilateral element, credits will have to be tradable to avoid discriminating against developing-country companies, and there seems no reason why the international community should try to set rules for distributing the credits any more than it would for any other asset within complex corporate contracts.

One issue that certainly cannot be avoided, however, is the need for robust procedures on the crediting process itself. The Protocol states that the first COP/MOP shall 'elaborate modalities and procedures' to ensure 'transparency, efficiency and accountability through independent auditing and verification of project activities'. The reference to 'independent auditing' shows awareness of the need to separate the various functions: those who invest, certify and verify all need to be independent of each other. It is already accepted that there is likely to be a substantial role for private and quasi-government agencies: the Swiss *Société Générale de Surveillance* and the *International Standards Organization* are already involved in AIJ projects and are likely to be drawn into the processes surrounding CDM projects.

In this area particularly, those negotiating the CDM would do well to study the experience of other international financial mechanisms. These include not just the multilateral banks, but the EU's structural funds and Common Agricultural Policy. If farmers in southern Europe could for years successfully claim financial support for crops which in fact never existed – or claim for the same crops several times over – the scale of the challenge facing the CDM is all too apparent.

In addition, if bilateral and portfolio structures are both operated, obviously the certification process needs to be compatible with them both, and adaptation and administration charges must be levied equally on them both.

One final aspect of crediting concerns timing. As argued in the Chapter 6, there are powerful reasons why credits should only be certified *ex-ante*, along with the assessed flow of emission savings arising from actual verified project performance. Any other approach multiplies tenfold the scope for fraud and for liability disputes in the event of project failure. However, investors will want to be able to judge the availability of CERs in negotiating project contracts. While the investors themselves – and not the CDM – should bear the risk of project non-performance, and of price risk concerning the value of CERs, it is not reasonable for them to be expected to second-guess the certification process itself. Consequently, there is likely to be a need for a pre-crediting approval process so that developers can proceed with some degree of assurance that their project will be deemed eligible for crediting when it is in operation.

7.5 D: distribution of activities under the CDM

African countries in particular have expressed fears that the CDM could reflect and even accentuate, rather than alleviate, the huge existing disparities in international investment flows *among* the developing countries. Just twelve developing countries have absorbed more than 80% of foreign private investment to the developing world since 1990, and the flows are almost entirely focused upon some 25–30 out of more than 120 developing countries. Panayotou notes that 'the CDM introduces yet another dimension of equity: equity between emerging economies and less developed countries ... because emerging economies have the better infrastructure, lower risk and largest greenhouse gas-saving potential, there is a major risk that the CDM could generate a financial flow towards those countries that are already receiving the bulk of private capital flows from industrialised countries.'[9]

But the CDM states as its guiding purpose 'sustainable development' and that by general acceptance includes a measure of equity. Sokona most eloquently voiced the concern of Africa in observing the pilot phase of AIJ:

> One mechanism designed to foster north–south cooperation, Actions Implemented Jointly, more or less passed Africa by – only one out of 75 ASIJ pilot projects current reported to the UNFCCC Secretariat is being implemented in Africa. Remarkably, an entire continent in the developing world has been effectively excluded from a process intended to strengthen the relationship between North and South on the basis of mutual interest.[10]

Spurred by such concerns, since Kyoto Africa has emerged as an increasingly active and coherent participant in the international negotiations. In the context of the CDM the Africa group has proposed that each group or region within the G77 be allocated a certain percentage of activities under the CDM. Like most quota systems, this may be quite problematic to implement as well as to negotiate, particularly if most activity under the CDM occurs on a relatively lightly regulated bilateral basis. However, if it does prove impractical, there could still be guidelines to the board or selection bodies to give preference to projects that meet sustainable development needs globally, taking account of geographical equity.

[9] T. Panayotou, 'Six Questions of Design and Governance', in Goldemberg, *Issues and Options*.
[10] Y. Sokona, 'What Prospects for Africa?', in Goldemberg, *Issues and Options*.

7.6 E: eligibility criteria

The Kyoto Protocol states that the co-purpose of the CDM shall be to 'assist' developing countries in achieving sustainable development and in contributing to the ultimate objective of the Convention. In addition to the requirements for additionality and voluntary participation, the other specific project criterion is for 'real, measurable and long-term benefits related to the mitigation of climate change'. All these considerations combine to form the criteria for project eligibility under the CDM. The fact that they go well beyond the criterion of least-cost abatement in the first commitment period is the other Achilles' heel of the conventional OECD view of the CDM as a tool for minimizing global abatement costs. This could be a serious cause of disputes.

Even apart from the questions of distributive equity noted above, the scope for dispute over what constitutes a desirable project that contributes to sustainable development can be judged by considering specific cases. Large dams are among the most controversial projects in the developing world; should they qualify? What would happen if China sought credits for extending its nuclear programme? What if a government wanted to promote a forestry project that was opposed by the people on the land in question, particularly if there were land disputes and perhaps even a political secession movement in the area in question? Even without nuclear power projects, World Bank lending offers a considerable history of such contentious projects. With the lure of credits, rows over CDM eligibility could be even worse. The fundamental fact is that there cannot be consensus because such disputes reflect clashes of values as well as interests.

Dessus focuses upon another aspect of eligibility, namely the need for *long-term* benefits.[11] Noting that CDM project emission savings on their own are unlikely to match the projected growth of developing-country emissions, he argues that criteria must take account of project replicability and contribution to capacity building and technology transfer; thus they must encompass, for example, infrastructure developments that would be hard to assess or justify under straightforward cost-benefit evaluations.

---

[11] B. Dessus, 'Equity, Sustainability and Solidarity Concerns', in Goldemberg, *Issues and Options*.

The debate over whether forestry should be included in the CDM is, of course, another facet of this argument, as well as of arguments over additionality and local sustainability.

There are two main approaches to this complex of issues. One would be to define lists of qualifying *project types*, which could be positive (assumed to qualify) or negative (automatically excluded). For example, it is widely accepted that certain kinds of renewable energy projects should qualify; and it is widely (though far from universally) argued that nuclear power should not be accepted as consistent with the 'sustainable development' objective. The omission of any specific reference to sinks under the CDM is sometimes used to support arguments that most sinks by definition would not meet the criteria of 'real, measurable and long-term benefits related to mitigation of climate change', by virtue both of the presumed huge uncertainties and the fact that most sinks would eventually decay and return their carbon to the atmosphere. But there are innumerable grey areas, in part because what is 'sustainable' may vary according to both perspective and context.

The other approach is to rely on *procedures* for determining eligibility on a case-by-case basis, set against much more specific criteria than those in the Protocol. This could be more comprehensive, but also risks being much slower and more bureaucratic. A two-stage approval procedure seems inevitable: one stage at the national level, based on what the host government regards as desirable; the second at the international level, reflecting the international community's interpretation of how the agreed criteria should be applied.

Overall, it seems hard to say which of the two approaches is preferable. Dessus[12] argues for an approach with an initial list of qualifying projects to kick-start the CDM; the list will then expand over time. This is probably the only way of expediting action, but it may be extremely difficult to negotiate because every country will have its own priority projects it would like to put forward for prompt crediting, and will be reluctant to have them set aside even on a temporary basis. The danger is that a prompt start with minimal controls might be agreed for a limited list,

---

[12] Ibid.

which will then expand under political pressures, leading to a system that is weak in both criteria *and* procedures.

The best way of avoiding this danger may be a hybrid system. Specifically, timing is very important in meeting many project tenders. Certain kinds of projects that clearly accord with many dimensions of sustainability, and that clearly would not be implemented in the absence of crediting, should be able to qualify much faster than those whose impacts are controversial or whose additionality is highly uncertain. It is important that the CDM establish a fast-track process for such projects, perhaps even with automatic crediting in certain cases. Other more controversial kinds of projects can hardly avoid a more extensive and complex process of evaluation and approval, or rejection.

Overall, it is plain that the eligibility criteria must and will be broader than just least-cost mitigation. This will both introduce further complexities, and further distance the CDM from the notion of just offering least-cost abatement in the first period.

7.7 F: financing sources

The sources of finance that the CDM taps will do much to define its nature. As noted, it explicitly allows for the participation of 'private and/or public entities'. One of the longest-standing concerns of developing countries in relation to financing of global environmental efforts has been to ensure 'financial additionality': that financing for global environmental objectives is genuinely additional to finance that would otherwise be obtained. They are extremely anxious to ensure that investments gaining credits under the CDM do indeed come from new and additional financial resources. Mere redirection of existing development aid towards abatement projects, they fear, could not only distort foreign aid, but also exacerbate inequities within the developing world and reduce abatement action within industrialized countries. As well as sharing these concerns, the EU fears that the CDM could become a direct competitor to the Global Environment Facility that it has done so much to create and build up. In mid-1998, US AID representatives in the OECD's Development Assistance Committee said that all US foreign aid programmes should be assessed for their impact on greenhouse gas emissions, sparking European fears not only that the United States would seek credits for its existing foreign aid programmes, but that the CDM would start to distort the fundamental purposes of foreign aid if donor attention is turned towards credit generation.

As long as direct government overseas development assistance can qualify for CERs there seems absolutely no way of ensuring financial additionality: it would indeed seem impossible to prevent foreign aid programmes becoming distorted. This is fundamentally an ethical issue. Governments have different criteria for foreign aid, but officially at least the principal aim is usually to help alleviate huge international inequities and to assist the poorest countries to develop. Notwithstanding the prevalence of 'tied aid', giving preferential treatment to industries from the donating country, and aid pursued for strategic reasons, it is *not* the defining purpose of ODA to help donor countries economically by alleviating their international environmental obligations – which is what crediting would do.

This suggests that government development assistance should generally not be eligible for crediting. Of course, publicly owned industrial corporations could not plausibly be excluded from gaining credits in the same way as private companies. Also, some government agencies themselves may be involved in CDM projects in some capacity. But they should not be the entities gaining credits, and in most cases it should not be hard to differentiate this from direct government development assistance.

The crediting mechanism could also help the CDM to encourage early action within industrialized countries. The driving incentive for private investment under the CDM would be to offset the pressure of domestic emissions-reduction legislation. Without such legislation, there is little incentive for private companies to engage in the CDM. Thus, ensuring that only companies can gain credits under the CDM might also help to accelerate domestic legislation on emissions control – without which the credits would have no value.

Excluding government agencies from acquiring CERs would, however, be controversial, and might be opposed particularly strongly by Japan, which faces some of the most severe constraints and has the world's largest public overseas aid and investment programmes. Some way will have to be found to accommodate these specific concerns. But the advantages seem far too great to lose. Restricting crediting to corporations would furthermore clarify the complementary economic roles of the GEF and the CDM. The GEF would remain the premier channel for OECD government finance to help developing countries adopt more sustainable practices, including the capacity and infrastructure to attract cleaner foreign investment, and to sponsor long-term developments (such as innovative technology) that would not attract private money. The CDM would in turn become the vehicle for the greening of foreign private investment that is likely to be so essential to achieving globally sustainable development in the twenty-first century.

7.8 G: governing the CDM

Finally, a creation as novel and important as the CDM inevitably raises profound questions of governance. The prime institution for this is intended to be the *executive board*. The composition of the executive board will define the balance of interests and expertise in the CDM, assuming it can be agreed at all. Yamin[13] notes that no limited-participation body has yet been set up under the UN FCCC, and both the proposed technology assessment panels, and the multilateral consultative process established under Article 13 of the Convention fell at this hurdle. The CDM may be an exception because all parties need it, but it will not be easy to achieve. Administratively, it would be desirable to keep the body small: Meira-Filho, who did more than anyone to conceive the CDM, suggests a board of nine representatives of Parties. The politics of participation are likely to dictate otherwise, as with the Global Environmental Facility which ended up with a 32-member Assembly.

This book cannot attempt to guess how the balance will be established between different regions, particularly G77 vs. OECD, on which most of the debate is likely to focus. It is more useful here to highlight the desirability of participation beyond these two groups. It is important that there should be active representation of the EITs, although they may not

---

[13] Yamin, 'Operational and Institutional Challenges'.

participate either as investors or recipients. This is because, as discussed in Chapter 6, there will be important issues concerning the relationship between the CDM and JI. It is also because these parties are the *only* ones with a direct economic stake in ensuring that the CDM has stringent rules on issues like additionality. As noted, one core problem is that it is in the economic interests of both investor and host countries to have lax criteria governing the generation of CERs. The only losers in the UN FCCC process would be the eastern European recipients of JI or trading resources, so they should be represented.

There is also a strong case for extending representation on the board beyond governments. It is not defined as a 'board of Parties', partly because the GEF made a strong case that it should not necessarily be excluded from representation on the CDM. Given that the CDM is likely also to be concerned primarily with private investment, there would also be a strong case for including an (elected) corporate representative familiar with the process of project tendering and financing. In this case, there could also be good grounds for including an (elected) representative of the environmental NGO community.

Such composition of the board could help to dilute the North–South politicization of the CDM's governance that otherwise seems inevitable. It would indeed make the CDM a manifestation of the emerging world structures that academics have highlighted during the 1990s in terms of the growing role of various transnational, non-governmental actors in the global economy. In turn, representation on the CDM would challenge these groups to establish more accountable procedures for choosing their representatives. Like the CDM itself, such an executive board would be a challenging move forward in international affairs, but one which may be essential if the CDM really is to be both effective and legitimate.

7.9 Conclusions

The CDM is a leap into *terra incognita* that is unlikely to work simply as a way to distribute abatement efforts globally at least cost. The difficulty of ensuring that emission savings are additional is compounded by the paradox that the most 'cost-effective' projects will be the least 'additional' and that strict project additionality would give perverse policy incentives; also, eligibility criteria are likely to introduce a number of factors other than just greenhouse gas limitation.

There are indeed powerful tensions over the basic objectives. The JUSSCANNZ countries may want to minimize costs globally, but the developing countries want to maximize resource and technology flows and to minimize interference with foreign aid, while eastern Europe will have an interest in minimizing use of the CDM overall. It is entirely appropriate that the developing countries, individually and collectively, should have a defining say in what could emerge as an instrument affecting the direction of development. The history of foreign investment and technology transfer has anyway repeatedly shown that investor-driven projects frequently fail if extensive local acceptance and participation are not achieved.

Design of the CDM faces dilemmas of integrity and size. Of course it is desirable to have a system that is rigorous, transparent, free of corruption, cost-efficient and quick in response, and which ensures an extensive flow of projects that generate additional emission savings and contribute to sustainable development. The fact that many of these objectives are not consonant – and some are in direct conflict with each other – is the reality over which negotiations will have to struggle. If the rules on additionality, contributions to sustainable development, etc., are rigorous, then govern-ance could be very complex and administration costs high, deterring investment. Weak rules and governance may attract more investment at the risk of corrupting and discrediting the system, as well as undermining the objectives of the Convention.

At best, the CDM could lever large flows of foreign investment towards sustainable and low-greenhouse-gas development. At worst, it could become not just a source of spurious emission credits, but a sink for the intellectual as well as some of the physical resources of the developing world, and a distraction from the fundamental goals of sustainable develop-ment. In either case, however, the CDM is likely to be constrained by the inherent limitations of credit-based, project-based systems. In terms purely of greenhouse gas control, crediting against an unknown baseline and meeting various other criteria will inevitably make the CDM somewhat cumbersome compared with a system based on allowance trading. As with emission crediting systems in the United States, it may be part of the pathway towards a broader emissions trading system.

Yet this could be in the distant future for many developing countries, and the CDM may perform a more fundamental and long-standing purpose. Ultimately, this chapter has argued that the defining purpose of the CDM should be to help direct foreign corporate investment towards goals of sustainable development, in its many forms and interpretations according in part to national preference. In this it should be complementary to the Global Environment Facility. Indeed, if appropriately designed, these institutions should correspond to the natural complementarity between the private and public sectors. State funding ('the GEF') is required for education, infrastructure and public-sector projects and to stimulate technical and institutional innovation ('capacity building'). Corporate finance, however, will provide the bulk of investment and innovation towards the collective societal goals – given appropriate context and incentives ('the CDM'). Such a structure is a natural concomitant to the globalization of the world economy at the end of the second millennium once recognition of a collective planetary constraint forces recognition of the need to cooperate in ensuring globally sustainable development.

Chapter 8

Prospects for the Kyoto regime

The Kyoto Protocol, together with the Climate Change Convention on which it rests, establishes the basic structure of commitments and mechanisms agreed by national representatives for addressing climate change. It represents the culmination of nine years of learning and institution building, domestically and internationally, following the establishment of the IPCC in 1988. As indicated by the analysis in the previous chapters, it still leaves many open questions and much to be resolved. This final chapter discusses the prospects for the Kyoto regime: the time scales established, the prospects for ratification and entry into force, and some of the bigger outstanding questions about the long-term evolution of the regime.

8.1 The Buenos Aires 'Plan of Action'

The first major political milestone after Kyoto was the next annual Conference of Parties to the Climate Change Convention, COP-4, held in Buenos Aires in November 1998. In many respects COP-4 was an epilogue to Kyoto, confirming the basic agreement and defining the follow-up process. It was widely described as 'dealing with Kyoto's unfinished business'.

Most countries needed several months even to start understanding what had been agreed at Kyoto, and the first meeting of the Subsidiary Bodies, held in June 1998 to start preparing for COP-4, was in part an opportunity to vent feelings and frustrations about the outcome, with some repetition of pre-Kyoto positions: opposition to trading by many developing countries, set against US efforts to reopen the question of developing-country commitments. There was some technical clarification on sinks, some acceptance that COP-4 would have to develop a work programme, and a lot of questions with few clear answers.

In some ways the most significant developments occurred before COP-4 even convened, as it became clear that all negotiators accepted the Protocol (together with other issues under the Convention) as the basis upon which they would work henceforth, and that the key agenda of COP-4 would be to set out a credible work programme. The specific result of the Conference was the Buenos Aires 'Plan of Action', which consisted of decisions in six specific areas:

- financial mechanism;
- development and transfer of technologies;
- addressing the specific needs and concerns of developing countries, including minimization of adverse impacts (Articles 4.8 and 4.9);
- activities implemented jointly under the pilot phase;
- the mechanisms of the Kyoto Protocol;
- preparations for the first session of the Conference of the Parties serving as the meeting of the Parties to the Kyoto Protocol.[1]

The decisions on the *financial mechanism* confirmed the Global Environment Facility as an 'entity entrusted with operation of the financial mechanism of the Convention' and broadened its scope. The shackles of the Convention's insistence on the concept of 'incremental costs' – a concept as inoperable as perfect additionality in the CDM – were relaxed, with reference to the need for 'flexibility to respond to changing circumstances'. The GEF's funding scope was formally expanded to include more wide-ranging support for building up the capacity of developing countries to address climate change issues, including full funding of their Second National Communications. The decision also established guidelines for the review of the financial mechanism every four years. Potentially all this paves the way for the GEF to develop a complementary role to the CDM, as suggested in Chapter 7.

The decision on the *development and transfer of technologies* also marked an important step forward in this long-running and tortuous

---

[1] Specifically: Decisions 2/CP.4 and 3/CP.4 on the financial mechanism, with others numbered sequentially in the order given here. In addition, a decision on sinks set out a work programme for decisions based on the IPCC Special Report on the topic.

debate. In place of the largely ineffective exhortations for developed countries to take all practicable steps to transfer technologies on special terms (e.g. see Chapter 2), that had marked previous exchanges, there was a much more focused agreement that addressed the roles of all Parties. It called for Annex II Parties (the OECD) to provide lists of environmentally sound technologies that were publicly owned, and for developing countries to submit prioritized technology needs, especially related to key technologies for addressing climate change. All Parties were urged to create an enabling environment to stimulate private-sector investment, and to identify projects and programmes on cooperative approaches to technology transfer. Most important, the agreement called for a consultative process to be established to consider a list of 19 specific issues and associated questions, set out in an Annex. It was hardly a radical breakthrough, but it embodied a considerable change of attitude in which the previous impasse melted away in favour of an approach which recognized that there were real problems that required real creative solutions in order to achieve global dissemination of environmentally sustainable technologies.

The decision on *Articles 4.8 and 4.9* of the Convention addressed the requirement to 'meet the specific needs and concerns of developing country parties' as set out in the Convention, and the corresponding requirement to minimize adverse impacts on vulnerable groups, including the impact of the response measures in the Protocol, as required under Articles 2.3 and 3.14 of the Kyoto Protocol. Building on Kyoto's implementing decision, it called for analysis of potential adverse impacts as input to workshops convened by the Secretariat, including specific proposals that could form the basis for specific decisions at the sixth Conference of Parties, COP-6.

The decision on *activities implemented jointly* agreed to continue and extend the AIJ pilot phase, and to continue capacity building, especially in the developing countries that lacked experience so far (notably in Africa). The review process called for in the Berlin decision will start at the tenth Subsidiary Bodies meeting in June 1999, and will be finished at or before COP-6. Informal consultation resulted in a text proposal for crediting of AIJ projects, but the G77 and China rejected this.

The decision on the Kyoto Protocol *mechanisms* established a work programme, undertaken 'with priority given to the CDM, to take a final decision on the mechanism (JI, the CDM and emissions trading) at COP-6'. This short decision invited the Parties to submit proposals for compilation as a miscellaneous document for input to technical workshops, with the Subsidiary Bodies charged subsequently to provide a synthesis of proposals by Parties. The Kyoto Protocol itself, of course, started life as a synthesis of proposals: COP-4 thus explicitly laid the foundations for a new negotiated agreement on implementing the mechanisms, to be agreed by COP-6. Attached to the decision was an inclusive list of possible topics for the work programme which illustrated just how much is outstanding: it lists 22 general topics, 50 on the CDM, 38 on JI and 32 on emissions trading, an astonishing total of 142 points that Parties wished to register as being of concern to them.

Finally, COP-4 initiated preparations for the first Conference of the Parties to serve as the Meeting of the Parties to the Kyoto Protocol at the first session after entry into force of the Kyoto Protocol. The work programme included a number of key issues including those relating to reporting, review of reports, and compliance procedures. These are likely to emerge as very important processes in implementing the regime.

Two items were notable for their absence from the Buenos Aires decisions. The Conference was mandated to conduct the Second Review of Adequacy of commitments under the Convention, but agreement on this proved impossible. All Parties agreed that the Convention's commitments were inadequate, as they had done in Berlin, but they could not agree why. The developing countries saw the issue as a way of increasing pressure on industrialized countries to control their emissions, but the United States in particular pursued the argument that the commitments were inadequate because developing countries did not have quantified obligations. The issue is likely to be considered again at COP-5.

The other 'missing' item – voluntary commitments by developing countries – gained by far the most publicity. Kazakhstan announced that it intended to apply to join Annex I and to take on commitments under the Protocol. Attempts by Argentina, the host country, to place the issue of voluntary commitments on the agenda were roundly rejected by the G77, most of which declined even to join in informal talks on the topic. Later in

the conference, when President Menem announced that Argentina intended to take on a binding commitment, it became clear why the country
was so interested in the question. Though no progress was made on the
issue at COP-4, the topic will almost certainly be forced onto the agenda
of COP-5, when both Kazakhstan and Argentina are likely to offer specific
commitments. The international community will then have to start working out whether and how to accept or negotiate what is entailed – as
discussed later in this chapter.

Thus Buenos Aires succeeded in 'maintaining the momentum' of
Kyoto, as many diplomats described it. In practice it did more than that: it
added considerable clarity about the time scales and politics of the
process, if little about the substance. Many important decisions are now
expected at the sixth Conference of Parties, due around the end of the year
2000. COP-6 could indeed rival Kyoto itself in the scale of the issues
scheduled to be resolved.

In terms of the politics, this was the first conference that was not
dominated by disputes within the OECD, though partly because the
United States and EU agreed to defer their differences on the trading
mechanisms. On the contrary, it was a conference at which some big
divisions among the developing countries were central at the outset,
emerging on the first day with the dispute over Argentina's efforts to put
voluntary commitments on the agenda; yet the conference ended as a
testament to their new-found strength. The developing countries achieved
long-sought advances on technology transfer; nothing was excluded from
the future debate on mechanisms for international transfer; they got almost
all they sought on the financial mechanism, including full funding for
their Second National Communications; and the energy exporters opened
a significant channel to further their concerns. This was achieved without
letting the OECD slip voluntary commitments back onto the formal
negotiating agenda through any back doors.

With hindsight the reasons are relatively simple. In climate change,
unlike most global negotiations, the developed countries are increasingly
asking the developing countries for action, rather than the other way
around. Several key developing countries and groups, including India,
Brazil and a newly coherent Africa group, had clarified their objectives

and concerns. Both the EU and the United States sought to engage the developing countries from the beginning, rather than expending most of their energy on trying to reach an EU–US compromise (though the EU continued to expend much of its energy negotiating within itself). The *de facto* defection of Argentina furthermore gave the G77 a sense that it had lost something – and the OECD a sense that it had gained something – before formal negotiations even started, so that the substance of the outcome was in some ways bound to offset this. The only major silent voice in Buenos Aires was eastern Europe. But this silence is unlikely to persist as the debate on implementing the transfer mechanisms moves forward, and competition between the different mechanisms becomes clearer.

All this suggests a more balanced and genuinely global nature to the various negotiations that have been effectively launched by the Buenos Aires Action Plan. It also means that they could become even more complex and contentious than Kyoto itself. Yet all these negotiations will be conducted in the shadow of one over-arching issue: whether and when the Kyoto Protocol will gain sufficient ratifications to enter into force, including ratification by the country that everyone knows just has to be 'on board', the United States.

8.2 Prospects for ratification and entry into force

Like all international agreements, the Kyoto Protocol states minimum conditions for ratification that must be met for the Protocol to come formally into legal operation ('entry into force'). For this, it must be ratified by at least 55 Parties, 'incorporating Annex I Parties which accounted in total for at least 55 per cent of the total CO_2 emissions for 1990' as reported by those Annex I Parties that had submitted their first communications by the time of Kyoto.[2] Table 1, appended to the Protocol, in Appendix 1, shows these data.

Ratification by 55 Parties is not a high hurdle. Failure on this criterion would require more than 120 Parties to fail to ratify – virtually all the G77

---

[2] Ukraine was the only significant Annex I country not to have submitted its first communication by the time of Kyoto; its emissions are thus not relevant to the 'entry into force' calculation.

group, or the vast majority of them plus the EU. This is hardly plausible and would anyway make the treaty largely irrelevant. The minimum fraction of CO_2 emissions is the real hurdle, though it is considerably lower than the 65 or 70% that Canada and Japan (and to some extent the United States and other JUSSCANNZ countries) sought. A combination of the United States with Russia and almost any other significant Annex I Party would prevent entry into force. So, though presumably of less relevance, could the EU in combination with the EU accession countries, Russia and Japan. Basically the clause means that although in practice the regime will require some common acceptance by the major emitters – the United States, the EU and Russia – it could come into force with only two of these three plus reasonable support from other Annex I Parties. This could be relevant particularly regarding the timing of entry into force, if ratification is blocked in any of these groups – perhaps by technicalities, or if the sentiment in the US becomes supportive enough to abide by the regime while the technical hurdle of a two-thirds Senate majority proves too high to reach.

In fact there are many countries for which ratification is not a foregone conclusion. Implementation debates are bound to be difficult in the Japanese Diet, and perhaps in the Russian Duma, for more complex political and procedural reasons. Some developing countries might also hesitate if they feared that somehow this would open them to being pressured subsequently into deleterious commitments (though it is hard to see why their ratification would make much difference to this).

Even in the EU there will be difficulties. Some powerful industrial voices in Germany have called for a renegotiation of the EU 'bubble' agreement under which Germany agreed a 21% reduction to allow for growth in poorer EU countries. If such voices were to dominate – perhaps with some of the German Greens deciding that a rapid nuclear phaseout was more important than climate change – it would destroy the basis of EU ratification, and thereby the Protocol. Some more serious voices in the EU have called for EU ratification to be made conditional upon US ratification. This would also wreck the Protocol, much as such conditionality undermined the European carbon tax proposals eight years earlier. It would link the incredibly complicated process of coordinating

15 national ratification procedures in the EU to the process of still more difficult and unpredictable debates in the US (in which EU hesitation would be portrayed as proof that the whole deal is unworkable).

Nevertheless, all this seems unlikely because both Japan and the EU (especially Germany) have too much invested in the Protocol, and have somewhat greater cohesion between administration and legislature than in the United States. This, emissions data and many other factors do serve to emphasize the centrality of US participation and ultimately ratification by the US legislature. Since the Protocol bears the imprint of US design so heavily, it is rather ironic that US ratification is seen as the acid test. That this is so is testament in part to the divide between the US administration and legislature, with the latter identified much more strongly with domestic preoccupations and Republican perspectives.

Even before the Buenos Aires conference, it was recognized that the US administration would not even submit the Protocol for ratification before the Presidential and Congressional elections in November 2000. Similarly, the Buenos Aires conference left almost all the key implementation decisions until COP-6 around the end of 2000.[3] COP-6 will determine the full package that will be put before the US, and probably other legislatures.

The biggest underlying problem is the inadequacy of US domestic action on its own emissions. It will be hard to make much headway internationally while emissions from the world's biggest economy, in absolute and (with some minor exceptions) per capita terms, are still growing. Successive administrations have made rather limited progress in curtailing America's appetite for fossil fuels, despite concerns that this dependency leaves the US economy more exposed and more vulnerable internationally. Of course, the United States is already legally committed to returning its greenhouse gas emissions to 1990 levels under the Convention, as signed and ratified by the Republican Bush administration. Opponents of the Kyoto regime have nevertheless opposed any measures that could be construed as consistent with implementing the Kyoto Protocol – so far, quite successfully. At this stage, the primary challenge does not concern additional developing countries' involvement – it is first to gain meaningful

---

[3] The exact timing of COP-6 in relation to the US election has yet to be determined at the time of going to press.

domestic participation in the United States and certain other major Annex I Parties in emissions limitation.

What then are the prospects for US ratification? There is little doubt that the Protocol would be rejected by the legislature of 1998–9. Equally, there is no doubt that the trend of US opinion, as it becomes better informed about climate change issues, is moving towards greater acceptance of the problem and greater appreciation of the US role in the global context. The United States will ultimately reach a point at which the Kyoto Protocol is recognized to be important and good for the country itself. But the longer effective action in the United States is deferred, the longer US emissions keep rising, the greater and more rapid will be the adjustment required. Given all the flexibilities, the United States could still comply with the Treaty if it started to take serious action shortly after 2000. If significant action were deferred beyond about 2004, the adjustment required would start to pose an insurmountable obstacle. US opponents of ratification do not need to win the argument; they simply need to delay losing it long enough to make the task of implementation too daunting. Ratification anyway is an extremely slow and ponderous procedure in the US legislature (and some others). Serious action towards ratification and implementation needs to be taken in the period between 2000 and 2004, even if formal ratification is not achieved until towards the end of this period (in which case, the Protocol could enter into force without the United States for some time).

This does not make the regime entirely contingent upon the year 2000 US Presidential election, as some have claimed. An incoming Republican president would soon be faced with the realities of the issue and of the international situation; if convinced of the need for ratification, a Republican president could be in a stronger position to push it through a reluctant Congress. As noted, the regime anyway could enter into force without formal US ratification if US sentiment was moving towards *de facto* compliance. But it does make the whole process heavily dependent upon the pace and evolution of the broader US debate on the Kyoto regime.

That debate will of course be influenced by developments in the scientific and economic analysis, but politically it is likely to be determined by two main factors: the attitudes of key non-governmental sectors, particularly (but not exclusively) US-based industry; and the evolution of the debate on developing-country commitments. This chapter now considers each in turn.

8.3 Business and public involvement in the Kyoto regime

In the early 1990s, much was written about the decline of the nation-state and the rise of non-governmental actors. To an extent, the Kyoto Protocol was an antidote to such theories: the agreement was struck in the face of strong opposition from powerful industries, particularly in the United States, and with a set of flexibilities that were opposed by almost all environmental NGOs. It was very much an agreement struck by governments, negotiating what they felt to be possible and appropriate. Its ratification and implementation will, however, undoubtedly depend heavily upon these other groups.

Developments in industry are among the most fascinating aspects of the climate change regime. In the early 1990s, the idea of significantly limiting greenhouse gas emissions would have been opposed by almost all the major industries on the planet: the fossil fuel industries, 'from production to the pump', along with most manufacturing, processing and automobile industries. The rest of the corporate world would at best have been indifferent, with the almost trivial exception of the nascent energy efficiency and non-fossil energy industries.

The corporate world has changed, and with good reason. Many insurance companies, particularly global reinsurance, have become worried by cumulative and correlated weather-related losses. Natural gas companies have recognized that climate change probably holds as much opportunity as threat, while electricity and auto companies may envisage only modest transitional problems, providing that action is well planned and not too hasty. The primary activities of oil and coal production, and their primary industrial customers, stand increasingly exposed in the corporate community.

Furthermore, strategies that at least cushion the apparent threat to some of these interests are readily apparent. British Petroleum, also with considerable natural gas interests, was the first 'oil major' to accept publicly that climate change was an issue of serious concern, and one that could be addressed. Its move, accepting these realities and starting to plan for them, acquired almost revered status in the climate change community. It embarked on a full–scale review of internal activities, and in autumn 1998 set a goal of reducing its own greenhouse gas emissions by 10% by 2005, to be implemented using an internal emissions trading programme that was already operating in pilot phase by then.[4] BP, which after its merger with Amoco is the largest oil company in the United States, also boosted its solar energy activities.

Royal Dutch/Shell moved similarly, if in a less publicized way. Shell had earlier in the decade published scenarios depicting a long-term transition to renewable energy with atmospheric stabilization well below 550 ppm. A company with long-standing positions on corporate ethics, it had already undergone a bruising experience over disposal of the Brent Spar buoy and in its Nigerian operations. In 1997 Shell restructured its global operations, established renewable energy as one of its five core businesses, and subsequently announced its own target: to reduce corporate emissions by 10% by 2002. Overall, in 1998 European business moved rapidly to accept the Kyoto Protocol, though in some cases with a certain scepticism as to whether governments really knew what they were taking on. No major European industry federation is formally opposed to the Kyoto Protocol.

Alongside these shifts in the attitudes of major corporations emerged a huge array of facilitating private-sector activities. Various companies have undertaken initiatives ranging from the purely voluntary to agreements negotiated with governments under the threat of more onerous regulation. These activities merit full-length study in their own right.[5]

One particular feature has been the emergence of 'emission offset' activities and services. Not waiting for formal intergovernmental legis-

---

[4] The system initially comprises ten business units around the world. The first trade was reported in November 1998, at a price of $17/tC: Rodney Chase, presentation at the Royal Institute of International Affairs, London, 27 November 1998.

[5] See for example T. ten Brink and M. Morere, *The Role of Voluntary Initiatives to Address Climate Change,* Report for ICC, WBCSD, USCID, Keidanren, WWF and UNEP, Oxford and Brussels: ECOTEC, 1998; and F. Szekely, *Voluntary Agreements,* Geneva: International Academy of Environment, 1998.

lation, some companies have sought to fund activities that 'offset' their emissions, or the emissions of new project activities, for example through forest protection or plantation programmes that absorb an equivalent amount of CO_2. Consultancies have emerged in the United States and elsewhere specializing in advising and organizing offset portfolios. In the UK, another entrepreneurial activity, the Carbon Trust, has sought not merely to offer offset projects but to develop these into the possibility of offering 'zero carbon' labels to industrial products whose CO_2 emissions have been offset in this way, as a tool for marketing advantage. In late 1998, the oil and gas company Amerada Hess started marketing 'zero carbon' gas in Britain's liberalized gas markets on this basis, and the Carbon Trust's director has espoused the clear aim of working with the oil companies to install 'zero carbon' petrol at Britain's filling stations.

The potential of such initiatives, if properly governed and accredited, is enormous. Furthermore, they serve to increase steadily the coalition of industries for which the Kyoto Protocol is perceived not primarily as a threat but also as an opportunity. These initiatives will acquire greater value if the value of restraining net CO_2 emissions, or of offsetting them with other activities, is recognized in legislation. In the United States, a draft Senate Bill to ensure that appropriate early action by companies is credited under any subsequent legislation has already been prepared, though this is probably but the first step in a long process.[6]

These various trends have already enabled many European industries to unite behind acceptance of the Kyoto Protocol, and increasing numbers of US companies are beginning to raise their head above the parapet and back the Protocol.[7] These trends may be reinforced in some cases by the growing US–European corporate mergers, not just BP-Amoco but also Daimler-Chrysler, and rapid expansion of cross-ownership in the electricity business.

---

[6] For analysis see R. Nordhaus, S. Fotis and V. N. Feldman, *Early Action and Global Climate Change*, Washington: Pew Center on Global Climate Change, 1998 (see note 7).

[7] Many companies already backed the International Climate Change Partnership (ICCP) and in 1988 about 20 companies backed the formation of the Pew Center, which exists to further action on climate change in the United States. BP, General Motors and Monsanto joined with the World Resource Institute for another initiative on 'Safe Climate, Sound Business' (WRI, Washington, 1998).

Another feature of the landscape of 'civil society' involvement with climate change is the continuing role of environmental NGOs, including an increasing number of alliances between NGOs and the more progressive companies.[8] In general, however (and with notable exceptions, as at Berlin), NGOs have not found it very easy to engage effectively with the climate change issue as it has become steadily more complex. Their attitude towards the Kyoto Protocol has been mixed, particularly in relation to the international transfer mechanisms. Their efforts to support the Berlin Mandate's focus on industrialized-country commitments have sometimes placed them in an awkward position regarding calls for stronger action by developing countries; more creative proposals on global equitable commitments have come mostly from outside the mainstream NGOs.

They have generally been most successful at raising or maintaining public concern, working intimately with the media. At one level – a very important level – this has been quite successful. Opinion polls, including those in the United States, show a generally continuing rise in concern about environmental issues including climate change. But the media have not found it easy to grapple with the complexities of the climate change negotiations. Environment correspondents tend to be scientists, and anyway work under the constraint of editors who expect to see simple scare stories on the inside pages, rather than the more complex treatment that might be accorded to a major economic agreement. Thus, there are real challenges to be faced in getting civil society effectively behind the Kyoto Protocol and its implementation.

Away from the international arena, the other big contribution of NGOs has been to foster action on emissions by governments and the range of civil society. Many US NGOs, for example, are active at the state and city level in support of legislation on energy efficiency and renewable energy, and some build bridges with relevant industries. NGOs in developing countries, while generally supporting their governments' line in the negotiations, lobby hard for more action on sustainable development at home. Such action obviously is critical to the whole regime.

---

[8] Ibid. Also, the Environmental Defence Fund, well known for its enthusiastic support for emissions trading and JI, is heavily involved in advising a number of companies for example about emissions trading systems.

There are, however, two problems facing this optimistic view that industry and wider 'civil society' will simply line up to support the Protocol. One is that the common theme behind which diverse industries have managed to unite is that of maximizing flexibility. As demonstrated earlier in this book, flexibility is a double-edged ˙sword: essential, but potentially problematic if taken to excess. The incentive for industry is that flexibility lowers costs, but if there is so much flexibility that core activities can continue almost unaffected, then it may simply act to defer the changes that are necessary for long-term resolution of the climate problem. A Kyoto Protocol ratified with so much flexibility that the core trends causing climate change continue almost unabated could be a Pyrrhic victory. The counter-argument is that the essential point is to get the regime in place; this will send the definitive signal that the world will tackle climate change, and weak commitments can be strengthened over time. There are, however, weaknesses in this argument, particularly given the long time scale already established by the first commitment period. The relevant question is not just whether the Protocol is ratified but in what form.

Second, although the stance of some industries in the United States is softening, overall attitudes there remain far more reticent. Major and powerful groups, both companies and employees, still fundamentally oppose the Protocol. Indeed whereas the unions in Europe have largely accepted that tackling climate change will generate at least as many jobs as it costs, the US Mineworkers' Union has provided leadership for widespread union opposition that is particularly powerful within the Democratic Party. Within the corporate world the alliances are also powerful and influential through systems of Congressional lobbying. Exxon in particular, together with the coal companies, has been at the forefront of US industrial opposition to action on climate change, and Mobil also undertook an aggressive advertising campaign against Kyoto. The proposed Exxon-Mobil merger would create the world's largest company, as an implacable opponent of US ratification. So much is perceived to be at stake that it could create a battle reminiscent of that between Rockefeller's Standard Oil and the US government almost a century earlier. That titanic struggle ended with the break-up of Rockefeller's empire. Exxon-Mobil and the coal companies would lose a similar war in the end; but they and their allies would only need to win the first few battles to do a lot of damage to the Kyoto regime. Their central weapon would be the reluctance of developing countries to take on more specific commitments. To this crucial topic we now turn.

8.4 First-period developing-country commitments: accession, expansion and inflation?

As the US domestic debate moves beyond denial, the central plank of the corporate opposition to the Kyoto regime, and a much wider feature of the debate particularly (but not only) in the United States, focuses upon the need for a global response. This is contrasted with the focus of the Kyoto Protocol upon industrialized-country commitments. As sketched at the end of Chapter 3, this issue has a long and difficult political history, and represents the most complex and divisive issue in the entire climate change regime.

The substantive source of the division, as noted in Chapter 3, Section 7, is easy to identify. From one standpoint (accepted in the Convention), it is quite inequitable and simply unrealistic to expect developing countries to take on specific commitments to solve a problem they have played hardly any part in creating to date, and before the industrialized countries have begun significantly to put their own house in order. From the other standpoint, the problem cannot be solved without their involvement, and indeed growth in the emissions from the biggest developing countries over coming decades could ultimately swamp any restraint in the OECD.

The intrinsic difficulty of this debate has been exacerbated by the deeply embedded institutional divisions between the OECD and the G77 and China group in the United Nations system, the excruciating internal politics of the G77, and the persistent political insensitivity of the OECD in relation to the G77, sometimes amounting to ineptitude. The repeated attempts to force the Kyoto negotiations beyond the agreed terms of the Berlin Mandate, and particularly the US proposal on universal evolution of commitments, only served to inflame resentments in the G77. The fall of the proposed article on voluntary accession – partly a consequence of this – makes the situation even more difficult. The issue has become so politicized and so institutionally messy that the way forward is now far from clear.

The deletion of the proposed article on voluntary commitments leaves as problematic the institutional process for any new country to adopt binding commitments under the Protocol, as explained in Chapter 4. Argentinian attempts to put this issue onto the agenda of COP-4 were rejected forcefully by the G77, and although the conference president stated defiantly that she would pursue the issue informally with a view to reporting back at the close of the conference, any such discussions made no progress worth reporting.

COP-4 did, however, inject a new dimension into the debate, with the announcements by Argentina and Kazakhstan that they wished to adopt commitments the following year, and that the latter country wished to join Annex I.

The incentive for Kazakhstan, if it joins on anything like the same basis as Russia and Ukraine, is obvious: its emissions by 1996 had declined to almost 50% below 1990 levels, so if it could take a commitment anywhere near 1990 levels and join the trading system, it could have a considerable surplus to sell. Argentina might need a significant 'growth target' to reap any such benefits, and would need to join the trading system, though its ability to fulfil the obligations for Annex I, including the indicative aim of Article 4.2a and b of the Convention, is uncertain.

The concept of 'growth targets' for developing countries that also join the trading system has been widely proposed.[9] But a host of questions then arise. On what criteria (if any) should commitments for new countries be accepted? Can countries take on quantified commitments under the Kyoto Protocol without joining Annex I of the Convention? If so, should such countries be allowed to trade? If not, what incentive do they have to join? If they are allowed to trade, should they then be excluded from being

---

[9] N. Helm, *Growth Baselines: Reducing Emissions and Increasing Investment in Developing Countries*, Washington: Center for Clean Air Policy, January 1998. A later publication from the World Resource Institute highlights the potential drawbacks, and argues that any such targets might best be defined in terms of improving aggregate emissions intensity (emissions per unit GDP) over time (K. Baumert, R. Bandari and N. Kete, *What Might a Developing Country Commitment Look Like?*, Washington: World Resources Institute, forthcoming June 1999).

recipients of the CDM, and would it be politically possible to define commitments that did not involve countries acquiring and then trading a large surplus?

It is certain that COP-5 will not be able to answer all these questions, but with two countries wishing to adopt specific commitments, it will be very hard for the developing countries collectively to keep these issues off the agenda. Starting to discuss these issues could be the most vital and most difficult task at COP-5.

The difficulties are quite formidable. One issue is that of 'inflation', if countries are allowed to take on commitments and join the trading system with a large surplus. In most OECD countries, the presumption is that such 'accession hot air' is the price for getting new countries into the system of specific commitments. But the inflation problem could be very serious. Our numerical studies presented in Chapter 5, extended to analyse the effect of just the accession of Kazakhstan, show that even this could have a substantial impact. Conceivably it could cause price collapse, if the collective commitments do prove to be quite weak, because in these circumstances prices are sensitive at the margin. Accession of larger countries could have considerably more impact.

This problem might be mitigated if an effective approach were developed towards tackling 'hot air', as discussed in Chapter 6, Section 4. It would then matter less if a country did take on a much inflated target, but it is a messy basis to give countries excess deliberately and then discount it. Another suggested approach is that developing countries might take on budgets against which they could sell, but they would not be held accountable if they overshot their assigned amount: they would have an incentive to limit emissions (so as to sell), but face no penalty if they did not.[10]

Another difficulty of these approaches is the further violation it would imply of the collective Article 3.1 commitment, for Annex I countries to reduce their emissions collectively to at least 5% below 1990 levels in the first commitment period.

The counter-argument again is that, even if countries do join with a surplus to trade, it is a 'necessary price' for establishing the regime and

---

[10] C. Philibert, *Four Proposals on Emissions Trading*, Paris: UN Environment Programme Energy and Industry Office, 1998 (in draft).

bringing new countries into it. The priority, according to this line of reasoning, is to gain entry into force with as many countries as possible in the framework, as early as possible, even if the price is a weak and inflated system. The system could be strengthened in subsequent periods.

However, it is far from clear that the central bloc of G77 countries would accede to first-period commitments on this ad-hoc basis. The participation of countries such as China and India is crucial. Voluntary accession as the main means of moving forward could in fact make it easier for them to stay outside than in more structured and collective negotiations, if they were unhappy with either the principles or terms of accession. As noted, the failure of the proposed article on voluntary accession anyway gives them a lot of power to block such accretion.

Furthermore country-by-country accession can be seen as a direct threat to the cohesion of the G77 group. If some countries can be tempted into the trading system, perhaps with surplus assigned amounts, the prospect for the G77 is one of slowly being nibbled away at the margins. The G77 countries know that ultimately they do need to be involved in more specific commitments if the climate problem is to be solved. Their real debate is when and how.

8.5 Longer-term approaches: an overview

What is lacking in all this, as highlighted by Jacoby et al., is a clear longer-term pathway either for domestic emissions control or for expansion of the commitments.[11]

In the climate change negotiations, developing countries have consistently refused to discuss specific global frameworks for quantified emission commitments. Their primary justification for this has been the moral and legal principle that action should be taken first by the developed world. In most international environmental treaties, developing-country commitments have followed in the wake of OECD action, as with phasing out CFCs under the Montreal Protocol (which deferred action by developing countries, and which major developing countries only joined in 1990,

---

[11] H. D. Jacoby, R.G. Prinn, and R. Schmalensee, 'Kyoto's Unfinished Business', *Foreign Affairs*, Vol. 77, No. 4, 1998.

after industrialized countries had begun implementing and strengthening their commitments). In the Climate Convention itself, indeed, developing countries are making progress on national reporting and accepting somewhat stronger guidelines, as OECD countries move forward in these areas. The failure of most OECD countries to get their emissions under control so far has given developing countries every cause to maintain their refusal even to discuss possible quantified commitments.

Whatever the economic arguments, these and other factors concerning the reality of developing-country politics make it unlikely and in fact unreasonable that many would join in commitments in the first period. The important question is whether and how they might be involved in subsequent periods. Mitchell notes that by the end of the first commitment period – assuming that Annex I Parties achieve their commitments without massive use of the CDM – then developing-country emissions could be roughly equal to those from Annex I. Those from Annex I would presumably be falling, those from developing countries rising fast. Progress even to a second commitment period under these conditions, Mitchell concludes, 'is difficult to imagine: the politics would be insufferable by the Annex I countries'.[12]

Three things could affect developing-country participation before negotiations on second-period commitments commence in 2005. First, more serious action is obviously needed in the developed countries to tackle their own emissions and fulfil their existing obligations under the Convention. It would help if the EU does achieve the Convention's goal, as it seems likely to; this can marginally but validly be attributed to the EU's internal efforts on climate change (see Chapter 5). Technically indeed this would mean that most Annex I Parties would have achieved the indicative aim in the Convention's Article 4.2 commitments.[13] Progress

---

[12] J. Mitchell, 'Post-Kyoto Implications for Energy Markets: the Geopolitical Perspective', in *Energy Markets, What's New? Proceedings of the 4th European Conference of the IAEE and the GEE,* Berlin: Gesellschaft für Energiewissenschaft und Politik, 1998.

[13] It would mean that the 15 countries of the EU, acting collectively, would have emissions in the year 2000 below 1990 levels, the 'strong' interpretation of the Convention's Article 4.2 commitments. In this they would be joined by the eleven Economies in Transition. Switzerland may also achieve the goal, leaving 'only' the six other JUSSCANNZ countries and Iceland missing the goal.

towards at least halting the growth of emissions in the United States in addition could greatly alter the tone of international discussions.

Second – and aided by the example of such action – the developing countries themselves may become less fearful of binding commitments, and indeed more positive towards the notion of limiting greenhouse gas emissions in the context of sustainable development. For example, studies conducted in the context of the IPCC's forthcoming Special Report on Emission Scenarios show that if projected coal consumption is actually realised, much of east Asia will be subject to intolerable levels of acidification, with severe consequences for agriculture and other activities. Finding efficient ways to limit coal consumption is likely to be intrinsic to the sustainable development of China and some other parts of east Asia, and commitments under the climate change regime could increasingly be seen in this context.

Finally, entry into force of the Kyoto Protocol, with its specific binding obligations including substantial assigned amount reductions by key Annex I Parties, could be plausibly considered as establishing commitments strong enough to legitimize opening discussions on developing-country commitments for subsequent periods. But what might emerge?

There is one possibility that might fit plausibly with the current structure of the Protocol, even concerning the first commitment period. It could occur if the CDM develops as a way to extend developing-country involvement in a more piecemeal way, and perhaps in ways that also link with the growing role of multinational corporations. If companies and countries become frustrated with all the requirements of demonstrating additionality and other eligibility criteria for the CDM, they may propose that their company, industry or sector adopts legally binding caps on its emissions in a given developing country or countries as a basis against which to trade. For example, a multinational chemical company could propose a cap on its emissions in one or more of the developing countries in which it operates. If it gains agreement from the countries concerned and the executive board of the CDM to use this as a legitimate basis against which reductions would automatically be regarded as generating CERs, then in effect one part of the industrial economy of the country or countries would have entered the structure of specific binding commitments, and the trading system, of the Kyoto Protocol.

Other basic approaches to the international allocation of future commitments, of less or greater specificity, are of course possible. Academic debate during the first half of the 1990s resulted in many proposals, and some discussion of the ethical basis upon which they rested.[14] One common suggestion is that countries should be required to take on commitments once they reach a certain level of per capita income. Others relate to proposals for 'contraction and convergence' discussed below.

The approach advanced as part of the Brazilian proposal put forward in the Kyoto negotiations (but deferred for later consideration) is seemingly very different but shares many features. It also proposes a specific, global and long-term basis for allocating emission allowances, emanating from an assessment of historical contributions towards realized global temperature increases. It would ultimately include all countries, albeit gradually, converging at least in terms of their assumption of responsibility for past emissions

Subsequent academic debate has refined analysis and added new specific proposals.[15] A more general approach is advanced by the Pew Center, in a report that proposes classifying countries according to an amalgam of three equity principles, namely responsibility for emissions (past, current and future), standard of living (based upon GDP), and opportunities for abatement (expressed in terms of emission intensities).[16] From this it proposes three groupings of countries: those that 'must act', those that 'should act, but differently', and those that 'could act' if they wished but

---

[14] An extensive review of the literature on criteria and proposals for international allocation is given in the IPCC Second Assessment Report, Working Group III, Chapter 4, 'Equity and Social Considerations', in *Economic and Social Dimensions of Climate Change*, CUP, 1996. A shortened review and analysis is given in M. Grubb, 'Seeking Fair Weather: Ethics and the International Debate on Climate Change', *International Affairs*, Vol. 71, No. 3, 1995.

[15] A recent summary and numerical economic analysis of different global distribution proposals is given in A. Rose, B. Stevens, J. Edmonds and M. Wise, 'International Equity and Differentiation in Global Warming Policy', *Environmental and Resource Economics*, Vol. 12, 1998, pp. 25–51.

[16] E. Claussen and L. McNeilly, *Equity and Global Climate Change*, Washington: Pew Center on Global Climate Change, 1998.

would be under no obligation to do so. Muller, in another recent contribution, focuses upon the relationship between different equity criteria and processes of negotiation.[17] He proposes that countries should rank different proposals for allocation according to their preferences, and that the different resultant rankings would help to determine the weight to be assigned to different proposals.

The major political obstacle to most such approaches, namely the deeply embedded structural division between the Annex I and the G77 groupings, has bedevilled the climate change negotiations since their inception. As noted in reviewing the outcome of the Kyoto negotiations (Chapter 3), the Protocol's biggest failings simply reflect the basic North–South divide in the UN system itself. It is an anachronistic division that bears ever less relationship to the emerging world of the twenty-first century, but for which it is not easy to develop any credible alternative (and considering proposals for reform would certainly take us beyond the scope of this book).

One principle has, however, succeeded in at least engaging the attention of many developing countries, though outside the negotiating process. In its general form this is the idea that allocations should be 'equitable'. The heads of state at the Non-Aligned Summit in September 1998, representing many developing countries, expressed their opposition to emissions trading by stating that it should proceed 'only after agreement on initial allocation on an equitable basis for all countries'. On some interpretations, this leads to rather more specific proposals consistent with the North–South divisions. We now consider them.

8.6 Emissions convergence: any room for compromise?

For the first commitment period, there is unlikely to be anything more than some piecemeal accession along the lines discussed above, combined perhaps with intensified activity in the CDM. Is there anything more coherent that might emerge for subsequent commitment periods? Indeed is there anything more promising than a series of ad-hoc Kyoto-type allocation negotiations writ ever larger and even more difficult?

---

[17] B. Muller, *Justice in Global Warming Negotiations: How to Obtain a Procedurally Fair Compromise*, Oxford: Oxford Institute of Energy Studies, December 1998.

One relatively simple proposal for allocation has emerged from the much more wide-ranging academic debate during the 1990s.[18] The ethical proposition that all human beings should have an equal entitlement to emit greenhouse gases leads to the proposal that emission allowances in the long term should be distributed in proportion to population. Most proponents of this approach recognize that the adjustment burdens of equal per capita emission allowances would make it politically impossible in the short term, and so the idea has taken the form of proposals that global emissions should over time *contract* in total, and *converge* towards equal per capita distribution of tradable emission allowances: contraction and convergence.

The model of contraction and convergence has for many people an intuitive appeal as a desirable and fair approach to the long-term problem of distribution, and it has emerged as the most politically prominent contender for any specific, global formula for long-term allocations, with increasing numbers of adherents in both developed and developing countries.[19] Politically, it has obvious appeal for developing countries, most (but not quite all) of which are low per capita emitters, and which would thereby receive an allowance for increased emissions under most formulations. By the same token, most formulations would lead over time to greatly reduced assigned amounts for the developed countries.

Figure 8.1 shows allocations that could be implied for Annex I and non-Annex I with per capita convergence over the century, for stabilizing concentrations at 450 ppm and 550 ppm CO_2. In the long term, allocations for the industrialized countries would be dramatically reduced from their current levels, since these per capita levels would clearly be wildly unsustainable if globalized.

---

[18] See note 14.

[19] A. Meyer and A. Cooper (Global Commons Institute) coined the term and have campaigned to promote the 'contraction and convergence' approach, backed with detailed and graphic numerical studies of what it might mean. Details may be found on the GCI web site, http://www.gci.org.uk, which includes access to a numerical model. The international parliamentarians group, Global Legislators for a Balanced Environment, has backed this approach; see Aubrey Meyer, 'Global Equity and Climate Change: A History of the UNFCCC Negotiations for a Global Solution', *GLOBE International*, Brussels, 1998; or Aubrey Meyer and Tony Cooper, 'Contraction and Convergence: A Global Solution to a Global Problem', in *Man Made Climate Change – Economic Aspects and Policy Options*, Proceedings of ZEW conference, Mannheim, Germany, March 1997.

Figure 8.1: Emissions convergence distributions under 450 ppm and 550 ppm

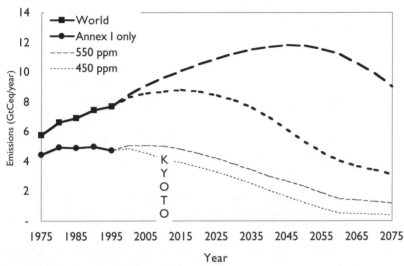

Source: Authors.

Note: The diagram shows emissions or 'assigned amounts' for industrialized and developing countries (Annex I and non-Annex I regions), under scenarios which stabilize atmospheric concentrations at 450ppm and 550ppm. The global trajectories are taken from the analysis of Haduong et al., depicted in Figure 5.2. In this diagram: historical data are simply pro rata +30% in recognition of (highly uncertain) LUCF changes and non-CO_2 gases, equalized to the 1997 base year estimates; up to 2010 developing countries follow IEA projections while industrialized countries undertake reductions; and assigned amounts converge linearly thereafter to equal per capita allowances in 2060. Population changes after 2040 do not change allowances.

Various academic objections to contraction and convergence have been raised. Some dispute the primacy of the egalitarian ethic that underpins it. Others point to the huge disparity in per capita wealth and emissions *within* developing countries and question the logic of giving entitlements to countries on egalitarian grounds as they might be used to support the lifestyle of an upper class which is every bit as extravagant as the West's, but masked by a larger sea of poverty. Other objections stem from the fact that it rests entirely on criteria of responsibilities, rather than any assessment of opportunities or equality of burdens, which has been the other main strand in the literature and a driving force in negotiations. Few other concrete proposals have emerged politically, however, apart from the

Brazilian proposal. Emerging interest in the possibility of using 'emissions intensity' as an indicator for defining voluntary commitments may also prove relevant, perhaps as a complementary indicator to per capita emissions. [20]

Since the G77 group has so far not entertained any debate about possible future global commitments, ideas of emissions convergence (which in the following discussion are taken to subsume the Brazilian and any other proposals for global schemes defining specific criteria for emission rights or commitments) have yet to be tested politically. Various issues emerge, however.

First, there are seemingly semantic but important fundamentals to clarify. Many developing-country representatives have insisted that they consider the issue as one of *entitlements*, not *commitments*. But the meaning of entitlements without commitments is unclear, and it smacks of seeking rights without responsibilities, always a questionable moral stance. The United States, for its part, has always been very wary of any international discussion couched in terms of economic 'rights', preferring more pragmatic language focused on actions and decisions.

Beyond these considerations, the most obvious political difficulty is that per capita emissions convergence would imply quite rapid and ultimately huge reductions in the allocations for industrialized countries, particularly the high per capita emitters in the New World countries that have already opposed any specific reference to the word 'equity' in emission allocations.[21] In fact, one striking observation, as illustrated in Figure 8.2, is that the Kyoto commitments do represent a considerable degree of convergence in per capita emissions, if not overall contraction.[22]

---

[20] See notes 9 and 18.

[21] The strength of opposition was brought home to the author in the final night at Kyoto. With the negotiations seemingly on the rocks, given Indian and Chinese opposition to emissions trading and voluntary accession, I asked one JUSSCANNZ representative (not from the United States) whether the impasse might be broken if a reference to the need for an 'equitable' basis for future commitments were to be inserted into the heading of the proposed article on voluntary commitments. At the mention of the word, he turned his back and walked away. The fear of any process that might lead from 'equitable' to 'equal' per capita principles was palpable.

[22] This may help to diffuse one potential danger to the Kyoto regime. The messy allocations of Kyoto have led some proponents of contraction and convergence to suggest that Kyoto should be scrapped in favour of something more rational. Here they join their bitterest

Only Russia, Ukraine and to some extent Australia represent serious divergence from the derived idea that per capita emission allowances should contract, at a rate roughly in proportion to the absolute historical level.[23]

Nevertheless, the processes by which these commitments were adopted point to a root political difficulty of formalized emissions convergence. This proposal seeks a logical, top-down and long-term resolution in the context of a political process that is inherently illogical, bottom-up and mostly concerned with the current or next round of commitments. The negotiating process is not logical (i.e. internally consistent in its application of principles) because Parties bring widely divergent principles and perceptions. It is bottom-up because it is built up from specific national positions and requires consensus which no country can be forced to join. It is short-term – compared with the time scale of the climate problem – because that is the time span directly relevant and feasible for negotiators at a given time, especially in view of the uncertainties. To a rational mind, formalized emissions convergence might well seem preferable to repeated and tortuous negotiations with ad-hoc outcomes along the lines of the Kyoto Protocol. But it faces the hurdle that no single rational mind can govern such a global process.

However, concepts of emission convergence could provide a focus for negotiations, perhaps as a starting point for skirmishing over commitments in later rounds. India's Tata Energy Research Institute suggests a 'convergence corridor' within which specific allocations would be negotiated.[24] Indeed if the G77 as an institution is to move beyond denial of the need for developing countries to engage in such debates – as it will have to do if

---

enemies, who would rather see no agreement at all. Indeed, it is conceivable that the Kyoto regime could be destroyed by an alliance of those for whom it is not good enough, with those for whom it is too good. But that seems unlikely and involves political speculation beyond the scope of this book.

[23] Arguably the Russian and Ukrainian anomalies can partly be justified by the observation that rapid contraction, implying import of allowances to offset the domestic burden, involves an element of assuming economic responsibility for emissions reductions abroad and should not apply to poorer countries. Anomalies within the EU 'bubble', that allows some poorer member countries to increase per capita emissions to well above levels sustainable in the long run, could be justified on similar grounds.

[24] R.K. Pachauri, in Tata Energy Research Institute, *Climate Change: Post-Kyoto Perspectives from the South*, New Delhi: TERI, 1998.

Figure 8.2: Kyoto commitments for industrialized countries

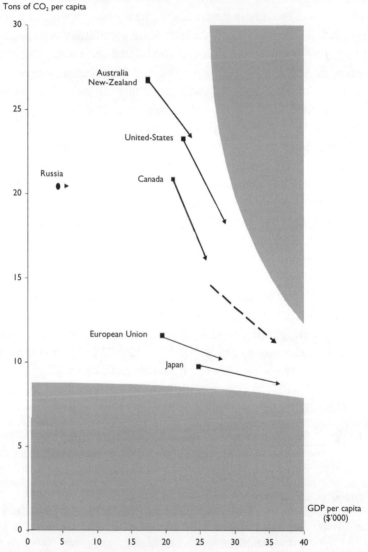

Note: The arrows connect points of 1990 per-capita emissions to the Kyoto commitment levels for 2010, given projections of GDP and population growth. The thick dotted line shows the Annex I average. All data CO_2 only without other gases, LUCF or the CDM.
Source: B. Dessus, 'Equity, sustainability and solidarity concerns', in J. Goldemberg et al., *Issues and Options: The Clean Development Mechanism*, New York: UNDP, 1998.

the climate problem is to be seriously addressed – then such ideas may constitute its best chance for maintaining some semblance of unity.

Insofar as more formal emissions convergence has been considered, it is widely assumed that it would split the negotiations straight along North–South lines. The reality may well be more subtle. Proposals for specific quantified convergence in some form would appeal not only to those countries that would stand to benefit directly, but also to those searching for rationality and order in a basically chaotic situation regarding the terms of global participation. It would tend to appeal to countries with strong intellectual and philosphical traditions oriented towards order and planning, such as France and Brazil. But it might well be anathema to those with a more libertarian tradition and intrinsic suspicion of grand plans. Specific quantified convergence proposals could just as easily divide the Francophone and the Anglo-Saxon worlds; other traditions are harder to place.

It is not easy to see room for compromise between such divergent world-views and interests, if and as the negotiations expand to include far more countries in far more widely divergent situations. But some willingness at least seriously to discuss globally equitable allocations is probably the price that the United States and allied countries will have to pay if the developing countries are to be persuaded to discuss global quantified emission commitments.

8.7   Conclusions

The Kyoto Protocol, together with the Convention on which it rests, forms the basis upon which future global efforts to tackle climate change will be built. The Buenos Aires conference confirmed that all the main country representatives accept it as such, and launched a complex set of inter-linked negotiations to put flesh on its foundational structure of specific commitments and mechanisms. Even this is only the next step in a long-term process. Many commentators have drawn analogies with the General Agreement on Tariffs and Trade which, as Jacoby et al. note, 'grew and evolved over time, adding countries and goods along the way, peacefully resolving conflicts between national economic interest, and contributing importantly to global economic growth ... but it took 50 years of hard work – even given an intelligent, forward looking design at the outset'.[25]

The negotiations launched under the Buenos Aires Plan of Action will culminate at COP-6 at the end of the year 2000, and will define the more detailed package that is likely to be presented for ratification in the United States and elsewhere. The battle for ratification will be long and difficult, perhaps in several countries including some European ones, but particularly in the United States. The Protocol could come into force without US ratification but the agreement could not be sustained without effective US participation achieved some time in the years 2000–4. This is likely to be influenced by three main factors. *Business* is growing steadily more positive towards action on climate change in general and, in Europe, towards the Kyoto Protocol in particular, but opposition remains deep in the United States. This is partly offset by rising popular concern as mediated by *NGOs and the media*, but these have great difficulty in grappling with and communicating the political and technical complexity of the agreement itself.

The third main factor will be the nature of future involvement by the *developing countries*, including whether, and if so how and when, they might start negotiating concerning specific quantified commitments. Some may choose to follow Argentina and adopt commitments unilaterally but this is problematic and will probably be limited. Sectoral commitments through the CDM might offer another possibility. But at some point, and contingent upon more adequate action by the industrialized countries and entry into force of the Protocol, the developing countries will have to start discussing quantified commitments, probably based around concepts of emissions convergence.

The political problem of tackling climate change has been popularly characterized as one of moving the US sceptics 'beyond denial' and gaining more 'meaningful participation' from developing countries. In a sense that can now be reversed. The core challenge is to gain meaningful participation of the United States and other major OECD countries *domestically* in their policies and politics. Morally and legally this should precede – but politically it may have to go in tandem with – a move

---

[25] See note 11.

beyond denial in developing countries about the need to start considering more specific and quantified commitments.

The fact that the Protocol itself does not contain *global* quantified commitments is being used as a main argument by its opponents. The idea that scrapping the Protocol is the way to attain global participation does not stand serious scrutiny, however: to the contrary, honouring principles and commitments agreed since the earliest stages, and implementing the Protocol's existing commitments and mechanisms, is the essential step required upon which more global discussions can then be based.

As described in this book, the Protocol, with the follow-up Buenos Aires plan, is the culmination of 10 years of intensive negotiations on climate change. Its inclusion of international economic instruments, including global reach through the CDM, is more sophisticated – despite the difficulties identified – than most political analysts ever considered possible. The many other elements, spanning processes on technology transfer and financial mechanisms, the global 'Article 4.1/Article 10' commitments, policies and measures, minimization of adverse impacts, sinks, and compliance mechanisms, represent a complex package designed to gain and sustain global participation. The central structure of successive five-year commitment periods, combined with reporting and review procedures, provides a natural basis for the evolution of the regime over time. For all its limitations, the Kyoto Protocol is a remarkable agreement, and history will judge it as one of the defining achievements of international diplomacy in the late twentieth century. With so much invested, so much at stake, and so much more to do, there is no turning back.

Part III

Appendices

Appendix 1

Kyoto Protocol to the United Nations Framework Convention on Climate Change

The Parties to this Protocol,
   Being Parties to the United Nations Framework Convention on Climate Change, hereinafter referred to as "the Convention",
   In pursuit of the ultimate objective of the Convention as stated in its Article 2,
   Recalling the provisions of the Convention,
   Being guided by Article 3 of the Convention,
   Pursuant to the Berlin Mandate adopted by decision 1/CP.1 of the Conference of the Parties to the Convention at its first session,
Have agreed as follows:

For the purposes of this Protocol, the definitions contained in Article 1 of the Convention shall apply. In addition:

1. "Conference of the Parties" means the Conference of the Parties to the Convention.
2. "Convention" means the United Nations Framework Convention on Climate Change, adopted in New York on 9 May 1992.
3. "Intergovernmental Panel on Climate Change" means the Intergovernmental Panel on Climate Change established in 1988 jointly by the World Meteorological Organization and the United Nations Environment Programme.
4. "Montreal Protocol" means the Montreal Protocol on Substances that Deplete the Ozone Layer, adopted in Montreal on 16 September 1987 and as subsequently adjusted and amended.
5. "Parties present and voting" means Parties present and casting an affirmative or negative vote.
6. "Party" means, unless the context otherwise indicates, a Party to this Protocol.
    7. "Party included in Annex I" means a Party included in Annex I to the Convention, as may be amended, or a Party which has made a notification under Article 4, paragraph 2(g), of the Convention.

Article 2

1. Each Party included in Annex I, in achieving its quantified emission limitation and reduction commitments under Article 3, in order to promote sustainable development, shall:

(a) Implement and/or further elaborate policies and measures in accordance with its national circumstances, such as:

  (i)   Enhancement of energy efficiency in relevant sectors of the national economy;

  (ii)   Protection and enhancement of sinks and reservoirs of greenhouse gases not controlled by the Montreal Protocol, taking into account its commitments under relevant international environmental agreements; promotion of sustainable forest management practices, afforestation and reforestation;

  (iii) Promotion of sustainable forms of agriculture in light of climate change considerations;

  (iv)   Research on, and promotion, development and increased use of, new and renewable forms of energy, of carbon dioxide sequestration technologies and of advanced and innovative environmentally sound technologies;

  (v)   Progressive reduction or phasing out of market imperfections, fiscal incentives, tax and duty exemptions and subsidies in all greenhouse gas emitting sectors that run counter to the objective of the Convention and application of market instruments;

  (vi)   Encouragement of appropriate reforms in relevant sectors aimed at promoting policies and measures which limit or reduce emissions of greenhouse gases not controlled by the Montreal Protocol;

  (vii) Measures to limit and/or reduce emissions of greenhouse gases not controlled by the Montreal Protocol in the transport sector;

  (viii) Limitation and/or reduction of methane emissions through recovery and use in waste management, as well as in the production, transport and distribution of energy;

(b) Cooperate with other such Parties to enhance the individual and combined effectiveness of their policies and measures adopted under this Article, pursuant to Article 4, paragraph 2(e)(i), of the Convention. To this end, these Parties shall take steps to share their experience and exchange information on such policies and measures, including developing ways of improving their comparability, transparency and effectiveness. The Conference of the Parties serving as the meeting of the Parties to this Protocol shall, at its first session or as soon as practicable thereafter, consider ways to facilitate such cooperation, taking into account all relevant information.

2. The Parties included in Annex I shall pursue limitation or reduction of emissions of greenhouse gases not controlled by the Montreal Protocol from aviation and marine bunker fuels, working through the International Civil Aviation Organization and the International Maritime Organization, respectively.

3. The Parties included in Annex I shall strive to implement policies and measures under this Article in such a way as to minimize adverse effects, including the adverse effects of climate change, effects on international trade, and social, environmental and economic impacts on other Parties, especially developing country Parties and in particular those identified in Article 4, paragraphs 8 and 9,

of the Convention, taking into account Article 3 of the Convention. The Conference of the Parties serving as the meeting of the Parties to this Protocol may take further action, as appropriate, to promote the implementation of the provisions of this paragraph.

4. The Conference of the Parties serving as the meeting of the Parties to this Protocol, if it decides that it would be beneficial to coordinate any of the policies and measures in paragraph 1(a) above, taking into account different national circumstances and potential effects, shall consider ways and means to elaborate the coordination of such policies and measures.

Article 3

1. The Parties included in Annex I shall, individually or jointly, ensure that their aggregate anthropogenic carbon dioxide equivalent emissions of the greenhouse gases listed in Annex A do not exceed their assigned amounts, calculated pursuant to their quantified emission limitation and reduction commitments inscribed in Annex B and in accordance with the provisions of this Article, with a view to reducing their overall emissions of such gases by at least 5 per cent below 1990 levels in the commitment period 2008 to 2012.

2. Each Party included in Annex I shall, by 2005, have made demonstrable progress in achieving its commitments under this Protocol.

3. The net changes in greenhouse gas emissions by sources and removals by sinks resulting from direct human-induced land-use change and forestry activities, limited to afforestation, reforestation and deforestation since 1990, measured as verifiable changes in carbon stocks in each commitment period, shall be used to meet the commitments under this Article of each Party included in Annex I. The greenhouse gas emissions by sources and removals by sinks associated with those activities shall be reported in a transparent and verifiable manner and reviewed in accordance with Articles 7 and 8.

4. Prior to the first session of the Conference of the Parties serving as the meeting of the Parties to this Protocol, each Party included in Annex I shall provide, for consideration by the Subsidiary Body for Scientific and Technological Advice, data to establish its level of carbon stocks in 1990 and to enable an estimate to be made of its changes in carbon stocks in subsequent years. The Conference of the Parties serving as the meeting of the Parties to this Protocol shall, at its first session or as soon as practicable thereafter, decide upon modalities, rules and guidelines as to how, and which, additional human-induced activities related to changes in greenhouse gas emissions by sources and removals by sinks in the agricultural soils and the land-use change and forestry categories shall be added to, or subtracted from, the assigned amounts for Parties included in Annex I, taking into account uncertainties, transparency in reporting, verifiability, the methodological work of the Intergovernmental Panel on Climate Change, the advice provided by the Subsidiary Body for Scientific and Technological Advice in accordance with Article 5 and the decisions of the Conference of the Parties.

Such a decision shall apply in the second and subsequent commitment periods. A Party may choose to apply such a decision on these additional human-induced activities for its first commitment period, provided that these activities have taken place since 1990.

5. The Parties included in Annex I undergoing the process of transition to a market economy whose base year or period was established pursuant to decision 9/CP.2 of the Conference of the Parties at its second session shall use that base year or period for the implementation of their commitments under this Article. Any other Party included in Annex I undergoing the process of transition to a market economy which has not yet submitted its first national communication under Article 12 of the Convention may also notify the Conference of the Parties serving as the meeting of the Parties to this Protocol that it intends to use an historical base year or period other than 1990 for the implementation of its commitments under this Article. The Conference of the Parties serving as the meeting of the Parties to this Protocol shall decide on the acceptance of such notification.

6. Taking into account Article 4, paragraph 6, of the Convention, in the implementation of their commitments under this Protocol other than those under this Article, a certain degree of flexibility shall be allowed by the Conference of the Parties serving as the meeting of the Parties to this Protocol to the Parties included in Annex I undergoing the process of transition to a market economy.

7. In the first quantified emission limitation and reduction commitment period, from 2008 to 2012, the assigned amount for each Party included in Annex I shall be equal to the percentage inscribed for it in Annex B of its aggregate anthropogenic carbon dioxide equivalent emissions of the greenhouse gases listed in Annex A in 1990, or the base year or period determined in accordance with paragraph 5 above, multiplied by five. Those Parties included in Annex I for whom land-use change and forestry constituted a net source of greenhouse gas emissions in 1990 shall include in their 1990 emissions base year or period the aggregate anthropogenic carbon dioxide equivalent emissions by sources minus removals by sinks in 1990 from land-use change for the purposes of calculating their assigned amount.

8. Any Party included in Annex I may use 1995 as its base year for hydrofluorocarbons, perfluorocarbons and sulphur hexafluoride, for the purposes of the calculation referred to in paragraph 7 above.

9. Commitments for subsequent periods for Parties included in Annex I shall be established in amendments to Annex B to this Protocol, which shall be adopted in accordance with the provisions of Article 21, paragraph 7. The Conference of the Parties serving as the meeting of the Parties to this Protocol shall initiate the consideration of such commitments at least seven years before the end of the first commitment period referred to in paragraph 1 above.

10. Any emission reduction units, or any part of an assigned amount, which a Party acquires from another Party in accordance with the provisions of Article 6 or of Article 17 shall be added to the assigned amount for the acquiring Party.

11. Any emission reduction units, or any part of an assigned amount, which a Party transfers to another Party in accordance with the provisions of Article 6 or of Article 17 shall be subtracted from the assigned amount for the transferring Party.

12. Any certified emission reductions which a Party acquires from another Party in accordance with the provisions of Article 12 shall be added to the assigned amount for the acquiring Party.

13. If the emissions of a Party included in Annex I in a commitment period are less than its assigned amount under this Article, this difference shall, on request of that Party, be added to the assigned amount for that Party for subsequent commitment periods.

14. Each Party included in Annex I shall strive to implement the commitments mentioned in paragraph 1 above in such a way as to minimize adverse social, environmental and economic impacts on developing country Parties, particularly those identified in Article 4, paragraphs 8 and 9, of the Convention. In line with relevant decisions of the Conference of the Parties on the implementation of those paragraphs, the Conference of the Parties serving as the meeting of the Parties to this Protocol shall, at its first session, consider what actions are necessary to minimize the adverse effects of climate change and/or the impacts of response measures on Parties referred to in those paragraphs. Among the issues to be considered shall be the establishment of funding, insurance and transfer of technology.

Article 4

1. Any Parties included in Annex I that have reached an agreement to fulfil their commitments under Article 3 jointly, shall be deemed to have met those commitments provided that their total combined aggregate anthropogenic carbon dioxide equivalent emissions of the greenhouse gases listed in Annex A do not exceed their assigned amounts calculated pursuant to their quantified emission limitation and reduction commitments inscribed in Annex B and in accordance with the provisions of Article 3. The respective emission level allocated to each of the Parties to the agreement shall be set out in that agreement.

2. The Parties to any such agreement shall notify the secretariat of the terms of the agreement on the date of deposit of their instruments of ratification, acceptance or approval of this Protocol, or accession thereto. The secretariat shall in turn inform the Parties and signatories to the Convention of the terms of the agreement.

3. Any such agreement shall remain in operation for the duration of the commitment period specified in Article 3, paragraph 7.

4. If Parties acting jointly do so in the framework of, and together with, a regional economic integration organization, any alteration in the composition of the organization after adoption of this Protocol shall not affect existing commitments under this Protocol. Any alteration in the composition of the organization shall only apply for the purposes of those commitments under Article 3 that are adopted subsequent to that alteration.

5. In the event of failure by the Parties to such an agreement to achieve their total combined level of emission reductions, each Party to that agreement shall be responsible for its own level of emissions set out in the agreement.
6. If Parties acting jointly do so in the framework of, and together with, a regional economic integration organization which is itself a Party to this Protocol, each member State of that regional economic integration organization individually, and together with the regional economic integration organization acting in accordance with Article 24, shall, in the event of failure to achieve the total combined level of emission reductions, be responsible for its level of emissions as notified in accordance with this Article.

Article 5

1. Each Party included in Annex I shall have in place, no later than one year prior to the start of the first commitment period, a national system for the estimation of anthropogenic emissions by sources and removals by sinks of all greenhouse gases not controlled by the Montreal Protocol. Guidelines for such national systems, which shall incorporate the methodologies specified in paragraph 2 below, shall be decided upon by the Conference of the Parties serving as the meeting of the Parties to this Protocol at its first session.
2. Methodologies for estimating anthropogenic emissions by sources and removals by sinks of all greenhouse gases not controlled by the Montreal Protocol shall be those accepted by the Intergovernmental Panel on Climate Change and agreed upon by the Conference of the Parties at its third session. Where such methodologies are not used, appropriate adjustments shall be applied according to methodologies agreed upon by the Conference of the Parties serving as the meeting of the Parties to this Protocol at its first session. Based on the work of, *inter alia*, the Intergovernmental Panel on Climate Change and advice provided by the Subsidiary Body for Scientific and Technological Advice, the Conference of the Parties serving as the meeting of the Parties to this Protocol shall regularly review and, as appropriate, revise such methodologies and adjustments, taking fully into account any relevant decisions by the Conference of the Parties. Any revision to methodologies or adjustments shall be used only for the purposes of ascertaining compliance with commitments under Article 3 in respect of any commitment period adopted subsequent to that revision.
3. The global warming potentials used to calculate the carbon dioxide equivalence of anthropogenic emissions by sources and removals by sinks of greenhouse gases listed in Annex A shall be those accepted by the Intergovernmental Panel on Climate Change and agreed upon by the Conference of the Parties at its third session. Based on the work of, *inter alia*, the Intergovernmental Panel on Climate Change and advice provided by the Subsidiary Body for Scientific and Technological Advice, the Conference of the Parties serving as the meeting of the Parties to this Protocol shall regularly review and, as appropriate, revise the global warming potential of each such greenhouse gas, taking fully into account any

relevant decisions by the Conference of the Parties. Any revision to a global warming potential shall apply only to commitments under Article 3 in respect of any commitment period adopted subsequent to that revision.

Article 6

1. For the purpose of meeting its commitments under Article 3, any Party included in Annex I may transfer to, or acquire from, any other such Party emission reduction units resulting from projects aimed at reducing anthropogenic emissions by sources or enhancing anthropogenic removals by sinks of greenhouse gases in any sector of the economy, provided that:
    (a) Any such project has the approval of the Parties involved;
    (b) Any such project provides a reduction in emissions by sources, or an enhancement of removals by sinks, that is additional to any that would otherwise occur;
    (c) It does not acquire any emission reduction units if it is not in compliance with its obligations under Articles 5 and 7; and
    (d) The acquisition of emission reduction units shall be supplemental to domestic actions for the purposes of meeting commitments under Article 3.
2. The Conference of the Parties serving as the meeting of the Parties to this Protocol may, at its first session or as soon as practicable thereafter, further elaborate guidelines for the implementation of this Article, including for verification and reporting.
3. A Party included in Annex I may authorize legal entities to participate, under its responsibility, in actions leading to the generation, transfer or acquisition under this Article of emission reduction units.
4. If a question of implementation by a Party included in Annex I of the requirements referred to in this Article is identified in accordance with the relevant provisions of Article 8, transfers and acquisitions of emission reduction units may continue to be made after the question has been identified, provided that any such units may not be used by a Party to meet its commitments under Article 3 until any issue of compliance is resolved.

Article 7

1. Each Party included in Annex I shall incorporate in its annual inventory of anthropogenic emissions by sources and removals by sinks of greenhouse gases not controlled by the Montreal Protocol, submitted in accordance with the relevant decisions of the Conference of the Parties, the necessary supplementary information for the purposes of ensuring compliance with Article 3, to be determined in accordance with paragraph 4 below.
2. Each Party included in Annex I shall incorporate in its national communication, submitted under Article 12 of the Convention, the supplementary information necessary to demonstrate compliance with its commitments under this Protocol, to be determined in accordance with paragraph 4 below.

3. Each Party included in Annex I shall submit the information required under paragraph 1 above annually, beginning with the first inventory due under the Convention for the first year of the commitment period after this Protocol has entered into force for that Party. Each such Party shall submit the information required under paragraph 2 above as part of the first national communication due under the Convention after this Protocol has entered into force for it and after the adoption of guidelines as provided for in paragraph 4 below. The frequency of subsequent submission of information required under this Article shall be determined by the Conference of the Parties serving as the meeting of the Parties to this Protocol, taking into account any timetable for the submission of national communications decided upon by the Conference of the Parties.

4. The Conference of the Parties serving as the meeting of the Parties to this Protocol shall adopt at its first session, and review periodically thereafter, guidelines for the preparation of the information required under this Article, taking into account guidelines for the preparation of national communications by Parties included in Annex I adopted by the Conference of the Parties. The Conference of the Parties serving as the meeting of the Parties to this Protocol shall also, prior to the first commitment period, decide upon modalities for the accounting of assigned amounts.

Article 8

1. The information submitted under Article 7 by each Party included in Annex I shall be reviewed by expert review teams pursuant to the relevant decisions of the Conference of the Parties and in accordance with guidelines adopted for this purpose by the Conference of the Parties serving as the meeting of the Parties to this Protocol under paragraph 4 below. The information submitted under Article 7, paragraph 1, by each Party included in Annex I shall be reviewed as part of the annual compilation and accounting of emissions inventories and assigned amounts. Additionally, the information submitted under Article 7, paragraph 2, by each Party included in Annex I shall be reviewed as part of the review of communications.

2. Expert review teams shall be coordinated by the secretariat and shall be composed of experts selected from those nominated by Parties to the Convention and, as appropriate, by intergovernmental organizations, in accordance with guidance provided for this purpose by the Conference of the Parties.

3. The review process shall provide a thorough and comprehensive technical assessment of all aspects of the implementation by a Party of this Protocol. The expert review teams shall prepare a report to the Conference of the Parties serving as the meeting of the Parties to this Protocol, assessing the implementation of the commitments of the Party and identifying any potential problems in, and factors influencing, the fulfilment of commitments. Such reports shall be circulated by the secretariat to all Parties to the Convention. The secretariat shall list those questions of implementation indicated in such reports for further consideration by

the Conference of the Parties serving as the meeting of the Parties to this Protocol.

4. The Conference of the Parties serving as the meeting of the Parties to this Protocol shall adopt at its first session, and review periodically thereafter, guidelines for the review of implementation of this Protocol by expert review teams taking into account the relevant decisions of the Conference of the Parties.

5. The Conference of the Parties serving as the meeting of the Parties to this Protocol shall, with the assistance of the Subsidiary Body for Implementation and, as appropriate, the Subsidiary Body for Scientific and Technological Advice, consider:

   (a) The information submitted by Parties under Article 7 and the reports of the expert reviews thereon conducted under this Article; and

   (b) Those questions of implementation listed by the secretariat under paragraph 3 above, as well as any questions raised by Parties.

6. Pursuant to its consideration of the information referred to in paragraph 5 above, the Conference of the Parties serving as the meeting of the Parties to this Protocol shall take decisions on any matter required for the implementation of this Protocol.

Article 9

1. The Conference of the Parties serving as the meeting of the Parties to this Protocol shall periodically review this Protocol in the light of the best available scientific information and assessments on climate change and its impacts, as well as relevant technical, social and economic information. Such reviews shall be coordinated with pertinent reviews under the Convention, in particular those required by Article 4, paragraph 2(d), and Article 7, paragraph 2(a), of the Convention. Based on these reviews, the Conference of the Parties serving as the meeting of the Parties to this Protocol shall take appropriate action.

2. The first review shall take place at the second session of the Conference of the Parties serving as the meeting of the Parties to this Protocol. Further reviews shall take place at regular intervals and in a timely manner.

Article 10

All Parties, taking into account their common but differentiated responsibilities and their specific national and regional development priorities, objectives and circumstances, without introducing any new commitments for Parties not included in Annex I, but reaffirming existing commitments under Article 4, paragraph 1, of the Convention, and continuing to advance the implementation of these commitments in order to achieve sustainable development, taking into account Article 4, paragraphs 3, 5 and 7, of the Convention, shall:

(a) Formulate, where relevant and to the extent possible, cost-effective national and, where appropriate, regional programmes to improve the quality of local emission factors, activity data and/or models which reflect the socio-economic conditions of each Party for the preparation and periodic updating of national inventories of anthropogenic emissions by sources and removals by sinks of all greenhouse gases not controlled by the Montreal Protocol, using comparable methodologies to be agreed upon by the Conference of the Parties, and consistent with the guidelines for the preparation of national communications adopted by the Conference of the Parties;

(b) Formulate, implement, publish and regularly update national and, where appropriate, regional programmes containing measures to mitigate climate change and measures to facilitate adequate adaptation to climate change:

    (i) Such programmes would, *inter alia*, concern the energy, transport and industry sectors as well as agriculture, forestry and waste management. Furthermore, adaptation technologies and methods for improving spatial planning would improve adaptation to climate change; and

    (ii) Parties included in Annex I shall submit information on action under this Protocol, including national programmes, in accordance with Article 7; and other Parties shall seek to include in their national communications, as appropriate, information on programmes which contain measures that the Party believes contribute to addressing climate change and its adverse impacts, including the abatement of increases in greenhouse gas emissions, and enhancement of and removals by sinks, capacity building and adaptation measures;

(c) Cooperate in the promotion of effective modalities for the development, application and diffusion of, and take all practicable steps to promote, facilitate and finance, as appropriate, the transfer of, or access to, environmentally sound technologies, know-how, practices and processes pertinent to climate change, in particular to developing countries, including the formulation of policies and programmes for the effective transfer of environmentally sound technologies that are publicly owned or in the public domain and the creation of an enabling environment for the private sector, to promote and enhance the transfer of, and access to, environmentally sound technologies;

(d) Cooperate in scientific and technical research and promote the maintenance and the development of systematic observation systems and development of data archives to reduce uncertainties related to the climate system, the adverse impacts of climate change and the economic and social consequences of various response strategies, and promote the development and strengthening of endogenous capacities and capabilities to participate in international and intergovernmental efforts, programmes and networks on research and systematic observation, taking into account Article 5 of the Convention;

(e) Cooperate in and promote at the international level, and, where appropriate, using existing bodies, the development and implementation of education and training programmes, including the strengthening of national capacity building, in particular human and institutional capacities and the exchange or secondment of personnel to train experts in this field, in particular for developing countries, and facilitate at the national level public awareness of, and public access to information on, climate change. Suitable modalities should be developed to implement these activities through the relevant bodies of the Convention, taking into account Article 6 of the Convention;

(f) Include in their national communications information on programmes and activities undertaken pursuant to this Article in accordance with relevant decisions of the Conference of the Parties; and

(g) Give full consideration, in implementing the commitments under this Article, to Article 4, paragraph 8, of the Convention.

Article 11

1. In the implementation of Article 10, Parties shall take into account the provisions of Article 4, paragraphs 4, 5, 7, 8 and 9, of the Convention.

2. In the context of the implementation of Article 4, paragraph 1, of the Convention, in accordance with the provisions of Article 4, paragraph 3, and Article 11 of the Convention, and through the entity or entities entrusted with the operation of the financial mechanism of the Convention, the developed country Parties and other developed Parties included in Annex II to the Convention shall:

   (a) Provide new and additional financial resources to meet the agreed full costs incurred by developing country Parties in advancing the implementation of existing commitments under Article 4, paragraph 1(a), of the Convention that are covered in Article 10, subparagraph (a); and

   (b) Also provide such financial resources, including for the transfer of technology, needed by the developing country Parties to meet the agreed full incremental costs of advancing the implementation of existing commitments under Article 4, paragraph 1, of the Convention that are covered by Article 10 and that are agreed between a developing country Party and the international entity or entities referred to in Article 11 of the Convention, in accordance with that Article.

   The implementation of these existing commitments shall take into account the need for adequacy and predictability in the flow of funds and the importance of appropriate burden sharing among developed country Parties. The guidance to the entity or entities entrusted with the operation of the financial mechanism of the Convention in relevant decisions of the Conference of the Parties, including those agreed before the adoption of this Protocol, shall apply *mutatis mutandis* to the provisions of this paragraph.

3. The developed country Parties and other developed Parties in Annex II to the Convention may also provide, and developing country Parties avail themselves of, financial resources for the implementation of Article 10, through bilateral, regional and other multilateral channels.

Article 12

1. A clean development mechanism is hereby defined.

2. The purpose of the clean development mechanism shall be to assist Parties not included in Annex I in achieving sustainable development and in contributing to the ultimate objective of the Convention, and to assist Parties included in Annex I

in achieving compliance with their quantified emission limitation and reduction commitments under Article 3.

3. Under the clean development mechanism:

   (a) Parties not included in Annex I will benefit from project activities resulting in certified emission reductions; and

   (b) Parties included in Annex I may use the certified emission reductions accruing from such project activities to contribute to compliance with part of their quantified emission limitation and reduction commitments under Article 3, as determined by the Conference of the Parties serving as the meeting of the Parties to this Protocol.

4. The clean development mechanism shall be subject to the authority and guidance of the Conference of the Parties serving as the meeting of the Parties to this Protocol and be supervised by an executive board of the clean development mechanism.

5. Emission reductions resulting from each project activity shall be certified by operational entities to be designated by the Conference of the Parties serving as the meeting of the Parties to this Protocol, on the basis of:

   (a) Voluntary participation approved by each Party involved;

   (b) Real, measurable, and long-term benefits related to the mitigation of climate change; and

   (c) Reductions in emissions that are additional to any that would occur in the absence of the certified project activity.

6. The clean development mechanism shall assist in arranging funding of certified project activities as necessary.

7. The Conference of the Parties serving as the meeting of the Parties to this Protocol shall, at its first session, elaborate modalities and procedures with the objective of ensuring transparency, efficiency and accountability through independent auditing and verification of project activities.

8. The Conference of the Parties serving as the meeting of the Parties to this Protocol shall ensure that a share of the proceeds from certified project activities is used to cover administrative expenses as well as to assist developing country Parties that are particularly vulnerable to the adverse effects of climate change to meet the costs of adaptation.

9. Participation under the clean development mechanism, including in activities mentioned in paragraph 3(a) above and in the acquisition of certified emission reductions, may involve private and/or public entities, and is to be subject to whatever guidance may be provided by the executive board of the clean development mechanism.

10. Certified emission reductions obtained during the period from the year 2000 up to the beginning of the first commitment period can be used to assist in achieving compliance in the first commitment period.

Article 13

1. The Conference of the Parties, the supreme body of the Convention, shall serve as the meeting of the Parties to this Protocol.
2. Parties to the Convention that are not Parties to this Protocol may participate as observers in the proceedings of any session of the Conference of the Parties serving as the meeting of the Parties to this Protocol. When the Conference of the Parties serves as the meeting of the Parties to this Protocol, decisions under this Protocol shall be taken only by those that are Parties to this Protocol.
3. When the Conference of the Parties serves as the meeting of the Parties to this Protocol, any member of the Bureau of the Conference of the Parties representing a Party to the Convention but, at that time, not a Party to this Protocol, shall be replaced by an additional member to be elected by and from amongst the Parties to this Protocol.
4. The Conference of the Parties serving as the meeting of the Parties to this Protocol shall keep under regular review the implementation of this Protocol and shall make, within its mandate, the decisions necessary to promote its effective implementation. It shall perform the functions assigned to it by this Protocol and shall:

   (a) Assess, on the basis of all information made available to it in accordance with the provisions of this Protocol, the implementation of this Protocol by the Parties, the overall effects of the measures taken pursuant to this Protocol, in particular environmental, economic and social effects as well as their cumulative impacts and the extent to which progress towards the objective of the Convention is being achieved;

   (b) Periodically examine the obligations of the Parties under this Protocol, giving due consideration to any reviews required by Article 4, paragraph 2(d), and Article 7, paragraph 2, of the Convention, in the light of the objective of the Convention, the experience gained in its implementation and the evolution of scientific and technological knowledge, and in this respect consider and adopt regular reports on the implementation of this Protocol;

   (c) Promote and facilitate the exchange of information on measures adopted by the Parties to address climate change and its effects, taking into account the differing circumstances, responsibilities and capabilities of the Parties and their respective commitments under this Protocol;

   (d) Facilitate, at the request of two or more Parties, the coordination of measures adopted by them to address climate change and its effects, taking into account the differing circumstances, responsibilities and capabilities of the Parties and their respective commitments under this Protocol;

   (e) Promote and guide, in accordance with the objective of the Convention and the provisions of this Protocol, and taking fully into account the relevant decisions by the Conference of the Parties, the development and periodic refinement of comparable methodologies for the effective implementation of this Protocol, to be agreed on by the Conference of the Parties serving as the meeting of the Parties to this Protocol;

(f) Make recommendations on any matters necessary for the implementation of this Protocol;

(g) Seek to mobilize additional financial resources in accordance with Article 11, paragraph 2;

(h) Establish such subsidiary bodies as are deemed necessary for the implementation of this Protocol;

(i) Seek and utilize, where appropriate, the services and cooperation of, and information provided by, competent international organizations and intergovernmental and non-governmental bodies; and

(j) Exercise such other functions as may be required for the implementation of this Protocol, and consider any assignment resulting from a decision by the Conference of the Parties.

5. The rules of procedure of the Conference of the Parties and financial procedures applied under the Convention shall be applied *mutatis mutandis* under this Protocol, except as may be otherwise decided by consensus by the Conference of the Parties serving as the meeting of the Parties to this Protocol.

6. The first session of the Conference of the Parties serving as the meeting of the Parties to this Protocol shall be convened by the secretariat in conjunction with the first session of the Conference of the Parties that is scheduled after the date of the entry into force of this Protocol. Subsequent ordinary sessions of the Conference of the Parties serving as the meeting of the Parties to this Protocol shall be held every year and in conjunction with ordinary sessions of the Conference of the Parties, unless otherwise decided by the Conference of the Parties serving as the meeting of the Parties to this Protocol.

7. Extraordinary sessions of the Conference of the Parties serving as the meeting of the Parties to this Protocol shall be held at such other times as may be deemed necessary by the Conference of the Parties serving as the meeting of the Parties to this Protocol, or at the written request of any Party, provided that, within six months of the request being communicated to the Parties by the secretariat, it is supported by at least one third of the Parties.

8. The United Nations, its specialized agencies and the International Atomic Energy Agency, as well as any State member thereof or observers thereto not party to the Convention, may be represented at sessions of the Conference of the Parties serving as the meeting of the Parties to this Protocol as observers. Any body or agency, whether national or international, governmental or non-governmental, which is qualified in matters covered by this Protocol and which has informed the secretariat of its wish to be represented at a session of the Conference of the Parties serving as the meeting of the Parties to this Protocol as an observer, may be so admitted unless at least one third of the Parties present object. The admission and participation of observers shall be subject to the rules of procedure, as referred to in paragraph 5 above.

Article 14

1. The secretariat established by Article 8 of the Convention shall serve as the secretariat of this Protocol.
2. Article 8, paragraph 2, of the Convention on the functions of the secretariat, and Article 8, paragraph 3, of the Convention on arrangements made for the functioning of the secretariat, shall apply *mutatis mutandis* to this Protocol. The secretariat shall, in addition, exercise the functions assigned to it under this Protocol.

Article 15

1. The Subsidiary Body for Scientific and Technological Advice and the Subsidiary Body for Implementation established by Articles 9 and 10 of the Convention shall serve as, respectively, the Subsidiary Body for Scientific and Technological Advice and the Subsidiary Body for Implementation of this Protocol. The provisions relating to the functioning of these two bodies under the Convention shall apply *mutatis mutandis* to this Protocol. Sessions of the meetings of the Subsidiary Body for Scientific and Technological Advice and the Subsidiary Body for Implementation of this Protocol shall be held in conjunction with the meetings of, respectively, the Subsidiary Body for Scientific and Technological Advice and the Subsidiary Body for Implementation of the Convention.
2. Parties to the Convention that are not Parties to this Protocol may participate as observers in the proceedings of any session of the subsidiary bodies. When the subsidiary bodies serve as the subsidiary bodies of this Protocol, decisions under this Protocol shall be taken only by those that are Parties to this Protocol.
3. When the subsidiary bodies established by Articles 9 and 10 of the Convention exercise their functions with regard to matters concerning this Protocol, any member of the Bureaux of those subsidiary bodies representing a Party to the Convention but, at that time, not a party to this Protocol, shall be replaced by an additional member to be elected by and from amongst the Parties to this Protocol.

Article 16

The Conference of the Parties serving as the meeting of the Parties to this Protocol shall, as soon as practicable, consider the application to this Protocol of, and modify as appropriate, the multilateral consultative process referred to in Article 13 of the Convention, in the light of any relevant decisions that may be taken by the Conference of the Parties. Any multilateral consultative process that may be applied to this Protocol shall operate without prejudice to the procedures and mechanisms established in accordance with Article 18.

Article 17

The Conference of the Parties shall define the relevant principles, modalities, rules and guidelines, in particular for verification, reporting and accountability for emissions trading. The Parties included in Annex B may participate in emissions trading for the purposes of fulfilling their commitments under Article 3. Any such trading shall be supplemental to domestic actions for the purpose of meeting quantified emission limitation and reduction commitments under that Article.

Article 18

The Conference of the Parties serving as the meeting of the Parties to this Protocol shall, at its first session, approve appropriate and effective procedures and mechanisms to determine and to address cases of non-compliance with the provisions of this Protocol, including through the development of an indicative list of consequences, taking into account the cause, type, degree and frequency of non-compliance. Any procedures and mechanisms under this Article entailing binding consequences shall be adopted by means of an amendment to this Protocol.

Article 19

The provisions of Article 14 of the Convention on settlement of disputes shall apply *mutatis mutandis* to this Protocol.

Article 20

1. Any Party may propose amendments to this Protocol.
2. Amendments to this Protocol shall be adopted at an ordinary session of the Conference of the Parties serving as the meeting of the Parties to this Protocol. The text of any proposed amendment to this Protocol shall be communicated to the Parties by the secretariat at least six months before the meeting at which it is proposed for adoption. The secretariat shall also communicate the text of any proposed amendments to the Parties and signatories to the Convention and, for information, to the Depositary.
3. The Parties shall make every effort to reach agreement on any proposed amendment to this Protocol by consensus. If all efforts at consensus have been exhausted, and no agreement reached, the amendment shall as a last resort be adopted by a three-fourths majority vote of the Parties present and voting at the meeting. The adopted amendment shall be communicated by the secretariat to the Depositary, who shall circulate it to all Parties for their acceptance.
4. Instruments of acceptance in respect of an amendment shall be deposited with the Depositary. An amendment adopted in accordance with paragraph 3 above shall enter into force for those Parties having accepted it on the ninetieth day after the date of receipt by the Depositary of an instrument of acceptance by at least three fourths of the Parties to this Protocol.
5. The amendment shall enter into force for any other Party on the ninetieth day after the date on which that Party deposits with the Depositary its instrument of acceptance of the said amendment.

Article 21

1. Annexes to this Protocol shall form an integral part thereof and, unless otherwise expressly provided, a reference to this Protocol constitutes at the same time a reference to any annexes thereto. Any annexes adopted after the entry into force of this Protocol shall be restricted to lists, forms and any other material of a descriptive nature that is of a scientific, technical, procedural or administrative character.

2. Any Party may make proposals for an annex to this Protocol and may propose amendments to annexes to this Protocol.

3. Annexes to this Protocol and amendments to annexes to this Protocol shall be adopted at an ordinary session of the Conference of the Parties serving as the meeting of the Parties to this Protocol. The text of any proposed annex or amendment to an annex shall be communicated to the Parties by the secretariat at least six months before the meeting at which it is proposed for adoption. The secretariat shall also communicate the text of any proposed annex or amendment to an annex to the Parties and signatories to the Convention and, for information, to the Depositary.

4. The Parties shall make every effort to reach agreement on any proposed annex or amendment to an annex by consensus. If all efforts at consensus have been exhausted, and no agreement reached, the annex or amendment to an annex shall as a last resort be adopted by a three-fourths majority vote of the Parties present and voting at the meeting. The adopted annex or amendment to an annex shall be communicated by the secretariat to the Depositary, who shall circulate it to all Parties for their acceptance.

5. An annex, or amendment to an annex other than Annex A or B, that has been adopted in accordance with paragraphs 3 and 4 above shall enter into force for all Parties to this Protocol six months after the date of the communication by the Depositary to such Parties of the adoption of the annex or adoption of the amendment to the annex, except for those Parties that have notified the Depositary, in writing, within that period of their non-acceptance of the annex or amendment to the annex. The annex or amendment to an annex shall enter into force for Parties which withdraw their notification of non-acceptance on the ninetieth day after the date on which withdrawal of such notification has been received by the Depositary.

6. If the adoption of an annex or an amendment to an annex involves an amendment to this Protocol, that annex or amendment to an annex shall not enter into force until such time as the amendment to this Protocol enters into force.

7. Amendments to Annexes A and B to this Protocol shall be adopted and enter into force in accordance with the procedure set out in Article 20, provided that any amendment to Annex B shall be adopted only with the written consent of the Party concerned.

Article 22

1. Each Party shall have one vote, except as provided for in paragraph 2 below.
2. Regional economic integration organizations, in matters within their competence, shall exercise their right to vote with a number of votes equal to the number of their member States that are Parties to this Protocol. Such an organization shall not exercise its right to vote if any of its member States exercises its right, and vice versa.

Article 23

The Secretary-General of the United Nations shall be the Depositary of this Protocol.

Article 24

1. This Protocol shall be open for signature and subject to ratification, acceptance or approval by States and regional economic integration organizations which are Parties to the Convention. It shall be open for signature at United Nations Headquarters in New York from 16 March 1998 to 15 March 1999. This Protocol shall be open for accession from the day after the date on which it is closed for signature. Instruments of ratification, acceptance, approval or accession shall be deposited with the Depositary.
2. Any regional economic integration organization which becomes a Party to this Protocol without any of its member States being a Party shall be bound by all the obligations under this Protocol. In the case of such organizations, one or more of whose member States is a Party to this Protocol, the organization and its member States shall decide on their respective responsibilities for the performance of their obligations under this Protocol. In such cases, the organization and the member States shall not be entitled to exercise rights under this Protocol concurrently.
3. In their instruments of ratification, acceptance, approval or accession, regional economic integration organizations shall declare the extent of their competence with respect to the matters governed by this Protocol. These organizations shall also inform the Depositary, who shall in turn inform the Parties, of any substantial modification in the extent of their competence.

Article 25

1. This Protocol shall enter into force on the ninetieth day after the date on which not less than 55 Parties to the Convention, incorporating Parties included in Annex I which accounted in total for at least 55 per cent of the total carbon dioxide emissions for 1990 of the Parties included in Annex I, have deposited their instruments of ratification, acceptance, approval or accession.
2. For the purposes of this Article, "the total carbon dioxide emissions for 1990 of the Parties included in Annex I" means the amount communicated on or before the date of adoption of this Protocol by the Parties included in Annex I in their first national communications submitted in accordance with Article 12 of the Convention.

3. For each State or regional economic integration organization that ratifies, accepts or approves this Protocol or accedes thereto after the conditions set out in paragraph 1 above for entry into force have been fulfilled, this Protocol shall enter into force on the ninetieth day following the date of deposit of its instrument of ratification, acceptance, approval or accession.
4. For the purposes of this Article, any instrument deposited by a regional economic integration organization shall not be counted as additional to those deposited by States members of the organization.

Article 26

No reservations may be made to this Protocol.

Article 27

1. At any time after three years from the date on which this Protocol has entered into force for a Party, that Party may withdraw from this Protocol by giving written notification to the Depositary.
2. Any such withdrawal shall take effect upon expiry of one year from the date of receipt by the Depositary of the notification of withdrawal, or on such later date as may be specified in the notification of withdrawal.
3. Any Party that withdraws from the Convention shall be considered as also having withdrawn from this Protocol.

Article 28

The original of this Protocol, of which the Arabic, Chinese, English, French, Russian and Spanish texts are equally authentic, shall be deposited with the Secretary-General of the United Nations.

DONE at Kyoto this eleventh day of December one thousand nine hundred and ninety-seven.

IN WITNESS WHEREOF the undersigned, being duly authorized to that effect, have affixed their signatures to this Protocol on the dates indicated.

Annex A

Greenhouse gases
Carbon dioxide (CO_2)
Methane (CH_4)
Nitrous oxide (N_2O)
Hydrofluorocarbons (HFCs)
Perfluorocarbons (PFCs)
Sulphur hexafluoride (SF_6)

Sectors/source categories
Energy
    Fuel combustion
        Energy industries
        Manufacturing industries and construction
        Transport
        Other sectors
    Other
        Fugitive emissions from fuels
        Solid fuels
        Oil and natural gas
        Other
Industrial processes
    Mineral products
    Chemical industry
    Metal production
    Other production
    Production of halocarbons and sulphur hexafluoride
    Consumption of halocarbons and sulphur hexafluoride
    Other

Solvent and other product use

Agriculture
    Enteric fermentation
    Manure management
    Rice cultivation
    Agricultural soils
    Prescribed burning of savannas
    Field burning of agricultural residues
    Other
Waste
    Solid waste disposal on land
    Wastewater handling
    Waste incineration
    Other

Annex B

Party	Quantified emission limitation or reduction commitment (percentage of base year or period)
Australia	108
Austria	92
Belgium	92
Bulgaria*	92
Canada	94
Croatia*	95
Czech Republic*	92
Denmark	92
Estonia*	92
European Community	92
Finland	92
France	92
Germany	92
Greece	92
Hungary*	94
Iceland	110
Ireland	92
Italy	92
Japan	94
Latvia*	92
Liechtenstein	92
Lithuania*	92
Luxembourg	92
Monaco	92
Netherlands	92
New Zealand	100
Norway	101
Poland*	94
Portugal	92
Romania*	92
Russian Federation*	100
Slovakia*	92
Slovenia*	92
Spain	92
Sweden	92
Switzerland	92
Ukraine*	100
United Kingdom of Great Britain and Northern Ireland	92
United States of America	93

Annex table: Total carbon dioxide emissions of Annex I Parties in 1990, for the purposes of Article 25 of the Kyoto Protocol (FCCC/CP/97/7/Add1)

Party	Emissions (Gg CO_2)	Percentage
Australia	288,965	2.1
Austria	59,200	0.4
Belgium	113,405	0.8
Bulgaria	82,990	0.6
Canada	457,441	3.3
Czech Republic	169,514	1.2
Denmark	52,100	0.4
Estonia	37,797	0.3
Finland	53,900	0.4
France	366,536	2.7
Germany	1,012,443	7.4
Greece	82,100	0.6
Hungary	71,673	0.5
Iceland	2,172	0.0
Ireland	30,719	0.2
Italy	428,941	3.1
Japan	1,173,360	8.5
Latvia	22,976	0.2
Liechtenstein	208	0.0
Luxembourg	11,343	0.1
Monaco	71	0.0
Netherlands	167,600	1.2
New Zealand	25,530	0.2
Norway	35,533	0.3
Poland	414,930	3.0
Portugal	42,148	0.3
Romania	171,103	1.2
Russian Federation	2,388,720	17.4
Slovakia	58,278	0.4
Spain	260,654	1.9
Sweden	61,256	0.4
Switzerland	43,600	0.3
United Kingdom of Great Britain and Northern Ireland	584,078	4.3
United States of America	4,957,022	36.1
Total	**13,728,306**	**100.0**

Appendix 2
Key themes in economic debates:
insights from the IPCC Second Assessment Report

The economics of climate change mitigation has been the subject of extensive study and debate particularly since the issue started rising to political prominence in the late 1980s. These debates have generally received less press attention than the scientific issues, and it is harder to identify clear progress towards consensus or to judge the extent to which analysis has influenced policy, partly because of the more disparate nature of the subject matter and the different communities of analysts involved. Nevertheless, the economic issues are clearly of great concern to governments and different economic arguments have been wielded by different interests.

Many of the economic debates gained a sharper edge in the context of discussions within Working Group III of the IPCC's Second Assessment Report (with which the author was closely involved).[1] Subsequent debates about the level of commitments to be adopted under the Kyoto Protocol focused more closely on the costs of abatement. This Appendix summarizes briefly some of the main economic debates in these instances and in the various lobbying efforts surrounding the making of the Kyoto Protocol. The focus is upon the aggregate economic debates, rather than specific national studies, or potential trade effects that are considered elsewhere.[2]

1 How costly is the problem and how valuable a life saved?

The natural instinct of most economists, faced with the problem of climate change, is to try to compare the costs of the problem – the impacts of climate change and adapting to it – against the costs of limiting emissions. The most prominent contributions of economists in the early 1990s in this area were the competing analyses by Nordhaus and Cline, who attempted to estimate likely damages arising from climate change and to compare them with limitation costs.[3]

Most estimates of damage have stemmed from studies of the potential economic impact of a doubling of atmospheric CO_2 concentrations on the United States,

[1] J.P. Bruce, H. Lee, and E. Haites (eds), *Climate Change 1995: Economic and Social Dimensions of Climate Change (Contribution of Working Group III to the IPCC Second Assessment Report)*, Cambridge: CUP, 1996.

[2] D. Brack et al., *International Trade and Climate Change Policies*, London: RIIA/Earthscan, forthcoming 1999.

[3] W.D. Nordhaus, 'Economic Approaches to Greenhouse Warming', in R. Dornbusch and J. Poterba (eds), *Global Warming: Economic Policy Responses*, Cambridge, Mass.: MIT Press, 1991; W.R. Cline, *The Economics of Global Warming*, Washington, DC: Institute for International Economics, 1992.

extrapolated in various ways to apply to other countries. These and related initial estimates were reviewed in the IPCC Second Assessment Report: the chapter authors concluded that 'best-guess central estimates of global damage, including non-market impacts, are in the order of 1.5–2.0% of world GNP for 2 x CO_2 concentrations and equilibrium climate change. These figures are best-guess results and several impact categories could not be assessed for lack of data.' The Policymakers' Summary acknowledged that 'aggregate estimates ... tend to be a few percent of world GDP, with in general, considerably higher estimates of damage to developing countries as a share of their GDP'. The Policymakers' Summary in this area was the most convoluted and awkward of all, and the most painfully disjointed from the underlying chapter; it is the only case in the history of the IPCC where authors sought formally to distance themselves from the governmentally negotiated Policymakers' Summary. Efforts to quantify the impacts of climate change have continued, with reviews claiming to discern signs of consensus being promptly rebuffed by others.[4]

The reasons for such disagreement are not hard to see. Trying to quantify economically the impacts of climate change is a nightmarish exercise in which huge uncertainties combine with several important underlying value judgments. The problems may be roughly grouped as follows.[5]

Intrinsic uncertainties and risk aversion

The large scientific uncertainties translate into large uncertainties about physical impacts. Economic studies have mostly focused on best-guess estimates for a doubling of atmospheric concentrations. But the reality may be substantially different, and better or worse outcomes do not simply cancel each other out, because aversion to more adverse-than-expected outcomes generally weighs more heavily. One of the more consistent findings of economic analysis is that this makes climate change a more serious problem economically than is suggested by best-guess estimates, a curious contrast to the political debate in which uncertainty is generally (but usually irrationally) used to justify inaction.[6]

---

[4] J.B. Smith, 'Standardized Estimates of Climate Change Damages for the United States', *Climatic Change*, Vol. 32, 1996, pp. 313–26; response by D. Demeritt and D. Rothman, *Climatic Change*, Vol. 40, 1998, pp. 313–26.

[5] Overviews are given in for example R.S.J. Tol, 'The Damage Costs of Climate Change: Towards More Comprehensive Calculations', *Environmental and Resource Economics*, 5, 1995, pp. 353–74; R.S.J. Tol, 'The Damage Costs of Climate Change: Towards a Dynamic Representation', *Ecological Economics*, 19, 1996, pp. 67–90. See also the various specific references in this section.

[6] W.D. Nordhaus, *Managing the Global Commons: the Economics of Climate Change*, Cambridge, MA: MIT Press. The exception to this is if there is reasonable expectation that learning will resolve the main uncertainties rapidly enough (and favourably enough) to justify the additional costs of delay. Issues of risk and uncertainty raise quite fundamental dilemmas in the economic appraisal of climate change: see for example J. Robinson, 'Risks, Prediction and Other Optical Illusions', *Policy Sciences*, Vol. 25, 1992, pp. 237–54.

Low-probability, high-impact surprises

A special but important example is the risk of major surprises. Impact studies have tended to assume not only that climate change is relatively smooth, but also that there are no major surprises. Perhaps the most credible 'surprise' is that ocean current patterns will change, maybe rather suddenly. This could result in rapid regional temperature changes of well over 5°C. The underlying issue is that greenhouse gas emissions are interfering with immensely complex and interactive global systems that are far from adequately understood.

Dynamics of change

Assessing the costs of being in a warmer world is not the same as assessing the costs of adapting to climate change. As a continuing dynamic process, climate change is unlikely to be smooth and average conditions will probably not be an adequate indicator. The most important physical impact may be changing probabilities of extreme events, such as floods, prolonged droughts or intense storms. These changes may also be extremely hard to predict, limiting ability to prepare by adaptation, and 'economically optimal' responses, such as migration, may be constrained by brute political realities.

Evaluation of non-market impacts

A fourth area of difficulty concerns how economics evaluates impacts that fall outside any context of market valuation, such as damage to ecosystems. Economics has developed various approaches to evaluate such impacts, but the core problem is that the different approaches do not give consistent answers. Notably, it is not uncommon to find that 'willingness to accept compensation' for an adverse impact differs wildly from 'willingness to pay' to prevent it – but there are no clear grounds for choosing between these measures.[7] The problems are even worse when there are no credible *international* estimates of either, and there are basic ethical debates about *who* should pay to avoid impacts in different regions. In the present stage of the debate, one can only guess at the amount of compensation the small island states or drylands of Africa might demand for suffering climate change. It is intrinsically hard to test the 'veracity' of very high claims for compensation, but they should not simply be ignored.[8]

---

[7] D. Pearce and A. Markandya, *Blueprint for a Green Economy*, London: Earthscan, 1989. Furthermore, personal valuation of environmental quality varies between individuals and collectively over time (S. Owens, 'Negotiated Environments? Needs, Demands and Values in an Age of Sustainability', *Environment and Planning*, A29:4, 1997, pp. 571–80).

[8] This is most obviously because people are constrained in how much they can pay to avoid damages, but can say they would require infinite compensation for suffering impacts for which money is no consolation: a claim for which, as Adams notes, 'there is no affordable test' (J. Adams, *Cost-benefit Analysis: Part of the Problem, Not the Solution*, Oxford: Green College Centre for Environmental Policy and Understanding, 1995).

The 'value of statistical life'

One particular aspect of these complexities led the IPCC into exceptionally hot water, through the need in any quantitative study to capture mortality impacts in terms of a 'value of statistical life', defined as 'the value people assigned to a change in the risk of death among a population'.[9] Economic calculations such as those of Nordhaus, Fankhauser[10] and others assume a 'value of life' roughly in proportion to national per capita GDP, derived essentially from 'willingness to pay' approaches. Reference to such calculations in the IPCC provoked a furious reaction, given the implication that lives lost in developing countries would count for far less than lives lost in rich countries.[11] Economists sought to defend themselves by pointing out that the observed value of statistical life does unquestionably differ enormously between countries: it would indeed be absurd for India to try to put the same resources into modern medical services as the United States, when its people suffer many more basic threats to life and health.

As long as health and the related decisions of nations remain essentially separate, therefore, there is nothing necessarily immoral in putting a different value on statistical life in different places; it is simply a reflection of reality. But climate change concerns impacts by some nations on others; and in aggregate, by rich nations on poorer ones. Whose value of statistical life should then be used in evaluating the damages? This pointed to the more general issue concerning valuation of non-market impacts, namely: whose valuation, and on what basis? It was tied in with the theoretical debate about 'willingness to pay' versus 'willingness to accept', and highly political and ethical debates about who should pay.[12] The original IPCC authors on this topic acknowledged some of these complexities and ultimately observed that the lack of any consistent global decision-making authority meant that economics should not be *expected* to come up with a single globally applicable answer to 'the cost of climate change'.[13]

[9] J.P. Bruce et al., *Climate Change 1995*, Policymakers' Summary, Footnote 7.

[10] S. Fankhauser, *Valuing Climate Change – the Economics of the Greenhouse*, London: Earthscan, 1995.

[11] The Global Commons Institute launched its first campaign on this issue. The theme was picked up by the Indian Environment Minister, Kamal Nath, who wrote to other heads of delegations at the first meeting of the Conference of Parties to the Convention, flatly rejecting: '... the absurd and discriminatory Global Cost/Benefit Analysis procedures propounded by economists in the work of IPCC WG-III ... we unequivocally reject the theory that the monetary value of people's lives around the world is different because the value imputed should be proportional to the disparate income levels of the potential victims ... it is impossible for us to accept that which is not ethically justifiable, technically accurate or politically conducive to the interests of poor people as well as the global common good.'

[12] M. Grubb, 'Seeking Fair Weather: Ethics and the International Debate on Climate Change', *International Affairs*, Vol. 71, No. 3, 1995, pp. 463–96.

[13] S. Fankhauser, R.S.J. Tol and D.W. Pearce, 'The Aggregation of Climate Change Damages: a Welfare-Theoretic Approach', *Environmental and Resource Economics*, Vol. 10, 1997, pp. 249–66.

Evaluating the very long term

A similar but even more difficult issue is how to evaluate the very long term and intergenerational equity. Since climate change is such a very long-term issue, the weight placed on the welfare of future generations is of central importance, one which may well override most other concerns and uncertainties. Without going into detail – rooms could be filled with the economic literature on time preference – again there are alternative approaches that yield very different results. A *prescriptive* approach derived from economic first principles tends to yield much higher weights on future welfare than a *descriptive* approach based upon the time preferences apparently revealed in current economic indicators such as the apparent market returns to long-term risk-free investments. The unique nature and time scale of climate change impacts leave plenty of room for debate about which is really appropriate.[14] Increasingly, indeed, the literature has challenged the fundamental assumption of most economists that a single 'discount rate' could be used to compare future impacts against the present.[15]

The overriding conclusion is that valuing the impacts of global climate change is fundamentally an ethical issue as much as it is a technical one. Furthermore, it is notable that on all three of the 'distributional' aspects noted above – valuation of non-market impacts, their aggregation across countries, and weighting of future impacts – paying greater explicit attention to the ethical issues indicates climate change to be a more serious issue than suggested by the simpler earlier calculations of most Western economists.[16]

The economic debate about the monetary value of climate change damages thus became extremely convoluted and at times politically highly contentious. The problem is bedevilled by the danger of 'counting the things that can be measured, not the things that count', and the whole process of valuation for such a global and very long-

---

[14] For an expansive economic discussion of discount rates, see the IPCC WG-III report, Bruce et al., *Climate Change 1995*, Chapter 4.

[15] The discount rate is of huge and fundamental importance in assessing how seriously we should take a long-term problem like climate change. Chichilnisky shows that standard discounting approaches ultimately involve dictatorship of the interests of present generations over those of future generations (G. Chichilnisky, 'An Axiomatic Approach to Sustainable Development', *Social Choice and Welfare*, 1996). Other critiques include C.L. Spash, 'Double CO_2 and Beyond: Benefits, Costs and Compensation', *Ecological Economics*, Vol. 10, 1994, pp. 27–36; and M.L. Weitzman, 'Why the Far-Distant Future Should Be Discounted at Its Lowest Possible Rate', *Journal of Environmental Economics and Management*, Vol. 36, 1998, p. 201–8. Another recent contribution argues that the discount rate should be not constant but varied as the logarithm of time t (G. Heal, 'Discounting and Climate Change – an Editorial Comment', *Climatic Change*, Vol. 37, 1997, pp. 335–43).

[16] The main counter-argument to this conclusion is that a more holistic ethical stance would identify tackling climate change as a far lesser priority than addressing global poverty directly. This is true, but its relevance is debatable. Governments show little sign of becoming more influenced by the ethical dimensions of development assistance, but they are actively engaged in international negotiations on climate change in which the international ethical issues have to be addressed in order to attract widespread participation. Or to put it more crudely, 'two wrongs don't make a right'.

term issue raises complex ethical issues on which economists – and philosophers – have so far proved unable to offer any consensus. Attempting to conduct cost-benefit analysis of climate change provides quite fundamental challenges to the frontiers of economic appraisal. It is thus not surprising that the political process focused on a more tangible framework – the objective of stabilizing the atmosphere at safe levels as specified in the Convention – and after the Convention, many analyses began to study pathways towards atmospheric stabilization.

2 How urgent is the problem and how costly to delay?

In moving the focus from apparently arcane arguments about the 'monetary-equi-valent costs' of climate change risks and damages, onto the practical issue of reducing risks by stabilizing atmospheric concentrations, proponents of more aggressive abatement assumed that the need for stronger action would become clearer. Initially at least, they were in for a shock. A paper by Wigley, Richels and Edmonds (hereafter WRE) applied a carbon cycle model to explore alternate emission pathways, and suggested that pathways in which abatement is deferred would be economically preferable.[17] They offered four main reasons:

- existing *capital stock* would make rapid change costly;
- *technical progress* would reduce the costs of abatement in the future;
- *discounting* would reduce the present value of actions if deferred (or, would require less resources to be put aside now to finance it);
- continuing *absorption* of CO_2 meant that earlier emissions allowed more to be emitted overall.

Thus, they argued, even if the atmosphere needed to be stabilized it would be cheaper to defer abatement action, while pursuing R&D. They attempted to inject this new literature into the IPCC Second Assessment Report even before their main paper was accepted for publication, and there followed vigorous debate which emphasized the extreme complexity of the subject.[18]

Capital stock turnover

WRE's observation that 'time is needed to re-optimize the capital stock' was double-edged as an argument for deferring action, since capital stock is continually being

---

[17] T.M.L. Wigley, R. Richels and J. Edmonds, 'Economic and Environmental Choices in the Stabilisation of Atmospheric CO_2 Concentrations', *Nature*, Vol. 379, 18 January 1996.

[18] A more detailed analysis is given in M. Grubb, 'Technologies, Energy Systems and the Timing of CO_2 Emissions Abatement: an Overview of Economic Issues', *Energy Policy*, Vol. 23, No. 2, 1997, pp. 159–72.

refurbished, retired or restructured and additional new investment is required to meet demand growth.[19] A key to economically efficient abatement is to make new capital stock less carbon-intensive than it otherwise would be. This will mean a steady reduction of emissions from the business-as-usual trajectory, starting as soon as climate change is recognized to be a potentially serious problem. The economic importance of getting this right was itself apparent from the scenarios in the WRE analysis. In their central case, their 'deferred abatement' trajectory showed emissions rising by more than 50% over the next 40 years, before dropping steadily. Thus the delay scenario involved constructing an *additional* 4 GtC per year of capital stock over and above that anyway required for replacement. The scenario would mean investing in at least as much new CO_2-based capital stock over the next few decades as is embodied in the world's entire energy systems today, and then dismantling it over the subsequent decades.

Especially when coupled with uncertainties about the actual objective (see below), this has powerful industrial implications. Inappropriate delay in constraining emissions is not necessarily in the interests of industry and could backfire: it could increase the exposure of industry to the risk that new, carbon-intensive investments will have to be prematurely retired at large cost and dislocation compared with the costs of avoiding such investments in the first place (e.g. coal-powered plants or mines left 'stranded', or frontier oil exploration and development left without sufficient high-price markets when they mature). The World Energy Council, in the conclusions to the Tokyo World Energy Congress, concurred that 'action postponed will be opportunity lost, guaranteeing that when action can no longer be avoided the ensuing costs will be higher; dislocations more severe; and the effects much less predictable, than if appropriate actions are taken today'.[20]

Building on this theme, a direct response to WRE by HaDuong, Grubb and Hourcade (hereafter referred to as HGH) explored the issue of inertia more closely.[21] They argued that the models underlying the WRE analysis greatly underestimated inertia, and showed that given a quantitatively more realistic value for inertia in the global energy system, deferring abatement until 2020 could prove extremely costly (Figure A2.1) for any stabilization objective much below about 550 ppm.

The HGH analysis went on to make a more general argument about uncertainty and inertia. If we delay action in the belief that we are aiming to stabilize concentrations at 550 ppm, for example, then after a couple of decades it may simply be too late to be able to stabilize at 450 ppm, however urgent the problem then turns

---

[19] Frequently the oldest capital stock is also the least efficient, with rising maintenance costs. The net costs of retiring such stock rather than refurbishing it for a longer (polluting) life may be small and may indeed result in net gain when other factors are considered. When the costs are finely balanced the economic issues are similar to those involved in new investment to meet demand growth.

[20] World Energy Council, *Energy for our Common World – What Will the Future Ask of Us?* Conclusions and Recommendations of the 16th WEC Congress, Tokyo and London: WEC, 1995.

[21] M. HaDuong, M. Grubb and J.C. Hourcade, 'Influence of Socioeconomic Inertia and Uncertainty on Optimal CO_2 Abatement', *Nature*, 390, 20 November 1997.

Figure A2.1: Costs of abatement after delay, for 450 ppm and 550 ppm CO_2-only concentrations

% of 1990 World Gross Product

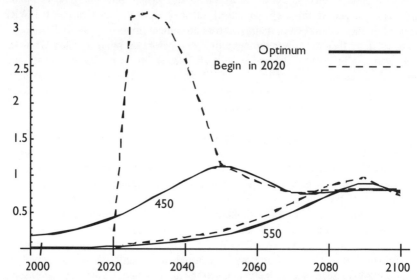

Source: M. HaDuong, M. Grubb and J.C. Hourcade, 'Influence of Socioeconomic Inertia and Uncertainty on Optimal CO_2 Abatement', *Nature*, 390, 20 November 1997.

out to be; and even stabilization at 500 ppm might by then involve radical changes of direction that could prove economically very disruptive. Also after such a delay, stabilization at a lower-than-expected level would require not only faster, but much deeper action – car-free cities, for example, rather than the low-emission, high-efficiency vehicles that might be consistent with a smoother abatement trajectory. Steady sustained pressure to limit emissions cannot expose us either economically or environmentally to the scale of risks that may be incurred by a long delay.

Technology development

The second line of argument for deferring abatement advanced by WRE was that technology development would reduce abatement costs in the future, so that it would be cheaper to defer abatement and invest in R&D instead. There are, however, two problems with this. First, not all abatement requires advanced technologies (at currently high but declining costs). In reality there is a whole continuum of abatement options. It is clearly a *non-sequitur* to imply that cheap abatement should be deferred on the grounds that the costs of more expensive action might decline in the future.

Second, this whole line of thinking assumed that all technological development occurs independently of emission abatement efforts. This envisages technology development as an 'exogenous' process occurring independently of market conditions. This may be true to the extent that technology development represents an automatic accumulation of knowledge, or is fostered primarily by government R&D. But the idea that new technologies develop autonomously, or arise primarily because governments pay to develop them, was abandoned decades ago by economists who work on technology issues.[22] Government R&D can help, but most effective technology development and dissemination is done by the private sector in pursuit of markets. In other words, much technology development is *induced* by market circumstances: market experience leads to cost reductions, and expectations about future market opportunities determine how industries deploy their R&D efforts.

This is not surprising, since in fact corporate R&D swamps government R&D in most countries. The energy sector itself, ranging from oil platforms to wind energy, has provided powerful examples of the fact that technology development depends strongly upon market conditions.

Induced technology development implies that the *act* of abatement would generate market opportunities, cash flows and expectations that enable industries to orient their efforts and learning in the direction of lower carbon technologies. Hence, on this model, action itself generates cheaper technological options arising out of accumulating experience.[23] In this case, deferring emission reductions simply delays the generation of options that can address the problem at low cost. The World Energy Council again rejected the nascent economic arguments offered by WRE; one of the Recommendations of the Tokyo World Energy Congress Statement urged 'governments, business decision-makers and energy consumers' to 'start taking action now to adapt to the needs of our long term future ... the next two or three decades represent the key period of opportunity for a transition to a more sustainable path of development for the long term. Research done and action taken now will begin the shift of direction required of "minimum regrets" action.'[24]

---

[22] K. Arrow, 'The Economic Implications of Learning-by-Doing', *Review of Economic Studies*, Vol. 29, 1962, pp. 155–73. A standard reference on induced technical change has become W.B. Arthur, 'Increasing Returns and Path Dependence in the Economy', Ann Arbor: University of Michigan Press, 1995. For studies of the evidence and implications concerning energy systems and climate change, see references in note 43.

[23] One objection that has been raised to this is the fact that stimulating innovation in one sector (e.g. energy) may reduce innovation in another sector. In fact, the evidence for this is rather slim; it is not at all obvious that challenges in one sector do in general cause them to draw 'innovation resources' from elsewhere, though the possibility deserves further empirical study.

[24] World Energy Council, *Energy for our Common World*.

Discounting and carbon cycle effects

The other two arguments advanced by WRE were that delay would lower the 'present value' of abatement costs by discounting (the costs were pushed further into the future), and that emitting more carbon earlier meant that more would be reabsorbed – hence more could be emitted in total – before the stabilization constraint was approached. In terms of modelling abatement costs under a stabilization constraint, both these were correct observations, but they had little relationship to the overall economic question of appropriate emission strategies.

First, considering damages expected from climate change, whether they are large or small, brings a new element into the calculations of time discounting. The argument about discounting would apply similarly to impacts: avoiding damages earlier also has a greater present economic value. Deferring emissions abatement defers abatement costs, but it brings impacts nearer. So even neglecting all other considerations, the implications of time discounting for the overall policy problem are not as clear-cut as implied in studies that considered only the question of stabilization without reference to damages (so that discounting applies only to abatement costs). WRE was in fact careful to note that the time paths of impacts would differ according to the time path of emissions, and presented calculations of how global average temperature and sea-level change varies between scenarios (though it did not draw the link with discounting, and a subsequent economic study suggested indeed that the monetary estimate of additional climate damage from the 'delay' scenarios might be comparable with the supposed economic benefits of deferring abatement).[25]

Also, the problem with the argument that carbon emitted earlier could be reabsorbed is that in fact it means greater overall peak rates of change in the atmosphere, more rapid change and more cumulative heat trapped. Since human societies are likely to have greater difficulty in adapting to rapid climate changes than to slow and smooth changes, again this is hardly a compelling line of argument.

To that extent, therefore, the use of economic arguments to justify deferring abatement backfired: the ensuing debates clarified the economic case for action. Analyses based on less overtly economic approaches, exploring 'tolerable windows'

---

[25] For WRE's central case, averaged over the next 50 years, the rate of temperature change appears to be more than 20% higher than is the case without deferral of emissions abatement. The Dutch IMAGE model has also been used to explore the implications of deferred abatement under a 450 ppm ceiling. Compared to the IPCC 450 ppm scenario, the scenario in which abatement is delayed until 2025 (followed by rapid reductions) leads to a 40% higher rate of global average temperature change over the first half of the next century, and a higher overall peak temperature later in the century; other indicators that are substantially affected throughout the next century by delayed action include maize yields and natural vegetation change (J. Alcamo, results presented at 3rd International Workshop on the IMAGE 2 Model, Delft, The Netherlands, 15–16 February 1996). Quantitative analysis of the effects of deferring abatement on climate change shows these impacts to be significant, and also to fall most heavily on developing countries; thus, deferring abatement also raises questions of equity (see for example R.S.J. Tol, 'On the Difference in Impact of Two Almost Identical Climate Change Scenarios', *Energy Policy*, Vol. 26, No. 1, pp. 13–20).

for emission trajectories towards certain climate-related goals, also emphasized the need for urgent action.[26] The debates did leave another important legacy, however. By focusing on the fact that it could take a long time to adjust and reorient the US energy system – an argument pressed very strongly by the US electricity industry, which feared the costs of early closure of existing coal-fired stock – the economic arguments reinforced pressures to ensure that if Kyoto did result in binding constraints, they should be delayed for as long as possible. The US administration adopted the position that action was necessary but that constraints should not become binding until 2010 at the earliest. The rest of the world did not succeed in shifting the US government from that position during the Kyoto negotiations.

3 How costly is abatement and what might be the 'hidden' benefits?[27]

Economists have for many years been studying the costs of limiting greenhouse gas emissions – particularly carbon dioxide from energy systems. These debates have been singularly confusing for the non-specialists, and while arguments about the economics of timing emissions abatement reached certain conclusions, those about the overall costs did not. Dissension among the scientists seemed modest compared to the wide-ranging estimates of the costs of limiting emissions: some analysts could not even agree on whether the costs would be positive or negative.

On costs and models

'Costs' is in fact a rather tenuous concept when analysed in detail; the IPCC has produced a whole report on cost methodology. In policy assessment, cost should be an indication of welfare loss. Many economic studies equate welfare loss directly with loss of GDP. This provided one source of disagreement. Even expressed in terms of the impact of emission constraints on GDP, however, and ignoring the presumed climate benefits of emissions control, studies still could not even agree on whether the costs were positive or negative. To understand why, it is useful to divide the 'economic cost' of limiting greenhouse gas emissions into three categories:

* measures that produce aggregate economic benefits irrespective of climate change considerations (e.g. promotion of economically beneficial energy efficiency measures)

---

[26] J. Alcamo and E. Kreilman, 'Emission Scenarios and Global Climate Protection', *Global Environmental Change: Human and Policy Dimensions*, Vol. 6, No. 4; J. Alcamo, 'The Safe Landing Approach: Risks and Trade-offs in Climate Change', in J. Alcamo, R. Leemans, and E. Kreileman (eds), *Global Change Scenarios for the 21st Century: Results from the Image 21 Model*, Oxford: Pergamon Press, 1998.

[27] This section is a shortened version of J. Fischer and M. Grubb, *The Use of Economic Models in Climate Change Policy Analysis*, Climate Change Briefing Paper No. 5, London: RIIA, October 1997.

– commonly, if unfortunately, known as 'no regret' measures;[28]
- short-run transitional or dislocation impacts arising from additional measures, for example from premature scrapping of existing capital and the limited flexibility of capital or labour markets to respond to structural changes;
- long-run continuing costs that arise from using resources that could have been used for other purposes in the absence of CO_2 constraints.

These cost components correspond fairly well to the three main classes of models employed in assessing the economic impact of CO_2 abatement:

- *bottom-up engineering* models that are specified in terms of the technologies available for supplying energy services, and rely on engineering assessments of the costs of constructing and running different technologies;
- *short- and medium-run macro-economic* models that are specified in terms of economic parameters such as investment, interest rates, labour markets, etc., and rely on historical evaluation of the relationships between the various economic activities;
- *long-run general equilibrium* (GE) models that estimate optimal use of resources within a given set of constraints, given various exogenous assumptions concerning price responses and costs, population growth and technological change.

Both macro-economic and general equilibrium models are usually known as 'top-down', though they differ radically in many respects. In addition, each main class of model has many different variants, and hybrids are possible and are increasingly applied (e.g. many general equilibrium models use partly bottom-up specification of technologies for electricity supply).[29] The vast majority, however, fall clearly into one of these three categories.

During 1996, partly to buy time and defuse some of the political debates surrounding the Kyoto negotiations, the US government embarked upon a major appraisal of the costs of limiting CO_2 emissions. The study was careful to use one model of each type. As explained below, the outcome was predictable, and the rest of this section explains why consensus was impossible – and may always be, when the 'cost' question is posed in such a general way.

---

[28] Many economists treat this category with some scepticism, on the grounds that if measures are really economically beneficial in their own right they should have been implemented anyway, and certainly should not be credited as benefits from CO_2 abatement. However, such measures might not have been implemented in the absence of climate change concerns for a variety of political, distributional or other considerations; they may be 'no regret' in terms of GDP impacts, but usually some group with political influence can find a reason to regret them.

[29] For more detailed classifications and some model types that do not fit into these categories, see Bruce et al., *Climate Change 1995*, and M. Grubb, J. Edmonds, P. ten Brink and M. Morrison, 'The Costs of Limiting Fossil Fuel CO_2 Emissions: a Survey and Analysis', *Annual Review of Energy and Environment*, 18, 1993, pp. 397–478.

Bottom-up models

Bottom-up models are based on detailed technical analyses and engineering studies of the life-cycle costs of currently available and emerging technologies for energy supply and end-use, including options to increase energy efficiency.[30]

These bottom-up models have been extensively applied for many years to estimate and cost the potential for reducing energy consumption and their associated emissions (of CO_2 etc.) through greater use of low-cost energy-efficiency and energy-supply technologies. The technical assessment by Working Group II of the Second Assessment Report of the Intergovernmental Panel on Climate Change (IPCC) relied exclusively on bottom-up assessments. The authors claimed that advanced technologies could supply energy needs with reduced CO_2 emissions at little extra cost. The IPCC's Working Group III relied primarily upon top-down frameworks but included reference to some bottom-up assessments. The two Working Groups managed to agree that improvements in energy efficiency could reduce CO_2 emissions by perhaps 10–30% over 2–3 decades at zero or negative costs (i.e. an economic benefit).[31] The potential improvements increase with time, since the new, more efficient technologies can then be integrated into normal plant replacement.

However, since the signing of the Framework Convention on Climate Change, there has been limited practical implementation of these potential improvements.

Limitations of bottom-up models

Bottom-up models have been refined and extended to allow for links between options and for various feedback effects and the effects of stock turnover.[32] However, they cannot allow for all feedback effects (e.g. effects on availability of capital and fuel markets). In addition, long-term projections rely upon price and performance projections for energy technologies which are highly uncertain.

Bottom-up models estimate the net resource costs of the energy-efficiency measures in terms of, for example, costs of labour, raw materials and capital equipment and savings in energy costs. However, they tend not to include the following less tangible (or hidden) but still important costs:

---

[30] The bottom-up models generate a marginal cost-effectiveness curve showing what additional levels of emissions reductions could be achieved at what costs by implementing specific control options – starting with the least costly first.

[31] IPCC (1996), *Climate Change 1995: Adaptation and Mitigation Technologies* (Working Group II), and *Climate Change 1995: Economic and Social Dimensions of Climate Change* (Working Group III), op. cit. note 1. There are many other examples of bottom-up studies including various works by US laboratories that culminated in a '5-lab' study concluding that the US could bring its emissions back to 1990 levels without net economic cost (Interlaboratory working group, 'Scenarios of US carbon reductions', Office of Efficiency and Renewable Energy, US Department of Energy, Washington DC). Other wide-ranging examples are the multi-volume works by F. Krause, *Cutting Carbon Emissions: Burden or Benefit,* International Project for Sustainable Energy Paths, California, 1993–97.

[32] The most widely used is the MARKAL model developed by the International Energy Agency, Paris, and now widely used for national studies.

- transaction costs of the (management) time involved in obtaining information about the more energy-efficient products and their applicability to an individual's circumstances;
- costs of disruption to the existing production lines as the new technology is introduced and tested, and costs of retraining workers to use the new technologies successfully so as not to impair existing production or product quality;
- any losses in consumers' welfare if the more energy-efficient products reduce comfort or convenience or are less reliable (e.g. welfare losses to consumers of moving to smaller, more fuel-efficient cars).

Bottom-up models, in other words, tend to give a relatively optimistic view of the potential for emissions limitation at modest (often negative) net cost, and neglect some important factors.

Macro-economic models

Medium-term macro-economic models have been developed to analyse the macro-economic implications of climate change policy options, mostly carbon taxes. These models have included greater disaggregation and refinement of the modelling of the energy and energy-using sectors (e.g. separation by fuel type).[33]

The macro-economic models set out the links among economic sectors. Thus a carbon tax would increase the price of energy and energy-using industries and hence inflation rates, which then affect wage rates. The resulting changes in costs for energy-using industries affect their investments and imports and exports, leading to changes in GDP levels. The model also allows for the carbon tax leading to increased output of some sectors. The links between industries are set out in an input/output matrix in the model. The model specifies elasticities which quantify how each sector would respond to a price change. The coefficients in the model are based on econometric analysis of past data. The models yield estimates of the impacts of policies on the standard economic indicators such as GDP, employment or unemployment, inflation and balance of payments.

Macro-economic models analyse the dislocation costs and structural changes that arise as an economy adjusts to price changes. They do not assume that the economy is at full employment and in equilibrium at all times, hence they can allow for persistent imbalances such as unemployment – unlike GE models which do not well capture such transitional costs.

Repetto and Austin[34] estimate that macro-economic models for the US economy generate cost estimates higher by about 1.4% of GDP than a GE model. However, the

---

[33] For example, see European Commission, *E3ME: an Energy-Environment-Economy Model for Europe*, Report for the Joule II Programme of DG XII of the European Commission, 1995.
[34] R. Repetto and D. Austin, *The Costs of Climate Protection: A Guide for the Perplexed*, Washington, DC: World Resources Institute.

implications of the difference between the models are often not clear-cut. In Europe, macro-economic models frequently give lower cost estimates than GE studies, possibly because they give more importance to the use of tax revenues to reduce existing distorting taxes, which lowers the costs of the policies (see below).

Limitations of macro-economic models include the following:

- their limited long-term properties constrain their suitability for analyses looking more than ten years ahead;
- they do not adequately incorporate the balancing factors that arise in market economies in response to (non-marginal) changes such as increased investment or prices;
- they tend to be highly complex and economy-specific.

These factors limit the capability of the macro-economic models to assess the economic impacts of abatement policies beyond the short or medium term (beyond 5–10 years) or to command consensus when applied internationally.

General equilibrium models

General equilibrium models are based on relationships between supply and demand for the various economic activities and factors of production that are mainly derived statistically from past data. They allow for (adjustments in) prices to attain an efficient allocation of resources in the economy so as to maximize consumers' welfare. GE models usually estimate changes in GDP levels which represent the continuing costs of constraints on CO_2 emissions, in terms of the opportunity costs of reallocating resources to a more costly outcome than would otherwise occur.[35]

General equilibrium models are appropriate for examining the impacts and any feedback effects associated with major non-marginal changes (e.g. in energy prices) to which macro-economic models are not well suited. They are also suitable for quantifying economic impacts and interdependencies at a global level.

Limitations of GE models include the following:

- they do not include estimates of the dislocation costs which arise as the economy moves to a new equilibrium in response to the climate change policy (see above);
- they assume perfect flexibility and resource utilization in all markets (e.g. in most

---

[35] Major GE models that have been published in peer-reviewed literature and applied to estimate the costs of abatement policies include: the Global 2100 / MERGE models of Alan Manne (Stanford) and Richard Richels (EPRI), and derivatives of this such as Peck and Tiesberg's CETA model and Rutherford's CRTM model; the OECD's GREEN model, developed into the EPPA model at MIT; the Edmonds-Reilly model and subsequent 'Second Generation Model' at the Battelle Pacific-Northwest Laboratories; the ABARE model developed at the Australian Bureau of Agricultural and Resource Economics. For further details see Bruce et al., *Climate Change 1995*.

such models, wages adjust to ensure full employment). They do not allow for the possibility that policies (e.g. reductions in distorting taxes) might lead to reductions in involuntary unemployment. This omission may not be appropriate where labour markets suffer from persistent unemployment.

The merits of both GE and macro-economic models are that they take full account of the diverse linked impacts on various economic activities of climate change policies, including feedback effects.

Limitations of both GE and macro-economic models are that:

• they explicitly do not allow for any potential 'no regret' regulatory measures;
• they depend upon particular theoretical representations of economic relationships between investment, prices, exchange rates, etc., which are sometimes contested;
• their estimates for key parameters such as the elasticities of demand are based largely on historical data. It is not clear whether these estimates are appropriate for looking far into the future for the major changes involved in climate change policies. In particular, they generally do not include specific technological options, which may be particularly important for large-scale changes that fall outside the range of historical price variations.

Key sensitivities

The results of economic modelling studies, of course, do not just depend on the models. They depend fundamentally upon assumptions. In fact, the number of critical assumptions turns out to be fairly limited – but variation across them can radically alter results.

Reference emissions and 'no regret' potentials

In the absence of abatement measures, energy demand and CO_2 emissions are expected to grow in most countries, at a rate that is highly uncertain and depends upon economic growth, structural change in the economy, and energy prices among other factors. The greater the business-as-usual growth, the greater the change required to achieve fixed emission goals, e.g. relative to 1990 levels. Because most top-down models analyse abatement costs in terms of the differences from a 'reference' case which is assumed to be optimal, the costs of achieving a given target depend heavily upon the assumed reference trajectory.

Frequently this optimal reference trajectory is assumed to be the same as business-as-usual emission projections, but bottom-up models highlight that the latter often appears not to be optimal as assessed on the basis of engineering costings.[36] To the

---

[36] See discussion in the IPCC Reports.

extent that cost-effective 'no-regret' policy measures do exist and can be implemented, top-down studies that analyse fixed targets (e.g. relative to 1990 emissions) will exaggerate the costs.

The use of carbon tax revenues and 'double dividends'

A carbon tax could raise substantial revenue. An energy tax proposed by the UK's Institute for Public Policy Research (IPPR) would generate £8.8 billion in revenue in 2005,[37] equivalent to approximately 3% of total UK government receipts or about 1.5% of GDP.

The early analyses of the costs of carbon taxes to control CO_2 emissions assumed that the revenue was redistributed to consumers in a lump-sum manner which would have no effect on incentives to employ labour or to work. However, this is not realistic. Subsequent studies have examined alternative ways of using the revenues.[38]

Using the revenue to reduce existing distorting taxes (e.g. employers' insurance contributions) could yield a double dividend in the form of higher GDP and employment because it would reduce the costs of existing distorting taxes, including the costs of hiring labour. Moreover, the carbon-intensive industries are highly capital-intensive, so that structural change from these to more labour-intensive sectors might increase employment. However, this depends upon whether unemployed people would take up the increased employment opportunities or whether the greater demand for labour would lead to higher wages and hence prices, which could offset deflationary macro-economic policies.

Claims for a 'strong' form of the double dividend, that a carbon tax could yield net economic benefits irrespective of environmental considerations, have not been widely supported by economists, and the potential for reducing unemployment through appropriate use of revenues is still a matter of considerable debate.[39] Nevertheless there is clearly scope for a weaker form of double dividend in which using the revenue to reduce existing distorting taxes could help significantly to offset the direct costs. The extent of this offset would depend on how the revenues are used.

---

[37] S. Tindale and G. Holtham, *Green Tax Reform: Pollution Payments and Labour Tax Cuts*, London: Institute for Public Policy Research, 1996.

[38] See T. Barker and N. Johnstone, 'International Competitiveness and Carbon Taxation', in T. Barker and J. Kohler (eds), *International Competitiveness and Environmental Policies*, Aldershot: Edward Elgar, 1997; and various papers in C. Carraro and D. Siniscalco (eds), *Environmental Fiscal Reform and Unemployment*, Dordrecht: Kluwer Academic Publishers, 1995.

[39] The distorting costs of a carbon tax would be higher than those of many existing taxes because a carbon tax would affect intermediate inputs in the energy and energy-using sectors. That, indeed, is part of their purpose. Therefore a carbon tax would entail some net costs which need to be compared with its environmental benefits in terms of contributing to lower climate change and its associated damage costs.

Elasticity of substitution and technology representation

Economic models represent energy technologies as ways of combining different inputs (e.g. labour and carbon) to produce useful energy. In most top-down models, 'elasticities of substitution' represent the way in which one input (e.g. labour) could substitute for another (e.g. carbon) as a continuous relationship across a wide range of possible technologies, implying that any combination of inputs in known technologies can be adopted. This overstates the substitutions that could actually be made in practice, and hence could underestimate the costs of climate change policies. Some models address this by presenting sensitivity analyses for lower elasticities of substitution.

The Australian ABARE model[40] adopts the different approach of defining a finite number of specific existing bundles of technologies. This is conceptually more plausible, but the limited bundle of established technologies employed means that in this respect the ABARE model tends to overestimate the costs of mitigation, especially for the medium and long terms.

Top-down models can combine substitution elasticities with a 'backstop technology' that would enable the generation of energy without emitting any greenhouse gases (e.g. renewables) at a certain cost threshold.[41] This can be an important factor in containing costs, depending upon the cost threshold assumed.[42] However, inevitably the models give a crude representation of the complex mix of technology options. For example, co-generation, wind energy and biomass energy technologies are now expected to play a significant role in limiting European CO_2 emissions in the period 2000–2020, but few top-down models explicitly include such technologies or resources.

Induced technological change

Most models allow for technical improvements that, for example, may reduce the costs of low-CO_2 technologies over time, but treat technological change as 'exogenous'. None of the three main modelling classes allow explicitly for the likely impact of policy changes (e.g. CO_2 abatement) on technological change and the costs of technologies. In computational terms this would be difficult to incorporate in GE models.

But studies have emphasized that much technical change is in fact induced by market circumstances and policies. Preliminary studies have stressed that making technical change endogenous (i.e. determined in the model according to market conditions and policies) may substantially reduce long-run abatement costs, alter emission trajectories,

---

[40] Brown et al., *The Economic Impact of International Climate Change Policy*, ABARE Research Report 97.4 by Australian Bureau of Agricultural and Resource Economics, Canberra, 1997.
[41] For example, Manne and Richels's Global 2100 and MERGE2 models; OECD's GREEN model; Edmonds-Reilly model.
[42] The IIAM model of Charles River Associates (Cambridge, MA) appears to include a non-fossil 'backstop' technology but at a cost that exceeds current costs of several renewable energy technologies, and which makes the backstop irrelevant except for long-run and deep reductions.

and increase the costs of deferring controls (which would also defer valuable technical change).[43]

Responding to this, other economists have emphasized that technical improvements in one sector (e.g. carbon-saving reductions) may be at the expense of improvements in other sectors; the net effect on total costs depends on how the returns from the induced changes compare with those from displaced changes in other sectors of the economy.[44] Some GE models (e.g. the ABARE model) assume that climate change policies can change the type and direction of technological change, but that the total amount of technological change is constant. Other non-modelling studies highlight the central role of innovation in technologies, systems and lifestyles in addressing climate change, and argue that economic models do not adequately capture such aspects.[45] At present these important debates are at an early stage of development.

Disaggregation

Greater disaggregation of fuel types and affected sectors enables models to reflect reality more closely and hence yield more accurate cost estimates. Thus backstop technologies may be more readily applicable for the electricity supply industry, but they appear more difficult and costly for transport, while some industrial sectors (e.g. aluminium) may be much more energy-intensive and vulnerable than the average. Aggregation of fuel types and the traded sectors affected by climate change policies may underestimate the impacts on competitiveness and 'leakage' of emissions.

Conclusions on models and abatement costs

Economic models have been widely used but there is little consensus about the overall economic impacts of limiting CO_2 emissions, let alone the additional complexities raised by including the other gases in the Kyoto Protocol. This is because of the variety and complexity of energy systems, technologies and markets, and the inherent uncertainties arising from the long time scales and wide geographical spread of the impacts. The three main classes of models used in the debates focus upon different cost issues and there is little common ground between them.

Bottom-up engineering models demonstrate that technologies exist that could reduce CO_2 emissions with net economic benefit if they were used in place of current techniques; they thus identify areas of potential 'no regret' measures that deserve

---

[43] M. Grubb, 'The Economics of Changing Course', *Energy Policy*, April 1995; A. Grubler and S. Messner, 'Technological Change and the Timing of Mitigation Measures', *Ecological Economics*, 1999; A. Grubler, *Technology and Global Change*, Cambridge: CUP, 1998.
[44] L.H. Goulder, S.H. Schneider, *Induced Technological Change, Crowding Out, and the Attractiveness of CO_2 Emissions Abatement*, Stanford University, CA.: Institute for International Studies.
[45] Giovanni Dosi et al., *Technical Change and Economic Theory*, London: Pinter, 1988.

public policy attention. However, they do not allow for the full hidden costs of implementing these technologies, so that the scope for such 'no regret' measures is less than these models suggest; top-down models, on the other hand, can be criticized for not allowing for any opportunities for 'no regret' measures, which means that they exaggerate the costs. Furthermore, macro-economic models vary in their representation of the energy sector and are economy-specific, and give widely differing results. General equilibrium models focus upon the optimal allocation of resources in an economy over time, generally (but not necessarily) assuming perfect foresight, an optimal 'baseline', and full employment of all available resources. This is hardly a convincing representation of economic realities. In the short and medium term at least, the real costs of emissions abatement almost certainly lie between the optimism of bottom-up studies and the pessimism of most top-down studies.

The US government analyses illustrated the scale of difficulties facing attempts to build consensus about economic impacts. To conduct its assessment in the run-up to Kyoto, the US administration chose one model from each of the three main classes. The bottom-up model suggested emissions could be returned to 1990 levels without economic loss; the GE model projected a loss of about 1% of GNP; and the macro-economic model projected a loss of 2.4% of GNP. After more than a year of effort, in July 1997 attempts to reach any consensus about the overall economic impacts of given emission targets were abandoned.

It is probably more useful to conclude that no single approach is better than another, and that there is no single answer to the over-generalized question of 'how costly?'. Rather, the different model types are appropriate to different specific policy questions and instruments.

Indeed, there was one such issue on which almost all economic studies could agree: emission reductions can be obtained more cheaply in some places than others, hence there are large cost savings to be obtained from international flexibility. Thus out of a generally frustrating economic debate, particularly in the United States, was born the consensus that whatever was done should be done as globally as possible, and as flexibly as possible.

Epilogue: the economic frontiers of climate change

Despite a decade of intense economic debate about climate change, we remain embarrassingly ignorant about many important aspects of the problem. Added to the issues noted above, one striking question in the energy dimension is where all the carbon required to reach large-scale atmospheric changes will come from. A concentration of 450 ppm CO_2 would require emissions totalling about 630–650 GtC over the next century; a concentration of 550 ppm CO_2 would require 870–990 GtC to be emitted. Global proven conventional oil and gas reserves (which contain about 80 GtC and 120 GtC respectively) are not sufficient to sustain demand growth beyond about the next twenty years, and new discoveries are nothing like sufficient to keep pace with global demand growth projections.[46] Coal resources are much bigger, though clean and easily accessible deposits may also be limited. There is no shortage of fossil carbon overall, but much of this lies in more difficult and costly deposits (such as deep ocean and tar sand deposits). Advanced economies have in fact been decarbonizing for a long time.

The long-term economics of climate change is thus intimately bound up with the overall economics of energy transitions in the century after the 'oil age', combined with continuing technological and structural change in our use of energy. From one perspective indeed the Kyoto targets could be seen as a useful discipline to help the world to start addressing this 'other' fundamental question in good time.

This is but one specialized facet of the economic issues surrounding climate change. As sketched in this chapter, economic debates are complex and not much has been resolved. The deeper one digs, the more searching are the questions that arise about the most basic assumptions embodied in conventional economic analyses. The 'global commons' scope of the issue brings to the fore fundamental assumptions about responsibility and evaluation in ways that do not apply within individual nation-states. The century-long time scales raise questions about intergenerational equity and processes of technical and behavioural change that go beyond the credible scope of conventional economics. Even the debates over 'no regret' abatement options are marked by quite basic debates about human rationality and social freedoms (the freedom to be unnecessarily wasteful in ways that may damage others) and the appropriate roles of government regulation. Finally, as discussed in the body of this book, the Kyoto Protocol itself introduces entirely novel international market instruments. Grappling with the more fundamental economic issues raised by climate change can feel like trying to understand modern cosmology with the tools of Newtonian mechanics. Progress is being made, but it remains unclear how much consensual guidance economics will be able to offer to the struggling policy-makers.

---

[46] All these data can be derived from the IPCC Second Assessment Report (Working Groups I and II), CUP, 1996. A summary and discussion is given in P. Kassler and M. Paterson, *Energy Exporters and Climate Change*, London: RIIA/Earthscan, 1997.

Appendix 3

Analysis of international trade in emission allowances

The mathematics of the ITEA model

The ITEA model has been kept simple to enhance transparency and interactiveness. New issues raised in the negotiations can be incorporated quickly and the implications explored. The model is most appropriate for exploring emission implications and relative costs. Like all such analysis, results are dependent on input assumptions.

The basis of ITEA is a straightforward marginal abatement cost function, defined relative to the BAU emissions in 2010, broken down into two separate components. The first component comprises reductions that incur negligible economic costs ('no regret' reductions), but cannot be achieved in the BAU scenario for political or other reasons. The second part of the curve represents linearly rising marginal costs after the 'no regret' reductions are exhausted. With x representing the percentage reduction below BAU, a the relative marginal abatement cost slope, and b the 'no regret' reduction percentage, the relative marginal cost (rmc) function is:

$$rmc = 0, \text{ for } x \leq b;$$
$$rmc = a \cdot (x{-}b), \text{ for } x > b. \tag{1}$$

When we fill in the assumptions for a and b (see next section) and the agreed allocation (in percentage below BAU, x) we find out the marginal costs for this country.

The formula for the marginal costs (mc) – with reductions in tonnes of carbon rather than percentage below BAU – is now easy to calculate. With $y = x.BAU$, the reductions in tonnes below BAU, $c = a/BAU$, and $d = b.BAU$, the 'no regret' reductions in tonnes, the marginal costs (mc) function is:

$$mc = 0, \text{ for } y \leq d;$$
$$mc = c \cdot (y{-}d), \text{ for } y > d. \tag{2}$$

The total costs (tc) for the reductions, without trading, resulting from this linear marginal cost curve, rise quadratically; they are:

$$tc = 0, \text{ for } y \leq d;$$
$$tc = c \cdot (y{-}d)^2, \text{ for } y > d. \tag{3}$$

Chapter 5 shows a graph of the standard relative marginal cost curve made up by the two components, and for the corresponding total cost for reductions.

ITEA includes the non-CO_2 greenhouse gases, which have very different abatement costs and 'no regret' options. Therefore we have created a second marginal cost curve for each country representing the other GHGs, grouped together according to their global warming potential (GWP). The allocations of the Kyoto Protocol are defined in CO_2-equivalents, therefore we have summed the two marginal cost curves of each country. As can be seen in Figure A3.1 the marginal abatement cost curve for all GHGs consists of three components, due to differences in marginal cost curves and emission projections for CO_2 on the one side and the other GHGs on the other. The formula used, however, represents only the first two, the third component being irrelevant for all reasonable solutions. The formula for the marginal abatement cost for all GHGs is:

$$mc = 0, \text{ for } y \leq d; \tag{4}$$
$$mc = c \cdot (y{-}d), \text{ for } y > d;$$
$$\text{with } 1/c = (1/c_{CO2}) + (1/c_{other\ GHGs}), \text{ and } d = d_{CO2} + d_{other\ GHGs}.$$

Because the market is assumed to be perfect in ITEA, we can sum the marginal cost curves of all Annex I Parties joining the trading regime to find the world marginal cost curve (*WMC*). The calculation and the answer are very similar to the formula for all GHGs:

$$WMC = 0, \text{ for } Y \leq D; \tag{5}$$
$$WMC = C \cdot (Y{-}D), \text{ for } Y > D;$$
$$\text{with } 1/C = \sum (1/c_i), \text{ and } D = \sum d_i, \text{ for all countries } i \text{ in Annex I.}$$

Knowing the world marginal cost we can deduce the other variables we want to know:

- domestic reductions of each country;
- choice of gases domestically;
- trade of emission quota between countries;
- the total costs of the Protocol for Annex I.

Assumptions

The main input assumptions are summarized in Chapter 5, Table 5.4. As indicated there, the data on business-as-usual projections are derived from national reports where available, with the exception of Japan and the EITs. The cost assumptions and their justification are as follows.

> *United States and Canada*
> Base: 'no regret' 10%, marginal cost slope 1.0.
> High: 'no regret' 6% (Canada 8%), marginal cost slope 1.0.

Figure A3.1: The marginal abatement cost curve for all GHGs (a), and an example of the curve for New Zealand (b)

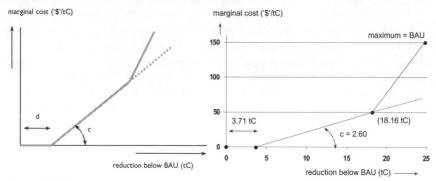

Source: ITEA.

A detailed study of technical options for emissions limitation conducted by US government laboratories (the '5-lab study') estimates that US CO_2 emissions could be returned roughly to 1990 levels without net economic costs, though various measures aimed primarily at improving energy efficiency.[1] Set against projected CO_2 growth of 26% above 1990 levels, this represents a 'no regret' potential of about 20% (20% of 1.26 is 0.25). However, this neglects some kinds of hidden costs and feedbacks, and not all the measures required are going to be politically feasible.[2] In our 'base case' we assume the realizable 'no regret' potential to be half of the 5-lab study estimate; in the 'high' case we assume that only just over a quarter can be realized. It is constrained by the impact of existing vehicle efficiency standards and widespread appliance standards, and recent downward revisions of emission projections that reflect some 'no regret' uptake in the base case.

Low energy prices, availability of natural gas and renewable sources, and the large scale of buildings, cars, etc. suggest lower marginal cost slope than in the EU. This assumption of lower marginal cost of emission reductions in the United States is supported by most detailed economic studies.

> *European Union*
> Base: 'no regret' 6%, marginal cost slope 1.2.
> High: 'no regret' 3%, marginal cost slope 1.2.

---

[1] Interlaboratory working group on energy-efficient and low-carbon technologies, 'Scenarios of US Carbon Reductions', US Department of Energy, Washington, DC, 1997.
[2] H.D. Jacoby, 'The Uses and Misuses of Technology Development as a Component of Climate Policy', MIT Joint Program on the Science and Policy of Global Change, Report No. 43, November 1998.

Table A3.1: Emission scenarios for EITs (from base year)

Country	Base year	1995 (%)	Projections (% from base year)		
			Reference	High	Other source comparison
Russia	1990	−29.2	−17.8	−4.7	−3.0
New EU					
Czech Republic	1990	−22.2	−7.2	+2.4	+0.3 (NC)
Estonia	1990	−43.3	−44.5	−35.5	−61.6 to −51.1 (NC), −34.1
Hungary	1985–7	−27.6	−14.5	−1.9	−19.3
Poland	1988	−22.4	−22.0	+15.8	−23.9
Slovenia	?	+6.1	+23.2	+42.8	+132.0
Other CEEC					
Bulgaria	1988	−40.4	−20.3	−12.1	−31.4
Croatia	?	−17.1	−3.7	+11.6	+37.5
Latvia	1990	−58.6	−22.2	−4.4	
Lithuania	1990	−63.5	−22.0	−13.9	−14.3 to −25.3 (NC), −16.7
Romania	1989	−39.0	−29.2	−17.9	−18.9
Slovak Republic	1990	−27.6	−17.8	−4.6	−14.2 to −4.1 (NC), −8.6
Ukraine	1990	−35.6	−17.7	−9.2	−15.9 (NC)

Source: The base and high case BAU projections are generated in ITEA, with respectively a 1% and 2% growth from 1995 emissions, or where available projections for the year 2000. Some EITs have emission projections for 2010 in their National Communication; in the right-hand column these projections are noted with (NC). The other projections are 'most likely emissions' collected by Missfeldt (personal communication).

'No regret' opportunities arise from the (removal of) continuing protection of the coal industry, slow development of energy efficiency standards, and opportunities arising from the growth in southern European countries. However, the scope is limited by the relatively high historical energy prices and relatively low emission projections, which suggest significant 'no regret' implementation, co-generation and renewables in the base projections. The high availability of low-cost natural gas constrains the marginal cost slope, which is 20% steeper than the reference defined for the US.

Japan
Base: 'no regret' 5%, marginal cost slope 1.4.
High: 'no regret' 3%, marginal cost slope 1.4.

Despite the high efficiency of the Japanese industrial sector and historically high energy prices, the rapid projected growth and relatively low efficiency of the building and service sectors suggest significant 'no regret' potential. The lack of readily accessible natural gas and limitations on indigenous renewable sources (with the possible but unproved exception of dry geothermal energy) imply a higher marginal abatement cost slope than the EU. The 'high' case involves higher BAU CO_2 emissions growth (20%) than in the reference (14.9%) owing to the large nuclear uncertainties (see Chapter 5).

Australia
Base: 'no regret' 12%, marginal cost slope 1.0.
High: 'no regret' 8%, marginal cost slope 1.0.

Low energy prices and the lack of significant standards for vehicles or appliances in Australia, also the continuing role of coal in power generation coupled with the availability of natural gas, imply 'no regret' opportunities somewhat larger than in the United States, especially given the very high projected growth in Australian emissions. The large land areas and gas availability, as in the United States, lower the marginal cost slope associated with displacing coal from power generation.

Other OECD
Base: 'no regret' 5% (Switzerland 0%), marginal cost slope 1.50.
High: 'no regret' 0%, marginal cost slope 1.50.

These countries are characterized by high energy prices and almost carbon-free electricity systems in the base year, giving little scope for 'no regret' measures (except perhaps in vehicle standards) and implying high marginal abatement costs. The low projection of growth in Switzerland implies uptake of almost all 'no regret' measures and more.

Economies in transition

Because of the great uncertainty in emissions from EITs two BAU scenarios are presented. The *base* scenario projects a 1% emissions growth per year from the most recent data (1995) or, if available, from the emission projection for the year 2000. A 1% growth is comparable to the annual BAU emissions growth in the United States. The *high* scenario projects a quick transition with 2% growth in emissions annually. The resulting BAU emissions in 2010 relative to their base year, which are used in ITEA, are also given in Chapter 5. Our projections are compared with the BAU scenarios of the National Communications for the Czech Republic, Estonia, Lithuania, the Slovak Republic and Ukraine, and scenarios from other sources for the other countries (see Table A3.1).

Our economic assumptions for the EITs are as follows.

Russia and Ukraine
Base: 'no regret' 14%, marginal cost slope 0.75.
High: 'no regret' 10%, marginal cost slope 0.75.

Widespread continuing inefficiencies in Russian and Ukrainian energy consumption, derived partly from non-payment, suggest a substantial potential for 'no regret' measures, particularly if external finance is utilized. These factors, coupled with availability of cheap natural gas (for Russia) and renewable resources, also keep the marginal cost of additional measures low.

New EU
Base: 'no regret' 10%, marginal cost slope 0.75.
High: 'no regret' 8%, marginal cost slope 0.75.
The conditions in some respects are similar to those in Russia, but most of these countries are much further along the road of economic transition and have far smaller non-payment problems, implying lower 'no regret' potentials.

Other CEEC (excluding Ukraine)
Base: 'no regret' 12%, marginal cost slope 0.75.
High: 'no regret' 8%, marginal cost slope 0.75.

Again the conditions are similar to those in Russia, but most of these countries are further along the road of economic transition and have lesser non-payment problems, implying lower 'no regret' potentials.

These economic assumptions, combined with the national reference projections and assumption concerning non-CO_2 gases set out in Chapter 5, form the main inputs for the quantified analysis of Chapter 5. The resulting cost estimates generally lie between those typical of 'bottom-up' and 'top-down' analyses. The lead author is currently involved in a more detailed international review of estimates of the cost of implementing the Kyoto commitments.

Appendix 4
Further reading and sources of information

The study of climate change and of the responses to it has grown rapidly, and the diversity of sources of information and analysis is bewildering to newcomers. This book has sought to reference items lightly and where directly relevant, without in any way attempting comprehensive coverage of the literature. Those seeking further information should find more than enough in the following sources.

The Secretariat of the UN Framework Convention on Climate Change is the repository for official information and documents relating to the UNFCCC and the Kyoto Protocol. The UNFCCC website, http://www. unfccc.de, is mostly organized according to meetings and the official documentation index, making it hard to navigate for the newcomer. The International Institute for Sustainable Development, http://www.iisd.ca, runs the Earth Negotiations Bulletin which covers all the main negotiating sessions.

The most authoritative sources of analytic information on both scientific and response aspects of climate change are reports by the Intergovernmental Panel on Climate Change (IPCC). Chapter 1 of this book gives the references for the volumes of the First (1990/1) and Second (1995/6) Assessment Report. The IPCC also publishes Technical Papers, which are intended to be analyses based upon material in the main Assessment Reports prepared at the request of the UNFCCC, and Special Reports on particular topics (*Regional Impacts of Climate Change*, 1998; *Aviation*, forthcoming 1999; *Technology Transfer*, forthcoming 2000). Their website is http://www.ipcc.ch.

The IPCC's formal task is to assess existing literature. Source scientific literature is published in the leading scientific journals *Nature* and *Science*, as well as the more specialist journals such as *Geophysical Research Letters* (American Geophysical Union, US). The World Climate Research Programme (Geneva) also publishes a newsletter with topical developments. The *New Scientist* regularly publishes stories on scientific and policy developments.

Various specialist newsletters cover climate change developments. The *Global Environmental Change Report* from Cutter Information Corp (Arlington, MA) reports on developments in the science and policy of climate change and ozone depletion, and various sister publications focus on some related international environmental topics (West European Environment; Business and Environment). The Bureau of National Affairs, Inc. (Washington, DC) publishes the broad-ranging *International Environment Reporter*. IISD run the 'ClimateL' email newsletter, which offers regular compilations of press reports on aspects of climate change; for details see their website (above). Environmental and business groups have their own networks, usually including email news services. Notable international groups among these are the Climate Action Network, the World Business Council on Sustainable Development, and the International Petroleum Industry Environmental Conservation Associate (IPIECA); these groups also publish important material around the climate change negotiations.

The UN Environment Programme also manages an information unit on climate change (see their website http://www.unep.ch). The *TIEMPO* newsletter, published from the climate research unit at the University of East Anglia (http://www.cru.uea.ac.uk) together with the International Institute for Environment and Development (http://www.oneworld.org/iied), focuses particularly on information for developing countries. A more broad-ranging international development information website that incorporates extensive information on climate change is at http://www.ids.ac.uk/eldis. Links from these sites to others around the world provide access to a global information resource.

Source material on analysis of responses to climate change is quite diverse. The journal *Climatic Change* (Dordrecht, NL: Kluwer) covers topics ranging from basic science to economics and analysis of the UNFCCC, and is particularly strong in areas linking impacts with response options. The *Pacific-Asian Journal of Energy* (Tata Energy Research Institute, New Delhi) is a leading journal of analysis from developing countries. *International Environmental Affairs* carried some of the most insightful source articles but was discontinued in 1998. *Global Environmental Change* (Butterworths) spans both natural science analysis of impacts and social science analysis of responses, while *Global Environmental Politics* focuses on political analysis. *Energy Policy* (Oxford, UK: Elsevier Science) is the leading international journal of energy policy and is strongest concerning (but is not confined to) energy-related aspects of response options, whilst *The Energy Journal* (International Association of Energy Economics, Cleveland, OH) and *Energy Economics* (Elsevier Science) focus more purely upon economic analysis. The *International Journal of Environment and Pollution* (Geneva: Inderscience Enterprises Ltd.) has strengthened its coverage of climate change with a double special issue 'EU Climate Policy: The European Commission Policy/Research Interface for Kyoto and Beyond' (Vol. 10, Nos. 3/4, 1998).

The legal literature on the climate change regime is growing more extensive. The *Yearbook of International Cooperation on Environment and Development* (Earthscan/ Fridtjof Nansen Institute), formerly the *Green Globe Yearbook*, carries excellent summaries of the status of all the major international environmental and developmental treaties, including regional conventions, together with some topical analyses. The *Review of European Community and International Environmental Law* (Blackwell/FIELD) carries research papers analysing agreements and responses to them, and published extensive analyses of the Kyoto Protocol in its special issue of July/August 1998 (Vol. 7, No. 2).

A large literature has emerged on specific topics. Joint Implementation (in its many variants) has attracted a particularly voluminous debate. In addition to the specific journal articles referred to in Chapter 3 (note 23), several substantial books have now been published on the topic. K. Chatterjee, *Activities Implemented Jointly to Mitigate Climate Change: Developing Country Perspectives* (Development Alternatives, New Delhi, 1997), offers a particularly extensive and thoughtful collection. Robert Dixon (ed.), *The UNFCCC Activities Implemented Jointly (AIJ) Pilot: Experiences and Lessons Learned* (Dordrecht: Kluwer, forthcoming 1999) will provide the most extensive analysis of the AIJ pilot phase and broader discussion of issues in joint implementation. The UNDP report, J. Goldemberg (ed.), *Issues and Options: The Clean Development Mechanism* (UNDP, New York NY, 1998), offers the most extensive analyses of the CDM. The *Joint Implementation Quarterly* (Foundation JIN, Paterswolde, the Netherlands: http://www.northsea.nl/jiq) carries news on projects and analysis.

Emissions trading has also attracted a large literature. The most comprehensive study relating specifically to the Protocol (but describing and drawing extensively upon experience with domestic emission trading systems) is probably that by Tom Tietenberg et al., *International Rules for Greenhouse Gas Emissions Trading* (Geneva: UNCTAD, 1999); (UNCTAD/GDS/GFSB/Misc.6). Other important and wide-ranging contributions on emissions trading include other works by Z. Zhang and A. Michaelowa (see UNCTAD study), by the OECD Environment Directorate (Paris: OECD), and by N. Matsuo (Institute for Global Environmental Strategies, Kanagawa, Japan). The published proceedings of a RIIA conference, Dean Anderson and Michael Grubb, *Controlling Carbon and Sulphur* (London: RIIA, 1997) present 16 papers covering domestic experience and the international debate in the formative period of the Kyoto negotiations.

Information on national climate change programmes is generally best obtained from websites of the relevant government departments. In-depth reviews by the Climate Change Secretariat are available on the UNFCCC website – generally at least a year after the initial government submissions, since the reviews have to be discussed by the governments concerned. The Climate Action Network publishes more prompt and critical reviews. The Financial Times Business Information published a series of three reports on *Climate Change and the Energy Sector: A Country-by-Country Analysis of National Programmes*, which set national programmes in their energy policy and political context (*Volume 1: The European Union*, 1997; *Volume 2: The non-EU OECD countries*, 1997; *Volume 3: The Economies in Transition*, 1998).

A few major research books have been published on broad themes of climate change policy. Analysis of the politics of negotiating the UNFCCC and its aftermath are given in I.M. Mintzer and J.A. Leonard, *Negotiating Climate Change: The Inside Story of the Rio Convention*, Cambridge: CUP, 1994), and M. Paterson, *Global Warming and Global Politics* (London: Routledge, 1996). Ian Rowlands, *The Politics of Global Atmospheric Change*, provides an international relations and regime analysis of the climate change and ozone regimes. The outcome of a 4-year research programme at Harvard University is summarized in Henry Lee, ed., *Shaping National Responses to Climate Change* (Washington, DC: Island Press, 1995). Many insights into developing country perspectives, and the constraints under which they negotiate, are to be found in Joyeeta Gupta, *The Climate Change Convention and Developing Countries – From Conflict to Consensus?* (Environment and Policy Series, Kluwer Academic Publishers, Dordrecht). Michael Grubb's own studies *Energy Policies and the Greenhouse Effect, Volume 1: Policy Appraisal* (RIIA, 1990) and *Volume II: Country Studies and Technical Options* (RIIA, 1991) still appear surprisingly relevant: the global institutions may have made huge strides during the 1990s, but in terms of the issues in energy policy, it seems, *plus ça change, plus c'est la même chose.*. Many more economics-oriented and modelling studies are referenced in Appendix 2.

The number of institutions engaged in climate change research grows steadily, and they are now far too numerous to review here. The OECD and its energy arm the International Energy Agency publish jointly on many areas of climate change economic and policy analysis, and offer the largest set of authoritative publications. The Tata Energy Research Institute (TERI) publishes the largest volume of research from the developing world. Insight particularly into the range of debates in the run-up to the Buenos Aires conference may be obtained from reports on a series of five workshops organized by the Geneva-based International Academy of Environment (closed down in June 1999). RIIA's Energy and Environmental Programme has published in the climate change field for some ten years; for details see its publications catalogue and the RIIA website (http://www.riia.org).

Index

The text of the Protocol itself (pp. 281–302) is not indexed. Figures in **bold** refer to tables, figures, and boxes.

abatement, cost and benefits, **312**, 313–22
  bottom-up models, 315–16
  general equilibrium models, 317–18
  macro-economic models, 316–17
activities implemented jointly (AIJ), 99–100, **101**
Ad Hoc Group on the Berlin Mandate (AGBM), 53
additionality, in CDM, 213–15, 227–31, 243
aerosols, 9, **10**, 11, 18, 22
African countries, 239
  and emissions trading, 95
  and the environment, 36
agriculture, effect of weather changes, 13
Alliance of Small Island States (AOSIS), 35, 38, 41, 47
  and emission targets, 53
  and UN FCCC, 49
Antarctic, changes in, 52
anthropogenic sources of greenhouse gases, 8–9, 11, 18
AOSIS, *see* Alliance of Small Island States
Argentina, 251–2, 263
Australia, xxxiv, 174
  'Australia Clause', 121
  carbon dioxide emissions, xxxvi, 161, 162
  commitment, 117n
  and the environment, 33
  and UN FCCC, 50
Austria, 30

base year (1995), 69–72, 116, 119
Belarus, 116
Berlin, first Conference of Parties 1995, 44–8
Berlin Mandate, box **48**, 53, 63, 65, 69, 124

bilateral or portfolio structure, of CDM, 232–7, **233**, box **235**
biomass energy, 19
Brazil, 268
  and emission targets, 58
  and environment, 36
  proposal on financial penalties, 101–2
British Petroleum, and climate change, 257–8
Btu tax, 44
'Bubble', xxvii, xxxv, 85–7, 122–4
  distribution, table **123**
Buenos Aires follow-up and Plan of Action, xlii, 106, 248–53
bunker fuels, xxxvi, 65, 67
business involvement in Kyoto Protocol, 257–62
  see also companies; corporate involvement

Canada, 127, 174
  carbon dioxide emissions, xxxvii
  commitment, 117n
  cutbacks, 162
  and environment, 33
  sinks, 187
capital stock turnover, 308–10
carbon cycle effects, 311–13
carbon dioxide (CO_2), xxxiv, 3–4, **73**, 119
  absorption, 24–5, **25**
  control, 10–11, 19
  emissions, xxxvi–xxxvii, **28**
    1990, in Protocol, **302**
    see also emissions
  increase, 8
  nature, 27
  optional global CO_2 trajectories, 158–60, **159**

projected increases and cutbacks, table **168**
Carbon Investment Fund, 100, 101
carbon sequestration projects, 98–9
   see also sinks
carbon taxes, 44, 67, 165, 196, 319
Carbon Trust, 259
certification, 204, 237
certified emission reductions (CERs), xxxviii,
   198–200, 202, 231–2, 238
China, 265
   emissions trading, 95–6, 111
   and environment, 35, 36
   and expansion of commitments, xl
   natural gas pipeline project, 235
   and technology transfer, 105
chlorofluorocarbons (CFCs), 9
Clean Air Act 1990, US, amendment, 9, 32
Clean Development Fund (Brazilian
   proposal), 102–3
Clean development mechanism (CDM), xxxv,
   xxxvii, xxxix, 103, 133–6, box **134**,
   195, 226–47
   additionality, 213–15, 227–32, 243
   administrative charges, 232
   bilateral or portfolio structure, 232–7, **233**,
      box **235**
   crediting, certification and corruption
      control, 237–8
   early crediting, 202–4
   distribution of activities, 238–9
   eligibility criteria, 240–2
   finance, 242–4
   government, 244–5
   potential impact, 186–90
   size, estimated, table **190**
   tradability of credits, 198, 200
climate change, 3–5, 6–7, 52
   commitment to, 65
   economics, 303–23 *passim*
   effects, 12–14, 305
   effect of Kyoto Protocol, 158
   impact on humans, 20, 23
   long-term, 307–8
   science of, 8–12
   socioeconomics, 15
Climate Change Convention, *see* UN
   Framework Climate Change
   Convention
Climate Technology Initiative (CTI), 105

Clinton, President Bill, 32, 44, 60
coal, 67, 70, 323
   support for, 165, 167
   US production, 31
commitment, table **118**
   developing countries, xli–xlii, 106–7, 107–
      11, 138–40, 265
   evolution of, 109
   Kyoto negotiations, 84–7
   and mechanisms for international transfer,
      194–5
   problems, 46
   review and strengthening, 149–50
   UN FCCC Article 4, 39–40
companies, and international transfer, 137
   see also business; corporate involvement
compensation, for lost revenues, 103–4
compensation fund, suggested by OPEC, 57
compliance and non-compliance with Kyoto
   Protocol, 102, 142–6, 210–11
'concrete ceiling', to supplementarity, 219–20
Conference of Parties, UN FCCC, 41–2, 146
   1st (COP–1), Berlin, 1995, 44–8
   2nd (COP–2), Geneva, 1996, 53–7, box 55
   3rd (COP–3), Kyoto, *see* Kyoto Protocol
   4th (COP–4), Buenos Aires, 1998, 106,
      248–53
   5th (COP–5), Bonn, 1999, 264
   6th (COP–6), 2000, xl, 252, 255 276
Congress, US, power over energy policy, 32
Convention on Climate Change, *see* UN
   Framework Convention
corporate involvement in JI and CDM, 195–6,
   207–8
corruption, control of, 237–8
Costa Rica, 67, 99, 100
crediting, in CDM, 202–4, **203**, 237–8
credits, xxviii
   tradability, 198–200
see also certified emission reductions;
   emission credits; emission
reduction units

deaths, consequent on weather, 14
   valuation, 306
deforestation, and CO_2, 8, 27, 32, 77–8, 229n
Dessus, B., 240–1
developed countries, 265
   and international flexibility, 97

and UN FCCC, 38
see also EU; economies in transition;
     JUSCANZ; and individual countries
developing countries, xxxv–xxxvi, xl, 113,
     138–42
  amendment procedures, 141–2
  'citizens worth less' accusation, 20, 306
  commitments, xlii, 106–7, 107–11, 138–40
  compensation, 103–4
  emission limitation, 158
  emissions, 108
  and environment, 35–6, 267
  financial aspects, 139
  and global frameworks, 265
  and international flexibility, 97–100
  minimizing adverse impacts, 140–1
  technology transfer, 104–6, 139–40, 246
  and UN FCCC, 38, 47
disasters, weather-related, 23–4
discounting, and carbon cycle effects, 311–13
dispute resolution, 148–9

early crediting, in CDM and JI, 202–4, **203**
economic consequences of Kyoto Protocol,
     160–71
  costs of Kyoto commitments, 163–5, **164,**
     176
economies in transition (EITs), xxxvi, xli,
     34–5, 113, 174
  central/eastern Europe, environment, 34
  commitment, 117n
  emission scenarios, **327**
  international flexibility, 87, 92–3
  and time scale, 71
ecosystems, under climate changes, 12–13, 305
El Niño, 21, 24
emission commitments, aggregate
     implications, table **178**
emission controls, xxxvii
emission credits, 195, 198–206
  early crediting, 202–4, **203**
  timing and liability, 204–6
  tradability, 198–200
  who should acquire them?, 200–2
emission distribution, Annex I, table **180**
emission permits, xxxii, 208–10
emission reduction units (ERUs), xxxviii,
     201–2, 203–4, 213
emission sources, 72–6, table **73**

emission targets, 58, 62
  assignment, 81–7
  definition, 68–80
  non-compliance, 71, 142–6, 210–11
  time aspects, 69–72
emissions:
  Annex I commitments, box **161**
  changes, table 1990–96 **82**
  convergence, 269–75
  developing countries, 108
  distribution, 17, table **118**
  gap between commitments and 1995
     levels, **162**
  under Kyoto Protocol, 115–22, table **118**
  by mid-1990s, 155
  projected 1990–2100, 156, **157**
  projections to 2010, table **166**
  projections and Kyoto commitments, 165–
     71, table **166**
  trends, 1990–95 and 1995–96, table **83**
  trends and cutbacks, 160–3
emissions trading, xxix, xxxvi–xxxviii, 59,
     89–96, 128–31, 195, 197
  accountability, **208**, 210–13
  corporate involvement, 206–7
  governance of, 206–17
  permits, 208–10
  principles, 206–7
  supplementarity, balancing and charges,
     217–24
energy equipment, 14
energy resources, change in USA, 31
energy supply, low-emitting energy systems,
     19
energy tax, 319
enforcement provisions, 146–9
environmental consequences, of Kyoto
     Protocol, 155–60
Environmental Groups, 51–2, 260
Environmentally Sound Technology (EST),
     104–5
environmentalism, in EU, 30
equity, in limitation of greenhouse gases, 17,
     269
Estrada-Oyuela, Raúl, 52, 56, 63n, 64, 107,
     111, 219
  and CDM, 135
  and developing countries, 110
  and emissions trading, 95–6

European Union (EU), 29–30
  carbon dioxide emissions, xxxvi
  commitments, 116, table **118**
  commitment bubble, xxvii, xxxv, 85–7,
    122–4
  and emission trading, 59
  pre-Kyoto, 56, 57–8
  and Kyoto Protocol, 65–8
  ratification issues, 254–5
  and sinks, 135n
  and supplementarity, 220
  and time-scale, 69, 70
  treatment, 174
  and UN FCCC, 49–50
  and USA, 65
Exxon-Mobil, 261–2

finance, for CDM, 242–4
financial responsibilities, under UN FCCC,
    41, 42
  developing countries, 139
financial penalties, 102–3, 148
Finland, 30
flexibility, *see* international flexibility
forestry:
  and CDM, 229, 241
  and sinks, 98–9, 121n, 143
Former Soviet Union (FSU), and
    environment, 34
fossil fuel exporters, *see* oil lobby
fossil fuels, 27, 77, 323
framework convention, *see* UN Framework
    Convention
France, 71, 94
fuel switching, 14

gas and sectoral coverage, 196–8
gas and source coverage, commitments, 117–19
General Agreement on Tariffs and Trade
    (GATT), 67
General Agreement on Trade in Services, 209
Germany, xxxvi, 45
global average temperature, 1860–1998, **22**
Global Climate Coalition, 18
Global Commons Institute, 270n
Global Environment Facility (GEF), xxxix,
    42, 45, 242, 244, 247, 249
  and AIJ projects, 100, 101
  funding, 106, 249

Global Remedy for the Environment and
    Energy Use, 105–6
global warming, xxxiii, 6, 7, 8, 11, 21, 305
  projections 1990–2100, 156, **157**
  see also greenhouse gases
global warming potential (GWP), 72, 73–4,
    73, 118
Gore, Vice-President Al, 32
government overseas development assistance,
    and CDM, 243
greenhouse gases (GHG):
  cause of climate change, 3, 5–6, 11
  emissions, xxxiv, 15, **28**
  emissions and commitments, table **118**
  impact, 23, 305
  inventories, 144
  list, in Kyoto Protocol, **73**, **300**
  measurement, 98
  rising, 7, 8–9
Greenpeace, 51, 76n
Group of 77 (G77) and China:
  emission cuts, 57, 58–9
  environment, 35–6
  expansion of commitments, xli–xlii, 113,
    262, 263, 265
  against North–South JI, 100
  and proposed Article 6, box **130**
  and UN FCCC 47, 49
Gulf Stream, 12, 23

halocarbons, increase, 9
headline commitments, 115–17
health, under climate change, 14
'hot air', xxx, 177, 181–2, 184, 189, 192,
    213–17, box **216**, 225, 264
Houghton, Sir John, 5, 8n
human influence on climate change, *see*
    anthropogenic sources of gases
Hungary, xxxiv, 117n
hydrofluorocarbons (HFCs), table **73**, 75

ice sheet, West Antarctic, 12
Iceland, xxxiv
India, 265
  carbon-saving projects, 100
  and emissions trading, 95–6, 111
  environment, 36
Indonesia, 36
industrial groups, and UN FCCC, 50–1

industrialized countries, Kyoto commitments, xxxiv–xxxv, table **118, 274**
   list, Annex B to Protocol, **301**
industry, and climate change, 257
inflation, xli, 192, 264
infrastructure, human, effect of weather change, 13–14
Intergovernmental Panel on Climate Change (IPCC), xxx, xxxiii, 3, 4–5
   First Assessment Report, 5–7
   Second Assessment Report (SAR), 7–8, 14–17, 24
      key debates and implications, 17–21
      key themes, 303–23
   Third Assessment Report, 19, 25
   Special Report on regional impacts, 24
   developments after SAR, 21–6
   see also Working Groups; Appendix 4
intergovernmental transfers, xxxviii
International Climate Change Partnership (ICCP), 259n
international emission transfers, xxxvi
   implementation architecture, xxxviii
   quantitative implications, xxxvi–xxxvii
International Energy Agency (IEA), 105
   projections on energy, 167–8, table **168**
international flexibility, xxxv, 112, 261
   aggregate results, 177–8, table **178**
   analytic framework, 171–3
   and Annex I countries, 87–96, 171–86
   caveats, 185–6
   composition and assumptions, 174–7
   cost and related implications, 182–5
   developing countries, 97–100
   implications for trading and emissions, 178–82
   stressed at Kyoto, 193
International Geophysical Year 1957, xxxiii, 4
International Standards Organization (ISO), 105
international trading in emissions allowances (ITEA), 173, **173**, 186, 324–9
   abatement cost modelling in, box **172**
international transfer:
   charges, 221–4
   clean development mechanism (CDM), 133–6
   emission credits, 195, 198–206
   see also emission credits

emissions trading, 128–31
   see also emissions trading
   governing international emissions trading, 206–17
   implementation, 194–225
   implications, 155–93
   joint implementation within Annex I, 131–2, box **132**
   mechanisms, 128–38
   principles, 194–8
   supplementarity, balancing and charges, 217–24
island states, 14, *and see* Alliance of Small Island States

Japan, 174
   carbon dioxide emissions, xxxvi, 32–3, 59
   commitment, 117n
   emissions trading, 94
   cutbacks, 162
   JI project investments, 132
   and Kyoto Protocol, 66
   and legally binding targets, 56
   and ratification, 254
   timescale, 71
   tribute to, 111
joint implementation (JI), of emission transfers, xxxi, xxxv, 45, box **46**, 175
   and developing countries, 97–100
   early crediting, 202–4
   estimating reductions, 98
   generation of credits from, 198, box **199**
   project-based, 88–9
   see also activities implemented jointly
JUSCANZ/JUSSCANNZ groups, 34, 35, 87, 246
   and a possible bubble, 124
   and international trading, 91–2
   and UN FCCC, 50

Kazakhstan, 251, 263
Kjellen, Bo, 107, 138n
Kyoto Protocol , 115–52
   Annex B, listing countries, 301
   bubbles, 85–7, 122–4
   Buenos Aires 'Plan of Action', 248–53
   business and public involvement, 257–62
   commitment, Annex I countries, 115–22, 138–40

commitment, developing countries, 107–11
compliance, revision etc, 142–50
conclusions, 111–14, 150–2
destruction, potential, 272n
developing countries, additional issues,
    138–42
economic consequences, 160–71
emission convergence, 269–75
emission targets, 68–80
emission target assignment, 81–7
environmental consequences, 155–60
evolution, xliii
first period, 262–5
future, xliv, 248–77
historical development, xxxiii–xxxiv,
    52–60
inflation, efficiency and competition
    between the mechanisms, 191–3
international flexibility (Annex I
    countries), 87–96
international flexibility (developing
    countries), 97–100
in longer term, 265–9
mechanisms for international transfer,
    128–38
negotiations, 61–114
negotiating process, 62–4
origin, 3
policies and measures, 65–8, 124–7, box
    125
political and legal foundations, 27–60
potential impact of flexibility within
    Annex I, 171–86
prospects, *see subheading* future
ratification and entry into force, 253–7
scientific basis, xxxiii–xxxiv
sectors/source categories, 300
signatures, xv, xxxiv
structure, xxiv–xxv
summary and conclusions, xxxiii–xlii
text, 281–302
time scale, xl
'Kyoto Surprise', 101–3

land loss, through sea level rise, 13–14
land-use change and forestry (LUCF), 119,
    120–2, 162, 186, 187
    emission, table, **78**
Latin America, and environment, 36

legislation on emissions, 196
life, human, 20, 306

methane (CH$_4$), xxxiv, 9, table **73**, 74, 119
Mexico, 109
Mitchell, J., 266
modelling, of greenhouse gas limitation, 15,
    16–17
Montreal Protocol, 9, 25, 104, 147, 265

National Energy Strategy, US, 32
natural gas, 31, 235
Netherlands, 57–8
New Zealand, 33, 66–7, 187
nitrous oxide (N$_2$O), 9, table **73**, 74, 119
'no regrets' measures, 8, 20, 318–19, 326
non-compliance, *see* compliance and non-
    compliance
non-governmental organizations (NGOs), 61,
    260
North–South division of countries, 29, 37, 269
    and JI, 97–100
Norway, 33, 34
nuclear fuel, 14

ocean currents, 12
oil:
    price collapse 1998, 167
    US production, 31
oil lobby, 67
    and environment, 35–6, 38, 41, 56–7
    and UN FCCC, 46–7
    see also Organization of Petroleum
        Exporting Countries
'open tissues', xiii–xiv
Organization of Petroleum Exporting
    Countries (OPEC), 35, 50–1, 56–7, 63,
    140–1
    and compensation, 103–4
Organization for Economic Cooperation and
    Development (OECD) countries, xli,
    113, 174
    and clean development mechanism, 226
    and emissions trading, 94–5, box **130**
    and fossil fuel CO$_2$ emissions, 27, **28**, 32–4
    reduced emissions, xxxiv
Oslo Protocol (Second Sulphur Protocol), 9–
    10
ozone depletion, 25–6

Panayotou, T., 239
penalties, 102–3, 148
perfluorocarbons (PFCs), table **73**, 75
permits, xxxii, 208–10
Pew Center, Washington, 268
Poland, 117n
political foundations, of UN FCCC, 49–52
portfolio fund model, 200, 232–7, **233**, box **235**
protocols, to the UN FCCC, 42–3

quantified emission limitation and reduction objectives (QELROs), 62, 219
see also emission targets

rain forest, Brazil, 24
reforestation, 188
research and development, *see* technology research and development
rice cultivation, 32
Rio Earth Summit 1992, 36, 37
Russia:
  allocations, xxxvi, 162
  and binding targets, 56
  commitment, 117n
  economics, 169, box **170**
  emissions, 93, box **170**
  and environment, 34–5
  ratification, 254

Saudi Arabia, 36, 56–7
Scandinavia, sinks, 187
Science Report of IPCC, 8–12
scientific developments, and UN FCCC, 52
scientific foundations, and IPCC, 5–7
sea-levels, rising, 6, 7, 8, 11–12, 13–14, 35
  projected 1990–2100, 156, **157**
Second Sulphur Protocol (Oslo), 9–10
Shell, 258
sinks, xxxiv–xxxv, xxxviii, 14, 15
  and Kyoto, 76–80, table **78**, 120–2
  potential impact, 186–90, table **188**
  role, 98–9
Sokona, Y., 239
source and sink flexibility, xxxiv–xxxv
South Korea, 109
storm intensity, 12
Subsidiary Body for Implementation (SBI), of UN FCCC, 42, 45

Subsidiary Body for Scientific and Technological Advice (SBSTA), of UN FCCC,
  42, 45, 105, 120n
Subsidies and Countervailing Measures (SCM) agreement, 209
sulphur, 9, 22
sulphur hexafluoride (SF$_6$), table **73**, 75
supplementarity, 195, 217–19
  'concrete ceilings', 219–20
  policy and measures criteria, 220–1
sustainable development, 133
Sweden, 30
Switzerland, 33, 34, 35, 45

Tata Energy Research Institute (TERI), 273
taxation, 65, 90
  see also carbon taxes
technological change, 320–1
technology assessment panels (TAPs), 105
technology research and development, 223, 310–11
technology transfer, 55, 98–9, 104–6, 139–40, 249–50
temperature:
  global averages 1860–1998, **22**
  trends 1860–1992, **10**
  see also global warming
Thailand, reforestation, 99n
timber, effect of weather changes, 13
time scales and horizons, 69–72
traded volumes, table **179**
trading, impact on cost distribution, **183**
  see also emissions trading
transitional economies, and environment, 34
Turkey, 109, 116

Ukraine, 174, 253n
  allocations, xxxvi, 162
  economics, 169, box **170**
  emissions, box **170**
  and environment, 35
Umbrella Group, 35, 124n
United Kingdom:
  carbon dioxide emissions, xxxvii,
  and UN FCCC, 49–50
UN Environmental Programme (UNEP), 4
UN Framework Convention on Climate Change (UN FCCC), 36–43, 93

elements, 37–43
financial responsibilities, 41, 42
institutional machinery, 41
origin, 7, 37
political changes, 49–52
ratification by countries, 43–4
structure, xxiv
and technology transfer, 104–5
United States, 53
carbon dioxide emissions, xxxvii, 161–2
and climate change, 31–2
commitment, 117n
emission control, 59–60, **181**
emissions trading, 90–3
and environment, 44
and the European Union, 65
international flexibility, 87
and the Kyoto Protocol, xxxvi, xl, 112,
124, 174
ratification, 255–7
and Russia, 34–5

and UN FCCC, 50
Uruguay Round, 209
US Initiative on Joint Implementation (USIJI),
99, 101

water shortage, 13
weather, extreme conditions, 11, 12, 23–4
Working Groups, IPCC:
I, 5–6, 18
II, 6, 12–14, 14–15, 19
III, 6–7, 15–16, box **16**, 20–1; Appendix 2
World Bank, Carbon Investment Fund, 100
World Climate Conference 1979, 4
World Climate Conference 1990, 7
World Climate Research Programme, 4
World Energy Council, 19, 309, 311
World Meteorological Organization (WMO), 4
Worth, Senator Tim, 53–4

Yamin, F., 232–4